HEAT CONDUCTION

HEAT CONDUCTION

Second Edition

Sadik Kakaç
Department of Mechanical Engineering
University of Miami, Coral Gables, Florida

Yaman Yener
Department of Mechanical Engineering
Northeastern University, Boston, Massachusetts

HEMISPHERE PUBLISHING CORPORATION
Washington New York London

DISTRIBUTION OUTSIDE NORTH AMERICA
SPRINGER-VERLAG
Berlin Heidelberg New York Tokyo

HEAT CONDUCTION: Second Edition

The first edition of this book was published under the title *Heat Conduction* by the Middle East Technical University.

1 2 3 4 5 6 7 8 9 0 B C B C 8 9 8 7 6 5

Library of Congress Cataloging in Publication Data

Kakaç, S. (Sadik)
Heat conduction.

Includes bibliographies and index.
1. Heat—Transmission. 2. Heat—Conduction.
I. Yener, Yaman, date. II. Title.
QC320.K35 1985 621.402'2 84-27975
ISBN 0-89116-391-3 Hemisphere Publishing Corporation

DISTRIBUTION OUTSIDE NORTH AMERICA:
ISBN 3-540-15244-X Springer-Verlag Berlin

TO FILIZ, DEMET, YASEMIN, AND ZEYNEP

True enlightenment in life is science-technology.

Mustafa Kemal Atatürk

Heat, like gravity, penetrates every substance of the universe; its rays occupy all parts of space. The theory of heat will hereafter form one of the most important branches of general physics.

J. B. Joseph Fourier, 1824

Contents

Preface

The study of heat transfer is one of the important fields of engineering science. Heat transfer problems are of great practical significance in many branches of engineering. Although it is generally regarded as most closely related to mechanical engineering, much work in this field has also been done in nuclear, chemical, metallurgical, and electrical engineering where heat transfer problems are equally important.

In this book, various problems of heat conduction in rectangular, cylindrical, and spherical coordinates are solved by the method of separation of variables, integral and Laplace transforms, numerical techniques, and the integral and variational methods. The material presented has evolved from a series of lecture notes developed by the authors when teaching a graduate course in heat conduction over a period of years.

This book is written for both engineering students and engineers practicing in areas involving the applications of heat diffusion problems. In general, the problems at the end of each chapter, in SI units, are designed to clarify the physically and/or theoretically important points, and to supplement the text. The authors have a strong conviction that this is essential for a clear understanding of the application of the theoretical results. Special attention has also been given to the derivation of basic equations and their solutions in sufficient detail to help the student clearly understand the subject matter.

In Chapter 1, the basic concepts and fundamentals of heat transfer are presented. Chapter 2 is devoted to the derivation of general forms of the heat conduction equation. Solutions of steady-state heat conduction problems in one-dimensional bodies with and without internal energy generation together with an extensive treatment of extended surfaces are given in Chapter 3. The mathematical techniques presented in this text involve the use of orthogonal functions, Fourier expansions, integral transforms, etc. Since these topics may be beyond the mathematical experience of some students, important aspects of these special subjects are introduced in Chapter 4. In Chapters 5 and 6, solutions of multi-dimensional steady- and unsteady-state heat conduction problems by the method of separation of variables are given, and in Chapters 7 and 8 the method of solution with integral and Laplace transforms are presented, respectively. Chapter 9 is included to introduce the reader to the finite-difference method. This method may provide the only means of obtaining solutions to some heat conduction problems. An emphasis is given to the finite-difference formulation of heat conduction problems, which can be adapted by an individual to his or her own computer facility. Chapter 10 includes miscellaneous topics of formulation and solution, such as the Duhamel method, the integral method, the variational method, and an introduction to phase-change problems.

In the second edition, the authors have retained the basic objective of the first edition. As before, not all the topics and solution methods of heat conduction could be covered. The topics chosen and the order and the depth of the coverage represent the personal judgement of the authors as to what is of first importance for, and what can be practically taught in a one semester course to the beginning engineering graduate students. The authors feel that a study of heat conduction in the logical sequence that is followed in this book provides the best way for a first-year graduate student to develop an understanding of heat conduction.

With few exceptions, no more engineering background than the usual undergraduate courses in thermodynamics, heat transfer, and advanced calculus is required of the reader.

The authors wish to express their sincere thanks to the Hemisphere Publishing Corporation, whose competent work made this publication possible.

The authors also wish to thank Ms. Donna Pressley for typing the manuscript and Mr. Ali Akgüneş who prepared the drawings at the Middle East Technical University, Ankara, Turkey.

We acknowledge with appreciation the suggestions and encouragement from many colleagues, who have been very helpful to us during the writing of both the first and the second editions. Our graduate students and assistants have also contributed to this book; their perceptive questions and comments have often forced us to review several portions of the manuscript.

Finally, this book could never have been written if it were not for the support of our wives. Their encouragement and patience has been an invaluable contribution.

Sadik Kakaç
Yaman Yener

Nomenclature

A	area, m^2
Bi	Biot number
c	specific heat, J/kg.K
c_p	specific heat at constant pressure, J/kg.K
c_v	specific heat at constant volume, J/kg.K
e	energy per unit mass, J/kg
E	energy, J
F_{ij}	radiation shape factor
Fo	Fourier number
g	gravitation acceleration, m/s^2
h	heat transfer coefficient, $W/m^2.K$
$\hat{i},\hat{j},\hat{k}$	unit vectors in the x-, y-, and z-directions
I_ν	modified Bessel function of the first kind of order ν
J_ν	Bessel function of the first kind of order ν
k	thermal conductivity, W/m.K
k_m	mean thermal conductivity, W/m.K
K	kernel
K_ν	modified Bessel function of the second kind of order ν
L	length, m
m	mass, kg
N	normalization integral, Eq. (4.23b)
p	pressure, N/m^2
P	perimeter, m
P_n	Legendre polynomial of order n
q	heat flux, W/m^2
$\dot{q}$	rate of internal energy generation per unit volume, W/m^3
Q	heat, quantity of heat, J

$\dot{Q}$	rate of heat transfer, W
Q_n	Legendre function of the second kind
r	radius
$\vec{r}$	position vector
R_t	thermal resistance, K/W
s	entropy per unit mass, J/kg.K
S	entropy, J/kg
t	time, s
T	thermodynamic temperature, K
u	internal energy per unit mass, J/kg
U	internal energy, J; overall heat transfer coefficient, $W/m^2.K$
v	specific volume, m^3/kg
$\mathcal{V}$	volume (element)
$\vec{V}$	velocity vector, m/s
W	work, J
$\dot{W}$	power, W
Y_ν	Bessel function of the second kind of order ν

Greek Letters

α	absorptivity; thermal diffusivity, m^2/s
δ	penetration depth, m; velocity boundary layer thickness, m
δ_t	thermal boundary layer thickness, m
ε	emissivity
η_f	fin efficiency
θ	polar angle, rad; temperature difference, K
λ	eigenvalue
ρ	mass density, kg/m^3; reflectivity

σ	Stefan-Boltzmann constant, $W/K^4 \cdot m^2$
τ	transmissivity
ϕ	azimuthal angle, rad; fin effectiveness
χ	reciprocal of the neutron diffusion length, 1/m

Coordinates

(x,y,z)	rectangular coordinates
(r,ϕ,z)	cylindrical coordinates
(r,ϕ,θ)	spherical coordinates

Subscripts

b	base of the fin
c	contact surface, center
cr	critical insulation thickness
e	electrical
f	fin conditions
i	initial condition
m	mean value
r	radiation
t	thermal
x	local condition
∞	ambient condition

1

Foundations of Heat Transfer

1.1 INTRODUCTORY REMARKS

Heat transfer is the study of energy transfer solely as a result of temperature differences. The laws which govern heat transfer are very important to the engineer in the design, construction, testing and operation of heat exchange devices. Heat transfer problems confront investigators in nearly every branch of engineering. Although it is generally regarded as most closely related to mechanical engineering, much work in this field has also been done in chemical engineering, nuclear engineering, metallurgical engineering, and electrical engineering, where heat transfer problems are equally important. It is probably this fundamental and widespread influence which has brought heat transfer into the category of an engineering science.

In thermodynamics, *heat* is defined as the form of energy that crosses the boundary of a thermodynamic system by virtue of a temperature difference existing between the system and its surroundings. That is, heat is a form of energy in transition and the temperature difference is the driving potential for its propagation. Since heat is energy in transit we can really talk about the transfer of heat. *Heat flow* is vectorial in the sense that heat is transferred in the direction of a negative temperature gradient; that is, from higher toward lower temperatures, which is, in fact, a statement of the second law of thermodynamics.

Thermodynamics is that branch of science which deals with the study of heat and work interactions of a system with its surroundings. The laws of thermodynamics may be used to predict the gross amount of heat transferred during a process in which a system goes from one equilibrium state (including mechanical, and chemical as well as thermal equilibriums) to another. In most instances, however, the overriding consideration may be the length of time over which the transfer of heat occurs or, simply, the time *rate* at which it takes place. The laws of thermodynamics alone are not sufficient to provide such information, neither can they explain the *mechanism* of heat transfer. The science of heat transfer, on the other hand, studies heat transfer mechanisms and extends thermodynamic analysis, through the development of necessary relations, to calculate heat transfer rates.

In the study of heat transfer in a medium, attention is directed towards the relations among the temperature distribution and heat flow in the medium, and geometry, dimensions and thermo-physical properties of the medium. Such studies are based upon the foundations comprising both theory and experiment. As in other branches of physical sciences, the theoretical part is constructed from one or more *physical (natural) laws*. Physical laws are statements in terms of concepts that have been found to be true through many years of experimental observations. A physical law is called a *general law* if the application of it is independent of the medium under consideration. Otherwise, it is called a *particular law*. There are, in fact, four general laws upon which all the analyses concerning heat transfer, either directly or indirectly, depend. These are:

(a) the law of conservation of mass,
(b) the first law of thermodynamics,
(c) the second law of thermodynamics, and
(d) Newton's second law of motion.

For most heat conduction problems, the use of the first and second laws of thermodynamics is sufficient. In addition to these general laws, it is usually necessary to bring certain particular laws into an analysis. There are three such particular laws that we will employ in the analysis of conduction heat transfer. These are:

(a) Fourier's law of heat conduction,
(b) Newton's law of cooling, and
(c) Stefan-Boltzmann's law of radiation.

1.2 MODES OF HEAT TRANSFER

The mechanism by which heat is transferred is, in fact, quite complex. There appear, however, to be three rather basic and distinct types or *modes* of heat transfer processes. These are:

(a) conduction,
(b) radiation, and
(c) convection.

Conduction is the process of heat transfer by molecular motion, supplemented in some cases by the flow of free electrons, through a medium (solid, liquid or gaseous) from a region of high temperature to a region of low temperature. When two bodies or materials at different temperatures are in direct contact, heat transfer by conduction also occurs across the interface between these bodies.

The mechanism of heat conduction in liquids and gases has been postulated as the transfer of kinetic energy of the molecular movement. Thermal energy stored in a fluid increases its internal energy by increasing the kinetic energy of its vibrating molecules and is measured by the increase of its temperature. A high temperature measurement would then indicate a high kinetic energy of the molecules, and the heat conduction would imply the transfer of kinetic energy by the more active molecules in the high temperature region to the molecules in the low molecular kinetic energy region by successive collisions. Heat conduction in solids with crystalline structures, on the other hand, depends on the energy transfer by molecular and lattice vibrations, and free electron drift. In general, energy transfer by molecular and lattice vibrations is not so large as the energy transfer by free electrons, and it is for this reason that good electrical conductors are almost as good heat conductors, while electrical insulators are usually good heat insulators. In the case of solids with amorphous structure, however, heat conduction depends upon the molecular energy transport only.

Radiation, on the other hand, is a process of heat transfer which takes place in the form of electromagnetic waves. Thermal energy in the form of electromagnetic waves or, in other words, thermal radiation can pass through certain types of substances and can also pass through a vacuum, whereas for heat conduction to take place a material medium is necessary.

Conduction is the only mechanism by which heat can flow in *opaque* solids. Through certain *transparent* or *semi-transparent* solids, such as glass and quartz, some energy is transmitted by radiation in addition to the energy that can be transmitted by conduction. With gases and liquids, if there is no observable fluid motion, the problem becomes one of simple conduction (and radiation). However, if there is observable fluid motion, energy is transported by temperature gradients (as in conduction) as well as in the form of internal energy due to the movement of the fluid particles. This process of energy transport by the combined action of heat conduction (and radiation), and the motion of fluid particles is referred to as *convection* or *convective heat transfer*. Although in the foregoing classification we have considered convection to be a mode of heat transfer, it is actually conduction (and radiation) in moving fluids.

In reality, temperature distribution in a medium is controlled by the combined effect of these three modes of heat transfer. Therefore, it is not actually possible to entirely isolate one mode from interactions with other modes. However, for simplicity in the analysis these three modes of heat transfer are almost always studied separately. In this book we will study conduction heat transfer only.

1.3 CONTINUUM CONCEPT

Matter, while seemingly continuous, is composed of molecules, atoms and electrons in constant motion and collisions. Since heat conduction is thought to come about through the exchange of kinetic energy among the particles, the most fundamental approach in analyzing the transfer of heat in a substance by conduction is therefore to apply the laws of motion to each individual particle or a statistical group of particles, subsequent to some initial state of affairs.

In most engineering problems our primary interest lies not in the molecular behavior of a substance, but rather in how the substance behaves as a continuous medium. In our study of heat conduction, we will therefore neglect the molecular structure of the substance and consider it to be a continuous medium-*continuum*, which is fortunately a valid approach to many practical problems where only macroscopic information is of interest. Such a model may be used provided that the size and the mean-free-path of molecules are small compared with other dimensions existing in the medium, so that a statistical average is meaningful. This approach, which is also known as the *phenomenological* approach to heat conduction, is simpler than microscopic approaches and usually gives the answers required in engineering. On the other hand, the information is not so complete with the continuum approach and certain parameters such as thermodynamic state and transport properties have to be introduced empirically. Parallel to the study of heat transfer processes by continuum approach, molecular considerations can also be used to obtain information on transport and thermodynamic properties. In this book we shall study phenomenological heat conduction theory.

1.4 DEFINITIONS AND CONCEPTS IN THERMODYNAMICS

In thermodynamics, a *system* is defined as an arbitrary collection of matter of fixed identity bounded by a closed surface which can be real or imaginary. All other systems outside the surface that interact with the system under consideration are known as *surroundings*. In the absence of any mass-energy conversions, the mass of a system not only remains constant, but the system must be made up of exactly the same submolecular particles. The four general laws listed in Section 1.1. are always stated in the first instance in terms of a system. In fact, one cannot meaningfully apply the general laws until a definite system is identified.

A *control volume* is any defined region in space across the boundaries of which matter, energy, and momentum may flow, within which changes of matter and energy, and momentum storage may take place, and on which external forces may act. The control volume may change its position or size with time. However, most often we deal with control volumes which are fixed in space and are of fixed size and shape.

The dimensions of a system or a control volume may be finite or infinitesimal. The complete definition of a system or a control volume must include at least the implicit definition of a coordinate system, since they may be moving or stationary.

The *thermodynamic state* of a system is its conditions, as described by a list of the values for all its *properties*. A property of a system is either a directly or an indirectly observable characteristic of that system which can in principle be evaluated quantitatively. Volume, mass, pressure, temperature, etc., are all properties of matter. When all the properties of a system are observed to be not changing, the system is said to be at an *equilibrium state*. A *process* is a change of state and is described in part by the series of states passed through by the system. A *cycle* is a process wherein the initial and final states of a system are the same.

If no transfer of energy as heat takes place between any two systems placed in contact with each other, they are said to be in *thermal equilibrium*. Any two systems are said to have the same *temperature* if they are in thermal equilibrium with each other. Two systems which are not in thermal equilibrium have different temperatures and energy transfer as heat may take place from one system to the other. Temperature is, therefore, that property of a system which measures the thermal level of the system.

The laws of thermodynamics deal with interactions between a system and its surroundings as they pass through equilibrium states. These interactions may be divided into two as (i) *work* and (ii) *heat* interactions. Heat has already been defined as the form of energy which is transferred across the boundary of a system due to temperature differences. Work, on the other hand, is a form of energy that is characterized as follows: When an energy form of one system (such as kinetic energy, potential energy, chemical energy, etc.) is transformed into an energy form of another system or the surroundings without the transfer of mass from the system and by means other than a temperature difference, the energy is said to have been transferred through the performance of work.

The variation of the temperature of a substance with the amount of energy stored within it is expressed in terms of the specific heat of the substance. Because of the different ways in which energy can be stored in a substance, the definition of specific heat depends upon the nature of energy addition. *Specific heat at constant volume* of a substance is the change in internal energy (see Section 1.5) for a unit mass per degree change of temperature between two equilibrium states of the same volume. *Specific heat at constant pressure* is the change in enthalpy (see Section 1.5) for a unit mass between two equilibrium states at the same pressure per degree change of temperature.

1.5 FIRST LAW OF THERMODYNAMICS

When a system undergoes a cyclic process, the first law of thermodynamics can be expressed as

$$\oint \delta Q = \oint \delta W \tag{1.1}$$

where the cyclic integral $\oint \delta Q$ represents the net heat transfer to the system and the cyclic integral $\oint \delta W$ is the net work done by the system during the cycle. By convention in thermodynamics heat transferred (added) to a system is taken to be positive and heat transferred (removed) from a system to be negative. Similarly, work done by a system is positive and work done on a system is negative. Both heat and work are *path* functions, *i.e.*, the amount of heat transferred or the amount of work done when a system undergoes a change of state depends on the path the system follows during the change of state. This is the reason the differentials of heat and work are inexact differentials denoted by the symbols δQ and δW.

Equation (1.1) states the first law of thermodynamics for a system during a cyclic process. For a process that involves an infinitesimal change of state during a time interval dt it can be expressed as

$$dE = \delta Q - \delta W \tag{1.2}$$

where δQ and δW are the small amounts of heat added to the system and the work done by the system respectively, and dE is the corresponding increase in the total energy of the system during the time interval dt. Energy, E, is a property of the system and like all other thermodynamic properties it is a *point* function. That is, dE depends upon the initial and final states only and not on the path followed between the two states. For a more complete discussion of point and path functions, the reader is referred to References [2,3]. The physical property E represents all energies of the system and is customarily separated into three parts as bulk kinetic energy, bulk potential energy, and internal energy; that is,

$$E = KE + PE + U$$

Internal energy, U, which includes all forms of energy in a system other than bulk kinetic and potential energies, represents the energy associated with molecular and atomic structure and behavior of the system.

If the system undergoes a process from an initial state 1 to a final state 2, integration of Eq. (1.2) yields

$$\Delta E = Q - W \tag{1.3}$$

where Q is the net heat transferred to the system, W is the work done by the system, and ΔE is the increase in the energy of the system during the process.

Equation (1.2) can also be written as a rate equation,

$$\frac{dE}{dt} = \frac{\delta Q}{dt} - \frac{\delta W}{dt} \tag{1.4a}$$

or

$$\frac{dE}{dt} = \dot{Q} - \dot{W} \tag{1.4b}$$

where $\dot{Q} = \delta Q/dt$ represents the rate of heat transfer to the system and $\dot{W} = \delta W/dt$ is the rate of work done (power) by the system.

We shall now proceed to develop the form of the first law of thermodynamics as it applies to a control volume. Consider an arbitrary control volume fixed in space as illustrated in Fig. 1.1. Matter flows across the boundaries of this control volume. Define a system whose boundary at time t happens to correspond exactly to that of the control volume. At some later time $t + \Delta t$ the system moves to another location and occupies a different volume in space. The first law for the system for this change is

$$\Delta E = \nabla Q - \nabla W \tag{1.5}$$

where ∇Q is the heat transferred to the system and ∇W is the work done by the system, and ΔE is the increase in the energy of the system during time interval Δt. Dividing Eq. (1.5) by Δt we get

$$\frac{\Delta E}{\Delta t} = \frac{\nabla Q}{\Delta t} - \frac{\nabla W}{\Delta t} \tag{1.6}$$

The left-hand side of this equation can be written as

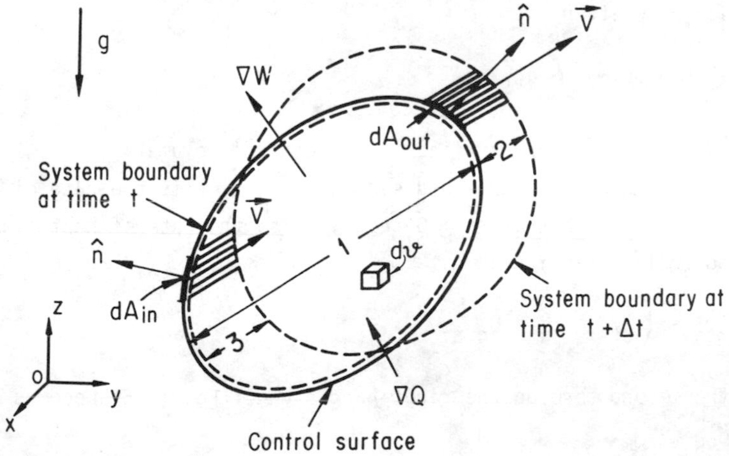

Fig. 1.1 First law of thermodynamics for a control volume.

$$\frac{\Delta E}{\Delta t} = \frac{E_1(t + \Delta t) + E_2(t + \Delta t) - E_3(t + \Delta t) - E_1(t)}{\Delta t}$$

$$= \frac{E_1(t + \Delta t) - E_1(t)}{\Delta t} + \frac{E_2(t + \Delta t)}{\Delta t} - \frac{E_3(t + \Delta t)}{\Delta t} \tag{1.7}$$

where E_1, E_2 and E_3 are the instantaneous values of E corresponding to three regions of space shown in Fig. 1.1.

As $\Delta t \to 0$, the first term on the right-hand side of Eq. (1.7) becomes

$$\lim_{\Delta t \to 0} \frac{E_1(t + \Delta t) - E_1(t)}{\Delta t} = \left(\frac{\partial E}{\partial t} \right)_{c.v.} = \frac{\partial}{\partial t} \int_{c.v.} e\rho dv \tag{1.8}$$

where dv is an element of control volume, ρ is the local density of that element, e is the energy per unit mass, and c.v. designates the control volume fixed in space and bounded by the control surface (c.s.). Further,

$$\lim_{\Delta t \to 0} \frac{E_2(t + \Delta t)}{\Delta t} = \int_{A_{out}} e\rho\vec{V}\cdot\hat{n}dA_{out} \tag{1.9}$$

and

$$\lim_{\Delta t \to 0} \frac{E_3(t + \Delta t)}{\Delta t} = - \int_{A_{in}} e\rho\vec{V}\cdot\hat{n}dA_{in} \tag{1.10}$$

where $\vec{V}$ denotes the velocity vector and $\hat{n}$ is the outward pointing unit vector normal to the control surface. Thus, as $\Delta t \to 0$, Eq. (1.7) becomes,

$$\frac{dE}{dt} = \lim_{\Delta t \to 0} \frac{\Delta E}{\Delta t} = \frac{\partial}{\partial t} \int_{c.v.} e\rho dv + \int_{A_{out}} e\rho\vec{V}\cdot\hat{n}dA_{out} + \int_{A_{in}} e\rho\vec{V}\cdot\hat{n}dA_{in} \tag{1.11}$$

which can also be written as

$$\frac{dE}{dt} = \frac{\partial}{\partial t} \int_{c.v.} e\rho dv + \int_{c.s.} e\rho\vec{V}\cdot\hat{n}dA \tag{1.12}$$

where dA is an element of the surface of the control volume.

The first term on the right-hand side of Eq. (1.6) is the rate of heat transfer to the system, but as $\Delta t \to 0$ this is also the rate of heat transfer across the control surface; that is,

$$\lim_{\Delta t \to 0} \frac{\nabla Q}{\Delta t} = \left(\frac{\delta Q}{dt} \right)_{c.s.} = \dot{Q}_{c.s.} \tag{1.13}$$

As $\Delta t \to 0$, the second term on the right-hand side of Eq. (1.6) becomes

$$\lim_{\Delta t \to 0} \frac{\nabla W}{\Delta t} = \frac{\delta W}{dt} = \dot{W} \tag{1.14}$$

where $\dot{W}$ is the rate of work done (power) by the fluid element in the control volume (system) on its surroundings at any time t. Hence, as $\Delta t \to 0$, Eq. (1.6) becomes

$$\frac{\partial}{\partial t}\int_{c.v.} e\rho dv + \int_{c.s.} e\rho\vec{V}\cdot\hat{n}dA = \dot{Q}_{c.s.} - \dot{W} \tag{1.15}$$

A final form for the first law expression may be obtained after further consideration of the power term $\dot{W}$. Work can be done by the system against its surroundings in a variety of ways. In this analysis we consider the work done against normal stresses (hydrostatic pressure), tangential stresses (shear), the work done by the system that could cause a shaft to rotate (shaft work), and the power drawn to the system from an external electric circuit. We neglect capillary and magnetic effects.

Consider first the work done against normal stresses at the boundary of the system, which is also called *flow work*. The work done by the system due to the normal stresses at an element of area dA_{out} during the time interval Δt is $p\,dA_{out}\Delta n$, where $\Delta n = (\vec{V}\cdot\hat{n})\Delta t$ is the component of the distance moved normal to dA_{out}, and p is the pressure on the surface element. Hence, the rate of work done by the system due to the normal stresses at the area element dA_{out} is $p\,dA_{out}\vec{V}\cdot\hat{n}$. Furthermore, the rate of work done by the system against normal stresses during the process as the mass $m_2(t + \Delta t)$ leaves the control volume is $\int_{A_{out}} p\vec{V}\cdot\hat{n}dA_{out}$. Similarly, the rate of work done by the normal stresses on the system as the mass $m_3(t + \Delta t)$ enters the control volume is $-\int_{A_{in}} p\vec{V}\cdot\hat{n}dA_{in}$. Thus, the net rate of work done by the system owing to the normal stresses may now be written as

$$\dot{W}_{normal} = \int_{A_{out}} p\vec{V}\cdot\hat{n}dA_{out} + \int_{A_{in}} p\vec{V}\cdot\hat{n}dA_{in} \tag{1.16a}$$

which can also be rewritten as

$$\dot{W}_{normal} = \int_{c.s.} p\vec{V}\cdot\hat{n}dA \tag{1.16b}$$

Let $\dot{W}_{shaft}$ represent the rate at which the system does shaft work and $\dot{W}_{shear}$ be the rate at which the system does work against shear stresses. Furthermore, the rate of work done on the system due to a power drawn to the system from an external electric circuit can be written as $\int_{c.v.}\dot{q}_e dv$, where $\dot{q}_e$ is the rate of internal energy generation per unit volume due to the power drawn to the system from the external circuit. Hence, as $\Delta t \to 0$, Eq. (1.15) becomes

$$\frac{\partial}{\partial t}\int_{c.v.} e\rho dv + \int_{c.s.} e\rho\vec{V}\cdot\hat{n}dA = \dot{Q}_{c.s.} - \int_{c.s.} p\vec{V}\cdot\hat{n}dA - \dot{W}_{shear} - \dot{W}_{shaft} + \int_{c.v.}\dot{q}_e dv \tag{1.17a}$$

which can also be written as

$$\frac{\partial}{\partial t}\int_{C.V.} e\rho dv + \int_{C.S.} (e + \frac{p}{\rho})\rho\vec{V}\cdot\hat{n}dA = \dot{Q}_{C.S.} - \dot{W}_{shear} - \dot{W}_{shaft} + \int_{C.V.} \dot{q}_e dv \tag{1.17b}$$

Since energy per unit mass may be written as

$$e = u + \tfrac{1}{2}V^2 + gz \tag{1.18}$$

where u is the internal energy per unit mass, $\frac{1}{2}V^2$ is the bulk kinetic energy per unit mass, and gz is the bulk potential energy per unit mass, then Eq. (1.17b) becomes

$$\frac{\partial}{\partial t}\int_{C.V.} e\rho dv + \int_{C.S.} (h + \tfrac{1}{2}V^2 + gz)\rho\vec{V}\cdot\hat{n}dA = \dot{Q}_{C.S.} - \dot{W}_{shear} - \dot{W}_{shaft} + \int_{C.V.} \dot{q}_e dv \tag{1.19}$$

where h is the *enthalpy* per unit mass defined as

$$h = u + \frac{p}{\rho} \tag{1.20}$$

Equations (1.17a,b) and (1.19) are the first law of thermodynamics written for a control volume.

1.6 SECOND LAW OF THERMODYNAMICS

The first law of thermodynamics, which embodies the idea of conservation of energy, gives us means for quantitative calculation of changes in state resulting from interactions between systems, but tells nothing about the direction that the process might take. In other words, the first law cannot explain physical observations such as the following: A cup of hot coffee placed in a cool room will always tend to cool to the temperature of the room, and once it is at room temperature it will never return spontaneously to its original hot state; air will rush into a vacuum chamber spontaneously; the conversion of heat into work cannot be carried out on a continuous basis with a conversion efficiency of 100 per cent; water and salt will mix spontaneously to form a solution, but the separation of such a solution cannot be made without some external means; a vibrating spring will eventually come to rest all by itself, etc. These observations concerning the one-wayedness of naturally occurring processes have led to the formulation of the second law of thermodynamics. Over the years, many statements of the second law have been made. Here we shall give the following *Clausius* statement: It is impossible for a self-acting system unaided by an external agency, to move heat from a body at low temperature to one at a higher temperature.

The second law of thermodynamics leads to a thermodynamic property--*entropy*--which is denoted normally by S. For any *reversible* process that a system undergoes during a time interval dt, the change in the entropy of the system is given by

$$dS = \left(\frac{\delta Q}{T} \right)_{rev.} \tag{1.21a}$$

For an *irreversible* process, the change in the entropy of the system is, however,

$$dS > \left(\frac{\delta Q}{T} \right)_{irr.} \tag{1.21b}$$

where δQ is the small amount of heat added to the system, and T is the temperature of the system at the time of heat transfer. Equations (1.21) may be taken as the mathematical statement of the second law, and can also be written in rate form as

$$\frac{dS}{dt} \geq \frac{1}{T} \frac{\delta Q}{dt} \tag{1.22}$$

By the same procedure as in the previous section, the control volume form of the second law can be obtained. Rather than going through the entire derivation, we shall give the result here, which is

$$\frac{\partial}{\partial t} \int_{c.v.} s\rho dv + \int_{c.s.} s\rho \vec{V} \cdot \hat{n} dA \geq \int_{c.s.} \frac{1}{T} \frac{\delta Q}{dt} \tag{1.23}$$

where s is the entropy per unit mass, and the equality applies to reversible processes and the inequality to irreversible processes.

1.7 TEMPERATURE DISTRIBUTION

Since heat transfer takes place whenever there is a temperature gradient in a medium, a knowledge of values of temperature at all points of the medium is essential in heat transfer studies. The instantaneous values of temperature at all points of the medium of interest are called the *temperature distribution* or *field*.

An *unsteady* (or *transient*) temperature distribution is one in which temperature not only varies from point to point in the medium, but also with time. It is to be noted that, when the temperature at various points in a medium changes, the internal energy also changes accordingly at these points. Following is a representation of an unsteady temperature distribution:

$$T = T(\vec{r}, t) \tag{1.24a}$$

where $\vec{r} = \hat{i}x + \hat{j}y + \hat{k}z$, and $\hat{i}$, $\hat{j}$ and $\hat{k}$ are the unit vectors in the x-, y- and z-directions, respectively.

A *steady* temperature distribution is one in which the temperature at a given point never varies with time; it is a function of space coordinates only.

Following is a representation of a steady temperature distribution:

$$T = T(\vec{r}), \qquad \frac{\partial T}{\partial t} = 0 \tag{1.24b}$$

The temperature distributions governed by Eqs. (1.24a) and (1.24b) are called *three dimensional*, since temperature is a function of three space coordinates ($\vec{r} = \hat{i}x + \hat{j}y + \hat{k}z$). When a temperature distribution is a function of two space coordinates, it is called *two dimensional*. For example, in the rectangular coordinate system,

$$T = T(x,y,t), \qquad \frac{\partial T}{\partial z} = 0 \tag{1.24c}$$

represents an unsteady two-dimensional temperature distribution.

When a temperature distribution is a function of one space coordinate only, the distribution is called *one dimensional*. For example, in the rectangular coordinate system,

$$T = T(x,t), \qquad \frac{\partial T}{\partial y} = \frac{\partial T}{\partial z} = 0 \tag{1.24d}$$

represents an unsteady one-dimensional temperature distribution.

If the points of a medium with equal temperatures are connected, then the resulting surfaces are called *isothermal surfaces*. Intersection of isothermal surfaces with a plane yields a family of *isotherms* on the plane surface. It is important to note that two isothermal surfaces never cut each other, since no part of the medium can have two different temperatures at the same time.

1.8 FOURIER'S LAW OF HEAT CONDUCTION

The basic law governing heat conduction may best be illustrated by considering a simple experiment: Consider a flat plate of thickness L such that the other two dimensions of the plate are very large compared to the thickness L as shown in Fig. 1.2. Let A be the surface area of the plate, and T_1, T_2 be temperatures of

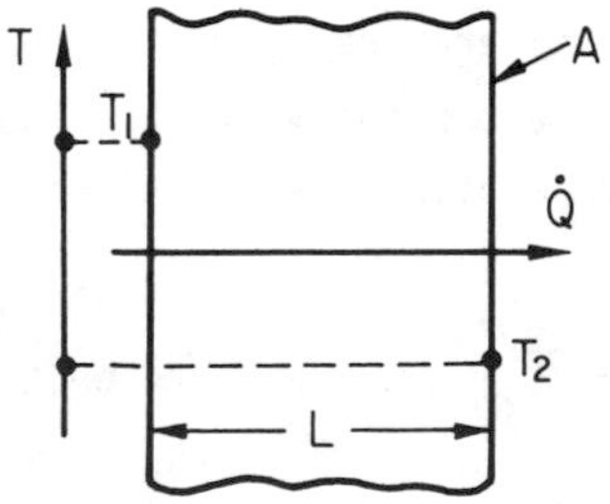

Fig. 1.2 Flat plate of thickness L.

the surfaces ($T_1 > T_2$). Since there is a temperature difference of ($T_1 - T_2$) between the surfaces of the plate, there will be a heat flow through the plate. From the second law of thermodynamics we know that the direction of this flow is from the higher temperature surface to the lower one. According to the first law of thermodynamics under steady conditions this flow of heat will be at a constant rate (see Problem 1.3). Experiments with different solids show that the rate of heat flow $\dot{Q}$ is directly proportional to the temperature difference ($T_1 - T_2$), the surface area A, and inversely proportional to the thickness L; that is,

$$\dot{Q} \sim A \frac{T_1 - T_2}{L} \tag{1.25}$$

When the proportionality constant is inserted, Eq. (1.25) becomes an equation:

$$\dot{Q} = kA \frac{T_1 - T_2}{L} \tag{1.26}$$

The positive proportionality constant k is called the *thermal conductivity* of the material of the plate. Equation (1.26) is known as the *Fourier's Law of Heat Conduction* after the well-known French scientist J. B. J. Fourier. It is to be noted that Eq. (1.26) is actually the definition of thermal conductivity; that is,

$$k \equiv \frac{(\dot{Q}/A)\cdot L}{T_1 - T_2} \tag{1.27}$$

Thermal conductivity is a thermo-physical property and units of this property are W/m•K in the SI system of units, kcal/hr•m•°C in the metric system, and Btu/hr•ft•°F in British thermal units.

A medium is said to be *homogeneous* if its thermal conductivity does not vary from point to point within the medium, and *heterogeneous* if there is such a variation. Further, a medium is said to be *isotropic* if its thermal conductivity is the same in all directions and *anisotropic* if there exists directional variation in thermal conductivity. Materials having porous structure, such as cork and glass wool, are examples of heterogeneous media, and those having fibrous structure, such as wood or asbestos, are examples of anisotropic media.

Consider now the same flat plate and let the temperature of the isothermal surface at location x be T(x) and at location x + Δx be T(x + Δx) as shown in Fig. 1.3. The rate of heat transfer through the layer shown in this figure can be written as

$$\dot{Q} = kA \frac{T(x) - T(x + \Delta x)}{\Delta x} \tag{1.28}$$

Here it should be noted that the material of the plate can be heterogeneous. In that case the layer which has a thickness Δx may be assumed to be locally

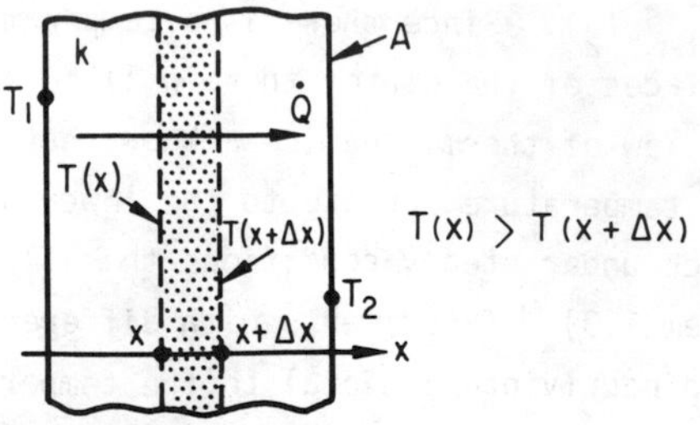

Fig. 1.3 One-dimensional heat conduction.

homogeneous as $\Delta x \to 0$.

If we rewrite Eq. (1.28) as $\Delta x \to 0$, we get

$$\dot{Q} = -kA \lim_{\Delta x \to 0} \frac{T(x + \Delta x) - T(x)}{\Delta x} \tag{1.29}$$

The limit in the above relation is by definition the derivative of temperature with respect to the coordinate axis x. Hence, Eq. (1.29) reduces to

$$\dot{Q} = -kA \frac{dT}{dx} \tag{1.30}$$

which is the Fourier's law of heat conduction stated in differential form for a one-dimensional system.

The quantity of heat transferred per unit time per unit area is called *heat flux*, q. Units of heat flux are W/m^2 in the SI system of units, $kcal/hr{\cdot}m^2$ in the metric system, and $Btu/hr{\cdot}ft^2$ in the British thermal units. Equation (1.30) can now be written in terms of heat flux as

$$q = \frac{\dot{Q}}{A} = -k \frac{dT}{dx} \tag{1.31}$$

Equation (1.31) can be interpreted as stating that if there is a negative temperature gradient dT/dx at location x, then there will be a heat flux in the positive x-direction across the isothermal surface at that location, and the magnitude of this flux is given by Eq. (1.31). Heat flux in the positive x-direction is considered positive; therefore, the minus sign in Eq. (1.31) is necessary in order to meet the requirement of the second law of thermodynamics that heat must flow from a higher to a lower temperature. That is, if the temperature gradient is negative then the heat flux is positive, and if, on the other hand, the gradient happens to be positive then the heat flux is negative (in the negative x-direction) as illustrated in Fig. 1.4.

The foregoing argument can now be extended to a medium with two- or three-dimensional temperature distributions. Figure 1.5 shows a set of isothermal

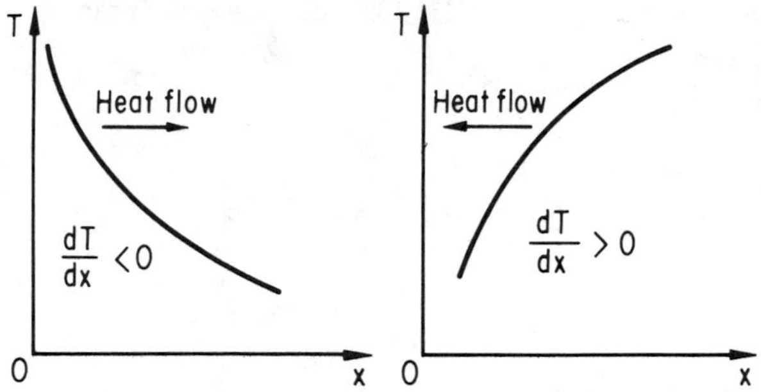

Fig. 1.4 Interpretation of Fourier's law.

surfaces in a solid body each differing in temperature by a small amount ΔT. Following the above argument, the heat flux across the isothermal surface at point P in the solid can be written as:

$$q_n = -k \frac{\partial T}{\partial n} \tag{1.32}$$

where $\partial/\partial n$ represents the differentiation along the normal to the isothermal surface, which is characterized by the unit direction vector $\hat{n}$ pointing in the direction of decreasing temperature. Equation (1.32) can also be written, for example, in the rectangular coordinate system as

$$q_n = -k \left(\frac{\partial T}{\partial x} \frac{\partial x}{\partial n} + \frac{\partial T}{\partial y} \frac{\partial y}{\partial n} + \frac{\partial T}{\partial z} \frac{\partial z}{\partial n} \right) \tag{1.33a}$$

or

$$q_n = -k \left(\frac{\partial T}{\partial x} \cos\alpha + \frac{\partial T}{\partial y} \cos\beta + \frac{\partial T}{\partial z} \cos\gamma \right) \tag{1.33b}$$

where, in fact, (α, β, γ) are the direction cosines of the normal vector $\hat{n}$; that is,

$$\hat{n} = \hat{i} \cos\alpha + \hat{j} \cos\beta + \hat{k} \cos\gamma \tag{1.34}$$

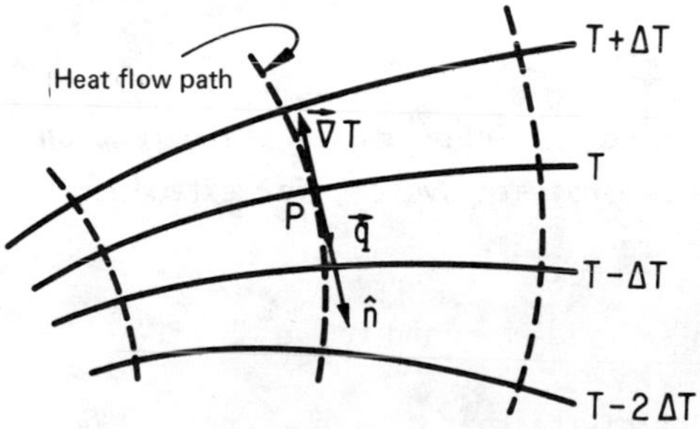

Fig. 1.5 Isothermal surfaces and heat flow paths in a solid body.

Therefore, using vector calculus Eq. (1.33b) can be rewritten as

$$q_n = -k\left(\hat{i}\,\frac{\partial T}{\partial x} + \hat{j}\,\frac{\partial T}{\partial y} + \hat{k}\,\frac{\partial T}{\partial z}\right)\cdot\hat{n} \tag{1.35a}$$

or

$$q_n = -k\,\vec{\nabla}T\cdot\hat{n} \tag{1.35b}$$

where

$$\vec{\nabla}T = \hat{i}\,\frac{\partial T}{\partial x} + \hat{j}\,\frac{\partial T}{\partial y} + \hat{k}\,\frac{\partial T}{\partial z} \tag{1.36}$$

is the *gradient* of the temperature distribution.

If we define a *heat flux vector* as $\vec{q} = \hat{n}q_n$, then it can be shown from Eq. (1.35b) that

$$\vec{q} = -k\vec{\nabla}T \tag{1.37}$$

Since in any coordinate system

$$\frac{\partial T}{\partial n} = \vec{\nabla}T\cdot\hat{n} \tag{1.38}$$

then the above development, *i.e.*, Eq. (1.37), is independent of the coordinate system used.

The magnitude of the heat flux across an arbitrary surface passing through the point P and having the unit direction vector $\hat{s}$ drawn normal to the surface will be equal to the magnitude of the component of $\vec{q}$ in s-direction; that is,

$$q_s = \vec{q}\cdot\hat{s} = q_n\hat{n}\cdot\hat{s} = q_n\cos\theta = -k\,\frac{\partial T}{\partial n}\cos\theta \tag{1.39}$$

where θ is the angle between the unit vectors $\hat{n}$ and $\hat{s}$. Since it is also true that

$$\frac{\partial T}{\partial n}\cos\theta = \frac{\partial T}{\partial s} \tag{1.40}$$

Eq. (1.39) reduces to

$$q_s = -k\,\frac{\partial T}{\partial s} \tag{1.41}$$

where $\partial/\partial s$ represents the differentiation in the direction of the normal $\hat{s}$.

In the rectangular coordinate system, for example, the three components of the heat flux vector $\vec{q}$ are

$$q_x = -k\,\frac{\partial T}{\partial x}, \qquad q_y = -k\,\frac{\partial T}{\partial y} \qquad \text{and} \qquad q_z = -k\,\frac{\partial T}{\partial z} \tag{1.42a,b,c}$$

which are the magnitudes of the heat fluxes at point P across the surfaces perpendicular to the directions x, y and z.

The gradient of the temperature distribution, $\vec{\nabla}T$, at point P is a vector pointing in the direction of increasing temperature normal to the isothermal surface passing through P. The heat flux vector $\vec{q}$ is also normal to the isothermal surface at point P, but points in the direction of decreasing temperature. Using Eq. (1.37) as the starting point of the definition of the Fourier's law of heat conduction, we can say that heat is transferred by conduction in the direction normal to isothermal surfaces from the higher temperature to the lower one. This law is well established for heat conduction in isotropic solids, and practical applications of it for various problems require the laboratory measurement of thermal conductivities of representative specimens. Fourier's law is also taken to be valid for unsteady problems as it has never been refuted.

So far we have considered Fourier's law of heat conduction in isotropic media only. In the case of anisotropic media, the heat flux vector is not necessarily parallel to the temperature gradient $\vec{\nabla}T$ and, therefore, the direction of the heat flux vector is not normal to isothermal surfaces. Fourier's law of heat conduction can be generalized for anisotropic media by assuming each component of the heat flux vector at a point to be linearly dependent on all components of the temperature gradient at that point. Thus, referred to a set of rectangular axes ox_1, ox_2, and ox_3, the components of the heat flux vector can be written as

$$q_1 = -k_{11}\frac{\partial T}{\partial x_1} - k_{12}\frac{\partial T}{\partial x_2} - k_{13}\frac{\partial T}{\partial x_3} \tag{1.43a}$$

$$q_2 = -k_{21}\frac{\partial T}{\partial x_1} - k_{22}\frac{\partial T}{\partial x_2} - k_{23}\frac{\partial T}{\partial x_3} \tag{1.43b}$$

$$q_3 = -k_{31}\frac{\partial T}{\partial x_1} - k_{32}\frac{\partial T}{\partial x_2} - k_{33}\frac{\partial T}{\partial x_3} \tag{1.43c}$$

where k_{ij} are called the *thermal conductivity coefficients*. They are the components of the *thermal conductivity tensor*,

$$[k_{ij}] = \begin{bmatrix} k_{11} & k_{12} & k_{13} \\ k_{21} & k_{22} & k_{23} \\ k_{31} & k_{32} & k_{33} \end{bmatrix} \tag{1.44}$$

Equations (1.43) can be written in a more compact form by using the cartesian tensor notation as

$$q_i = -k_{ij}\frac{\partial T}{\partial x_j}, \qquad i,j = 1,2,3 \tag{1.45}$$

In this book we will not study conduction heat transfer in anisotropic media. Interested readers may refer to references such as [5].

1.9 THERMAL CONDUCTIVITY

Thermal conductivity of a substance is a thermo-physical property of that substance. It can be interpreted from Eq. (1.27) as being equal to the heat transfer rate across a unit area through a unit thickness for a unit temperature difference. The magnitude of the thermal conductivity of materials varies over wide ranges; for example, from 0.0152 W/m·K for carbon dioxide at 300°K to 429 W/m·K for pure silver at 300°K. Thermal conductivity of a material depends on its chemical composition, physical structure, and state of it. It also depends on the temperature and pressure to which the material is subjected. In most cases, however, thermal conductivity is much less dependent on pressure than on temperature, so that the dependence on pressure may be neglected and the thermal conductivity is tabulated as a function of temperature for a given substance. In some cases, thermal conductivity may also vary with the direction of heat flow.

The variation of thermal conductivity with temperature may be neglected when the temperature range under consideration is not too large or the dependence of the thermal conductivity on temperature is not too severe. For numerous materials, especially within a small temperature range, the variation of thermal conductivity with temperature can be represented by the linear function

$$k(T) = k_0[1 + \gamma(T - T_0)] \tag{1.46}$$

where $k_0 = k(T_0)$, T_0 is a reference temperature and γ is a constant called the *temperature coefficient of thermal conductivity*.

Heat conduction in gases and vapors depends mainly on the molecular transfer of the kinetic energy of the molecular movement. That is, heat conduction is the transmission of kinetic energy by the more active molecules in the high temperature regions to the molecules in the low molecular kinetic energy regions by successive collisions. According to the kinetic theory of gases, the temperature of an element of gas is proportional to the mean kinetic energy of its constituent molecules. Clearly, the faster the molecules move, the faster they will transfer energy. This implies that the thermal conductivity of a gas should be dependent on its temperature. For gases at moderately low temperatures the kinetic theory of gases may be used to accurately predict the experimentally observed values. A very simple version of the kinetic theory (traffic-model) leads to the following approximate relation for gases:

$$k = \frac{\rho c_v \overline{V} \lambda}{3} \tag{1.47}$$

where c_v is the specific heat at constant volume, $\overline{V}$ is the mean velocity of the molecules, λ is the molecular mean-free-path between collisions, and ρ is the gas density. A more accurate analysis gives a numerical constant in Eq. (1.47) slightly different from 1/3.

In liquids the molecules are more closely spaced and therefore molecular force fields exert a strong influence on the energy exchange in the collisions. Because of this fact, liquids have much higher values of thermal conductivities than gases. The thermal conductivity of nonmetallic liquids, generally, decreases with increasing temperature, exceptions being water and glycerine. Liquid metals are useful as heat transfer media in nuclear reactors where high heat removal rates are essential.

Solid materials may have solely crystalline structures, such as quartz, may be in amorphous solid state, such as glass, a mixture of the two, or may be somewhat porous in structure with air or other gases in the pores. Heat conduction in solids with crystalline structures depends on the energy transfer by molecular and lattice vibrations, and free electrons. In general, the energy transfer by molecular and lattice vibrations is not as large as by the electron transport. It is for this reason that good electrical conductors are almost good heat conductors while electrical insulators are usually good heat insulators. Materials having high thermal conductivities are called *conductors*, while materials of low thermal conductivity are referred to as *insulators*.

In the case of amorphous solids, heat conduction depends on the molecular energy transport. It is for this reason that thermal conductivities of such solids are of the same order of magnitude as those observed for liquids. That is, amorphous solids have smaller thermal conductivities than solids with crystalline structure.

For pure crystalline metals the ratio of the thermal conductivity, k, to electrical conductivity, k_e, is found to be nearly proportional to the absolute temperature. A modified Lorenz equation expressing this relation is $k/k_e = 783 \times 10^{-9}\ T(°R)$. This equation does not hold for amorphous materials or alloys of metals. Thermal conductivities of alloys may be less than that of any constituent; for example, constantan is an alloy of 55 percent copper (Cu) and 45 percent nickel (Ni) and has k = 23W/m·K, while for pure copper k = 401W/m·K, and for nickel k = 90.7W/m·K. The thermal conductivities of amorphous solids increase with temperature. The thermal conductivity data of selected typical solids are given in Appendix A.

It is to be expected that solid bodies with pores filled with gases will have rather low values of thermal conductivity compared to more dense non-porous materials. For any particular solid the thermal conductivity increases with density. It also increases with moisture content. The moisture in solids presents a special problem in the presence of a temperature gradient. It usually migrates towards the colder regions, thus changing the thermal conductivity and, perhaps, damaging the material or the surrounding structure. For solids with porous structure or loosely packed fibrous materials (cork, glass wool, etc.) we can only talk about an "apparent" thermal conductivity and the apparent thermal conductivities of such materials usually reach minimum values as the "apparent" density is decreased. At very low densities, the gas spaces may be so large that an internal convection flow may result which increases heat transfer rates and, therefore, thermal conductivity values. In cellular or porous type materials internal radiation may also become important. If internal radiation is very significant the curve of k versus T will be concave upward.

1.10 NEWTON'S COOLING LAW

Convection has already been defined as the process of heat transport in a fluid by the combined action of heat conduction (and radiation) and fluid motion. As a mechanism of heat transfer, convection is most important between a solid surface and a fluid.

Consider a solid surface and a fluid flowing over this surface as shown in Fig. 1.6. It is an experimental observation that the fluid particles adjacent to the surface stick to it and have zero velocity relative to the surface. Other fluid particles attempting to slide over the stationary fluid particles on the surface are retarded as a result of viscous forces among the fluid particles. These effects of viscous forces originating at the surface extend into the body of the fluid. The velocity of the fluid particles asymptotically approaches that of the undisturbed free stream in a short distance δ (velocity boundary layer thickness) from the surface. The velocity of the flow will appear as shown in Fig. 1.6a.

If the solid surface is at a temperature T_w, different from T_∞ of the free

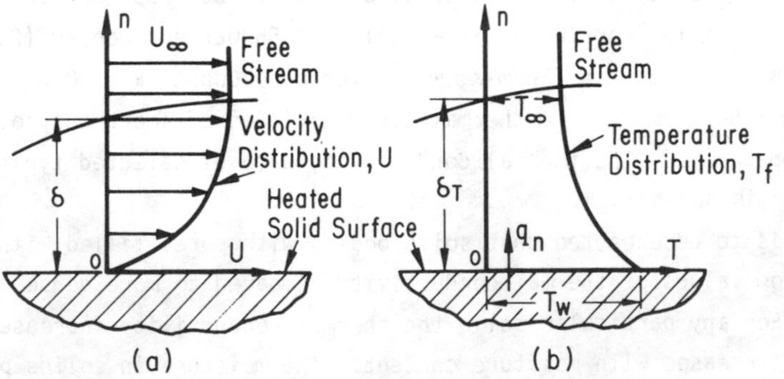

Fig. 1.6 Velocity and thermal boundary layers along a solid surface.

stream, then heat will flow by conduction either from the particles of the solid to the stationary fluid particles on the surface or vice versa. If $T_w > T_\infty$ as shown in Fig. 1.6b, then heat will flow from the solid to fluid particles at the surface. The energy thus transmitted increases the internal energy of the fluid particles and is carried away by the motion of the fluid. The temperature distribution in the fluid adjacent to the surface will appear as shown in Fig. 1.6b. That is, the temperature of the fluid particles will be T_w on the surface of the solid, but approaches the free stream value T_∞ in a short distance δ_T (thermal boundary layer thickness) from the surface. The temperature distribution in the fluid adjacent to the surface will appear as shown in Fig. 1.6b. It is certainly dependent on the velocity distribution shown in Fig. 1.6a.

Since fluid particles are stationary at the surface, the rate of heat transfer from the surface to the fluid per unit area of the surface is given by

$$q_n = -k_f\left(\frac{\partial T_f}{\partial n}\right)_w \tag{1.48}$$

where k_f represents the thermal conductivity of the fluid, T_f is the temperature distribution in the fluid, the subscript w means the derivative is evaluated at the surface, and n denotes the normal direction to the surface.

In 1701, Newton expressed the rate of heat transfer from a solid surface to a fluid per unit area of the surface by the equation

$$q_n = h(T_w - T_\infty) \tag{1.49}$$

where the quantity h is called the *heat transfer coefficient* or *film coefficient*, and T_w and T_∞ are the surface and fluid temperatures, respectively. In literature, Eq. (1.49) is known as *Newton's Law of Cooling*, which states that the rate of heat transfer per unit area from a surface to a fluid flowing along it is equal to the product of the temperature difference between the surface and the stream far from the surface and a quantity h. In fact, Newton's law of cooling, Eq. (1.49), is not a law, but the defining equation for the heat transfer coefficient; that is,

$$h \equiv \frac{q_n}{T_w - T_\infty} = \frac{-k_f(\partial T_f/\partial n)_w}{T_w - T_\infty} \tag{1.50}$$

Units of heat transfer coefficient h are $W/m^2{\cdot}K$ in the SI system of units, $kcal/hr{\cdot}m^2{\cdot}°C$ in the metric system, and $Btu/hr{\cdot}ft^2{\cdot}°F$ in British thermal units.

It should be noted that h is also given by

$$h = \frac{-k_s(\partial T_s/\partial n)_w}{T_w - T_\infty} \tag{1.51}$$

where k_s represents the thermal conductivity of the solid and T_s is the temperature distribution in the solid.

If the fluid motion involved in the process is induced by some external means such as a pump, blower, or fan then we speak of *forced convection*. If the fluid motion is caused by any body force within the system, such as a result of the density gradients near the surface, then the process is called *free convection*.

Heat transfer coefficient is actually a complicated function of the flow conditions, transport and thermo-physical properties (viscosity, thermal conductivity, specific heat, density) of the fluid, and geometry and dimensions of the surface. Its numerical value, in general, is not uniform over the surface. Table 1.1 gives the order of magnitude of the range of values of heat transfer coefficients under several conditions.

1.11 STEFAN-BOLTZMANN'S LAW OF RADIATION

As mentioned in Section 1.2, in contrast to the mechanisms of conduction and convection, where energy transfer through a material medium takes place, experimental observations show that energy may also be transported in the absence of a physical medium. This transfer takes place in the form of *electromagnetic waves*, and is known as *thermal radiation*.

All substances, solid bodies as well as liquids and gases, at normal and especially at elevated temperatures emit energy in the form of electromagnetic waves and are also capable of absorbing such energy. When radiation is incident on the surface of a body, part of it is reflected. The remainder penetrates into the body, which may then be absorbed as it travels through the body as illustrated in Fig. 1.7.

If the material of the body is a strong absorber of thermal radiation, the

Table 1.1 Approximate values of h, $W/m^2 \cdot K$.

Fluid	Free Convection	Forced Convection
Gases	5-30	30-300
Water	30-300	300-10,000
Viscous oils	5-100	30-3,000
Liquid metals	50-500	500-20,000
Boiling water	2,000-20,000	3,000-100,000
Condensing water vapor	3,000-30,000	3,000-200,000

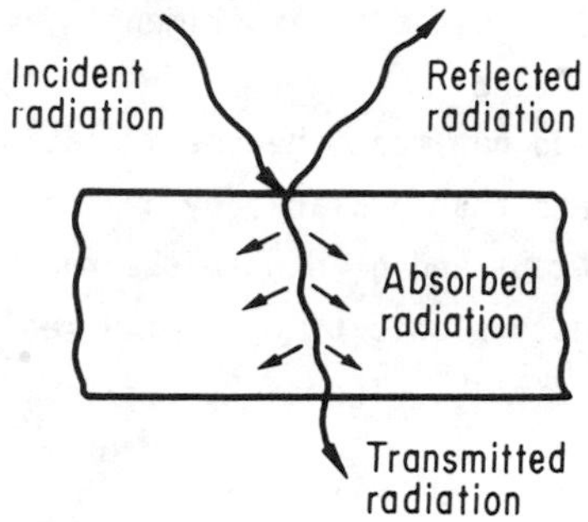

Fig. 1.7 Absorption, reflection and transmission of incident radiation.

radiation that penetrates into the body will be absorbed and converted into internal energy within a very thin layer adjacent to the surface. Such a body is called *opaque*. If the thickness of the material of the body required to substantially absorb radiation is large compared to the thickness dimension of the body, or if the material is *transparent*, then most of the radiation will be transmitted entirely through the body.

When radiation impinges on a surface, we define the fraction reflected as *reflectivity* ρ, the fraction absorbed as *absorptivity* α, and the fraction transmitted as *transmissivity* τ. Thus,

$$\rho + \alpha + \tau = 1 \tag{1.52}$$

For opaque materials $\tau = 0$. Therefore, Eq. (1.52) reduces to

$$\rho + \alpha = 1 \tag{1.53}$$

An ideal body which absorbs all the impinging radiation energy without reflection and transmission is called a *blackbody*. Therefore, for a blackbody Eq. (1.52) reduces to $\alpha = 1$. Only a few surfaces such as carbon black and platinum black approach the blackbody in their ability to absorb radiation energy.

A blackbody also emits the maximum possible amount of radiation [10,11]. The total emission of radiation per unit surface area per unit time from a blackbody is related to the fourth power of the absolute temperature according to the *Stefan-Boltzmann's Law of Radiation* as

$$q_b^r = \sigma T^4 \tag{1.54}$$

where T is the absolute temperature of the surface of the body, and σ is the *Stefan-Boltzmann constant* with the value 5.6697×10^{-8} W/m^2·K^4 in the SI system of units, and 0.1714×10^{-8} Btu/hr·ft^2·°R^4 in British thermal units.

The basic equation (1.54) for the total thermal radiation from a blackbody was proposed by Stefan in 1879, based on experimental evidence, and derived theoretically by Boltzmann in 1884.

Real bodies (surfaces) do not meet the specifications of a blackbody, but emit radiation at a lower rate than a blackbody of the same size and shape, and at the same temperature. Let q^r and q_b^r denote the radiation flux (*i.e.*, radiation emitted per unit surface area per unit time) from a real and a black surface of the same temperature, respectively. Then, the *emissivity* ε of the real surface is defined as

$$\varepsilon = \frac{q^r}{q_b^r} \tag{1.55}$$

Thus, for a black surface $\varepsilon = 1$. For a real body exchanging radiation energy only with other bodies at the same temperature, *i.e.*, for thermal equilibrium, it can be shown that $\alpha = \varepsilon$, which is a statement of one of the Kirchhoff's laws in thermal radiation [11]. The value of ε depends upon the material, its state, and the condition of the surface. Emissivities of various materials are given in Appendix A.

Consider now two isothermal surfaces A_1 and A_2 having emissivities ε_1 and ε_2, and the absolute temperatures T_1 and T_2, respectively. Let us assume that these two surfaces exchange heat with each other only. For example, A_1 and A_2 may be two large parallel surfaces having negligible losses from the ends as shown in Fig. 1.8a, or A_1 may be a surface completely enclosed by the surface A_2 as shown in Fig. 1.8b.

Assuming that the Kirchhoff's law is valid, it can be shown that the net radiation heat exchange between the surfaces A_1 and A_2 is given by

$$\dot{Q}_{12} = \sigma A_1 \mathcal{F}_{12}(T_1^4 - T_2^4) \tag{1.56}$$

where the factor $\mathcal{F}_{12}$ is defined as

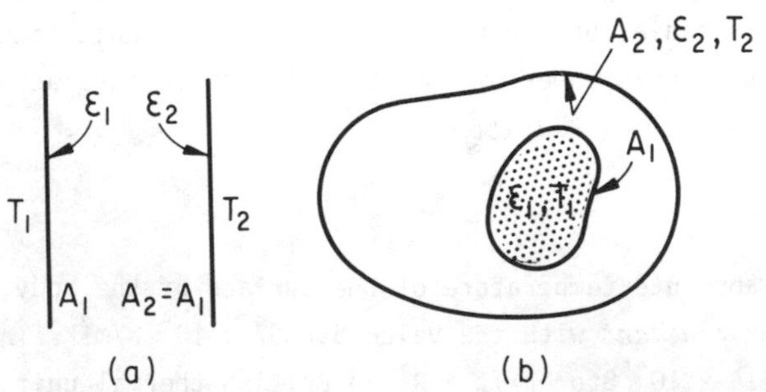

Fig. 1.8 Two isothermal surfaces A_1 and A_2 exchanging radiation energy.

$$\frac{1}{F_{12}} = \frac{1}{\varepsilon_1} + \frac{1}{\varepsilon_2} - 1 \tag{1.57}$$

for the parallel surfaces of Fig. 1.8a and

$$\frac{1}{\mathcal{F}_{12}} = \frac{1}{F_{12}} + \frac{1}{\varepsilon_1} - 1 + \frac{A_1}{A_2}\left(\frac{1}{\varepsilon_2} - 1\right) \tag{1.58}$$

for the surfaces of Fig. 1.8b. Here F_{12} is a purely geometrical factor called *radiation shape factor* or *configuration factor* between the surfaces A_1 and A_2, which is equal to the fraction of radiation leaving surface A_1 that reaches surface A_2. Radiation shape factors, in the form of equations and charts, are given for several configurations in literature [10,11]. For surface A_1 it is clear that

$$F_{11} + F_{12} = 1 \tag{1.59a}$$

and for surface A_2

$$F_{21} + F_{22} = 1 \tag{1.59b}$$

Obviously, if the surface A_1 is a completely convex surface then $F_{11} = 0$, and Eq. (1.58) reduces to

$$\frac{1}{\mathcal{F}_{12}} = \frac{1}{\varepsilon_1} + \frac{A_1}{A_2}\left(\frac{1}{\varepsilon_2} - 1\right) \tag{1.60}$$

In an enclosure with N number of surfaces, for the surface A_i, it is clear that

$$\sum_{j=1}^{N} F_{ij} = 1 \tag{1.61}$$

For flat or convex surfaces in the enclosure which cannot see themselves $F_{ij} = 0$.

For configurations including more than two surfaces, the evaluation of heat transfer by radiation becomes involved and interested readers may refer to books on thermal radiation such as References [10,11] for more information.

REFERENCES

1. Shapiro, A. H., *The Dynamics and Thermodynamics of Compressible Fluid Flow*, Vol. 1, The Ronald Press Co., New York, 1953.
2. Keenan, J. H., *Thermodynamics*, John Wiley and Sons, Inc., New York, 1941.
3. Van Wylen, G. J., and Sonntag, R. E., *Fundamentals of Classical Thermodynamics*, 2nd ed., Revised Printing, SI Version, John Wiley and Sons, Inc., New York, 1978.
4. Arpaci, V. S., *Conduction Heat Transfer*, Addison-Wesley Publishing Co., Inc.,

Reading, Massachusetts, 1966.

5. Özışık, M. N., *Heat Conduction*, John Wiley and Sons, Inc., New York, 1980.
6. Özışık, M. N., *Basic Heat Transfer*, McGraw-Hill Book Co., New York, 1977.
7. Holman, J. P., *Heat Transfer*, 5th ed., McGraw-Hill Book Co., New York, 1981.
8. Chapman, J. A., *Heat Transfer*, 3rd ed., MacMillan Publishing Co., Inc., New York, 1974.
9. Kakaç, S., *Isi Transferine Giriş I: Isi Iletimi* (in Turkish), Middle East Technical University Publications, No. 52, Ankara, 1976.
10. Özışık, M. N., *Radiative Transfer and Interactions with Conduction and Convection*, John Wiley and Sons, Inc., New York, 1973.
11. Siegel, R., and Howell, J. R., *Thermal Radiation Heat Transfer*, 2nd ed., Hemisphere Publishing Co., Washington, D. C., 1980.
12. Luikov, A. V., *Analytical Heat Diffusion Theory*, Academic Press, New York, 1968.
13. Kennard, E. H., *Kinetic Theory of Gases*, McGraw-Hill Book Co., New York, 1938.

PROBLEMS

1.1 In most of the commonly encountered engineering problems involving steady flows through devices, the inlet and outlet flows to and from the device are usually regarded as one dimensional. Develop the first law of thermodynamics (the steady-state steady-flow energy equation) for such a device as illustrated in Fig. 1.9.

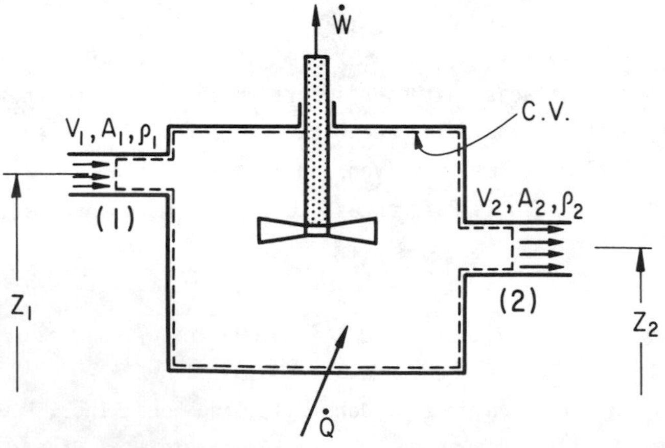

Fig. 1.9 Figure for Problem 1.1.

1.2 Write the first law of thermodynamics, Eq. (1.19), for a volume element in a stationary solid.

1.3 Prove that under steady-state conditons the rate of heat transfer by conduction through a flat plate whose surfaces are maintained at constant and uniform temperatures T_1 and T_2 is constant.

1.4 Find an expression for the steady temperature distribution in the flat plate of Problem 1.3 if the thickness of the plate is L and its thermal conductivity k is constant. Take the surface at temperature T_1 as the origin of the x-axis.

1.5 The plane wall shown in Fig. 1.10 has one surface maintained at T_1 and the other at T_2. The temperature at the center plane of the wall is measured to be T_3, and the rate of heat flow through the wall is $\dot{Q}$. Assuming that the thermal conductivity of the wall varies linearly with temperature find an expression for the thermal conductivity of the wall as a function of temperature and the rate of heat flow through the wall.

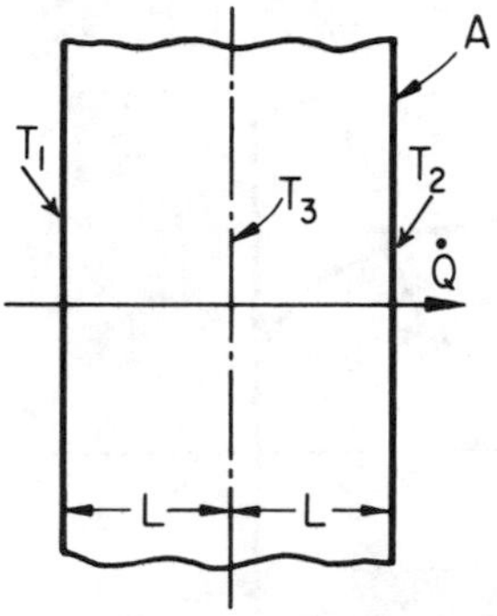

Fig. 1.10 Figure for Problem 1.5.

1.6 The temperature profile at a section in water flowing over a flat surface is experimentally measured to be

$$T(°C) = 20 + 80\ e^{-800y}$$

where y is the distance in m measured normal to the surface with y = 0 corresponding to the surface. What is the value of heat transfer coefficient at this location? Assume that the thermal conductivity of water is k = 0.62 W/m·K.

1.7 The steady-state temperature distribution in a flat plate is given by

$$T(°C) = 150 - 400\ x^2$$

where x is the distance measured in m along the width of the plate from the surface at 150°C. Determine the heat fluxes at the two surfaces of the plate. The thermal conductivity of the wall material is 40 W/m·K and the thickness of the wall is 0.25m.

1.8 Estimate the equilibrium temperature of a long rotating cylinder of diameter D, oriented in space with its axis normal to the sun rays. The cylinder is at a point in space where the irradiation of the sun (*i.e.*, energy incident on a surface perpendicular to the sun rays per unit time per unit area) is 1500 W/m^2. Assume that the absorptivity of the surface of the cylinder to solar radiation, α_s, is equal to its emissivity, ε, and the outer space is a blackbody at 0 K.

1.9 The temperature distribution in a plane wall during a transient heat transfer process is shown in Fig. 1.11 at a specific time t_1. Is the wall being heated or cooled? Explain.

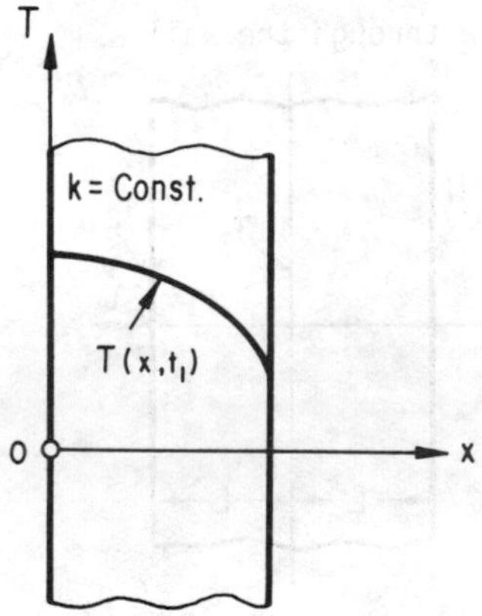

Fig. 1.11 Figure for Problem 1.9.

2

General Heat Conduction Equation

2.1 INTRODUCTION

The main objective of the study of heat conduction is to determine the temperature distribution and heat flow rates in a medium in the presence of a nonuniform temperature distribution or to determine the rate of heat flow between two media in contact due to a temperature difference existing between them. As we have discussed in Chapter 1, the mechanism of heat conduction in the molecular level is visualized as the exchange of kinetic energy between the microparticles in the high and low temperature regions. In the phenomenological heat conduction theory, however, the molecular structure of a medium is disregarded and the medium is taken to be a continuum. Analytical investigations into heat conduction based on the concept of continuum start with the derivation of the heat conduction equation. The heat conduction equation is a mathematical relation, expressed by a differential equation, between temperature, and time and space coordinates.

2.2 DERIVATION OF GENERAL HEAT CONDUCTION EQUATION

To derive the general heat conduction equation we consider a stationary isotropic and opaque solid as shown in Fig. 2.1a. Let $T(\vec{r},t)$ represent the tempera-

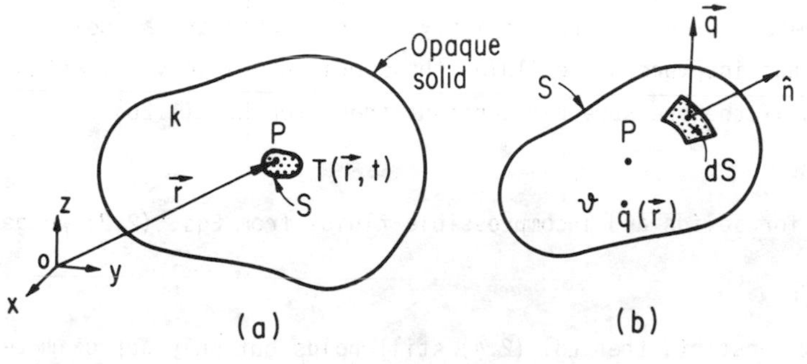

Fig. 2.1 An opaque solid and the control volume.

ture at a point P at time t, and let k be the thermal conductivity which may be a function of space coordinates and/or of temperature T. The density and specific heat of the solid are assumed to be constant, independent of temperature. Suppose that the point P is enclosed by any surface S lying entirely within the solid. Let $\hat{n}$ be the outward-drawn unit vector normal to the closed surface S, and υ be the volume of the space enclosed by S as illustrated in Fig. 2.1b. Let us further assume that internal energy is generated in the solid due to power drawn from an external electric circuit at a rate $\dot{q}_e = \dot{q}_e(\vec{r},t)$ per unit volume. Since

$$\vec{V} = 0, \ \dot{W}_{shaft} = 0, \text{ and } \dot{W}_{shear} = 0$$

then the first law of thermodynamics given by Eq. (1.19) reduces to

$$\frac{\partial}{\partial t}\int_{\upsilon} e\rho d\upsilon = \dot{Q}_s + \int_{\upsilon} \dot{q}_e d\upsilon \tag{2.1a}$$

where e is the energy per unit mass of the solid. Equation (2.1a) can also be written as

$$\int_{\upsilon} \rho \frac{\partial u}{\partial t} d\upsilon = \dot{Q}_s + \int_{\upsilon} \dot{q}_e d\upsilon \tag{2.1b}$$

where u is the internal energy per unit mass of the solid.

In general, for a substance which is homogeneous and invariable in composition

$$du = \left(\frac{\partial u}{\partial v}\right)_T dv + c_v dT \tag{2.2a}$$

and

$$dh = \left(\frac{\partial h}{\partial p}\right)_T dp + c_p dT \tag{2.2b}$$

where $v(=1/\rho)$ denotes specific volume, and c_v and c_p are specific heats at constant volume and pressure, respectively. Other terms are as defined in Chapter 1. For solids and incompressible fluids the specific volume v = constant. On the other hand, if the pressure p = constant then from Eq. (1.20) we have

$$dh = du \tag{2.3}$$

Therefore, for solids and incompressible fluids from Eqs. (2.2) we get

$$c_v = c_p = c \tag{2.4}$$

If p is not constant, then Eq. (2.4) still holds but only approximately since the difference $c_p - c_v$ for solids and incompressible fluids is negligibly small.

Thus, introducing $du = cdT$ in Eq. (2.1b) we get

$$\int_V \rho c \frac{\partial T}{\partial t}\, dv = \dot{Q}_s + \int_V \dot{q}_e dv \tag{2.5}$$

In a fissionable material internal energy is generated as a result of fission reactions which consist of continuous changes in the composition of the material as fissionable material is turned into internal energy. Since these composition changes are generally small, the thermo-physical properties of such a material may be assumed constant. Although this internal energy generation cannot be identified as a power input to the system from an external power source, the right-hand side of Eq. (2.5) can be modified as

$$\int_V \rho c \frac{\partial T}{\partial t}\, dv = \dot{Q}_s + \int_V \dot{q}_e dv + \int_V \dot{q}_n dv \tag{2.6}$$

where $\dot{q}_n = \dot{q}_n(\vec{r},t)$ represents the rate of internal energy generation per unit volume due to nuclear reactions.

A similar argument can also be given for the case of internal energy sources or sinks resulting from exothermic and endothermic chemical reactions. Hence, in general, Eq. (2.6) can be written as

$$\int_V \rho c \frac{\partial T}{\partial t}\, dv = \dot{Q}_s + \int_V \dot{q} dv \tag{2.7}$$

where $\dot{q} = \dot{q}(\vec{r},t)$ represents the rate of internal energy generation in the solid per unit volume and this generation may be due to electrical, nuclear, chemical or even infrared sources, etc.

The term $\dot{Q}_s$ in Eq. (2.7) represents the net rate of heat entering the volume across its bounding surface S. Integrating the rate of heat transfer $\vec{q}\cdot\hat{n}\, dS$ from a surface element dS over the entire control surface S we get

$$\dot{Q}_s = -\int_S \vec{q}\cdot\hat{n}\, dS \tag{2.8}$$

where the negative sign is in accordance with the sign convention for heat flow. Hence, the substitution of Eq. (2.8) into Eq. (2.7) yields

$$\int_V \rho c \frac{\partial T}{\partial t}\, dv = -\int_S \vec{q}\cdot\hat{n}\, dS + \int_V \dot{q}\, dv \tag{2.9}$$

The surface integral in the above equation can be converted into a volume integral by using the *divergence* theorem as

$$\int_S \vec{q}\cdot\hat{n}\, dS = \int_v \vec{\nabla}\cdot\vec{q}\, dv \tag{2.10}$$

Substituting Eq. (2.10) into Eq. (2.9) we get

$$\int_v \rho c \frac{\partial T}{\partial t}\, dv = -\int_v \vec{\nabla}\cdot\vec{q}\, dv + \int_v \dot{q}\, dv \tag{2.11a}$$

or

$$\int_v \left[\rho c \frac{\partial T}{\partial t} + \vec{\nabla}\,\vec{q} - \dot{q}\right] dv = 0 \tag{2.11b}$$

Since the integral in the above relation vanishes for every volume element v, its integrand must vanish everywhere, thus yielding

$$-\vec{\nabla}\cdot\vec{q} + \dot{q} = \rho c \frac{\partial T}{\partial t} \tag{2.12}$$

When we introduce Fourier's law of heat conduction into the above relation we get the differential equation for the temperature distribution. For an isotropic medium the Fourier's law is

$$\vec{q} = -k\vec{\nabla}T \tag{2.13}$$

Therefore, Eq. (2.12) becomes

$$\vec{\nabla}\,(k\vec{\nabla}T) + \dot{q} = \rho c \frac{\partial T}{\partial t} \tag{2.14}$$

This is the *general heat conduction equation* for heterogeneous isotropic solids. Equation (2.14) may be rearranged to give

$$k\nabla^2 T + \vec{\nabla}k\cdot\vec{\nabla}T + \dot{q} = \rho c \frac{\partial T}{\partial t} \tag{2.15}$$

where ∇^2 is the well-known *Laplacian operator.*

If k is a function of space coordinates only, then Eq. (2.15) is a *linear* partial differential equation. On the other hand, when k depends on temperature Eq. (2.15) becomes a *non-linear* partial differential equation.

For homogeneous isotropic substances k is constant and the general heat conduction equation reduces to

$$\nabla^2 T + \frac{\dot{q}}{k} = \frac{1}{\alpha}\frac{\partial T}{\partial t} \tag{2.16}$$

where $\alpha = k/\rho c$ is called the *thermal diffusivity* of the medium. Equation (2.16) is also called the *Fourier-Biot Equation.*

Thermal diffusivity of a medium is a thermo-physical property. Its units are m^2/s in SI system of units, m^2/hr in metric system and ft^2/hr in British thermal units. A high value of thermal diffusivity can result either from a high value of thermal conductivity k, which indicates a rapid rate of energy transfer, or from a low value of thermal capacity ρc, which means that less thermal energy moving through the medium will be absorbed and used to raise the temperature, thus more energy would be available for further transfers. Therefore, the larger the value of α, the faster will the heat diffuse through the medium.

In the absence of internal energy sources and power drawn to the system from an external electric circuit (both will also be named as internal heat sources) the heat conduction equation (2.16) takes the form

$$\nabla^2 T = \frac{1}{\alpha}\frac{\partial T}{\partial t} \tag{2.17}$$

which is called *heat diffusion equation*.

For steady-state conditions, in the presence of internal heat sources, we get

$$\nabla^2 T + \frac{\dot{q}}{k} = 0 \tag{2.18}$$

which is called *Poisson equation*.

Under steady-state conditions, for a stationary medium without internal heat sources, the heat conduction equation (2.16) becomes

$$\nabla^2 T = 0 \tag{2.19}$$

which is the *Laplace equation*.

From the Laplace equation (2.19) we conclude that under steady-state conditions the temperature distribution in a stationary medium with constant thermal conductivity and without internal heat sources does not depend on the thermophysical properties of the medium, but is determined only by the shape of the body and the temperature distribution along its boundaries.

A summary of the special cases of Eq. (2.16) along with the conditions that apply to each is presented in Table 2.1.

In the above relations $\nabla^2 T$ represents the Laplacian of the temperature distribution. In rectangular coordinates it is given by

$$\nabla^2 T \equiv \frac{\partial^2 T}{\partial x^2} + \frac{\partial^2 T}{\partial y^2} + \frac{\partial^2 T}{\partial z^2} \tag{2.20}$$

where $T = T(x, y, z, t)$.

In cylindrical coordinates for $\nabla^2 T$ we have

$$\nabla^2 T \equiv \frac{\partial^2 T}{\partial r^2} + \frac{1}{r}\frac{\partial T}{\partial r} + \frac{1}{r^2}\frac{\partial^2 T}{\partial \phi^2} + \frac{\partial^2 T}{\partial z^2} \tag{2.21}$$

where $T = T(r, \phi, z, t)$. The relations between the rectangular and cylindrical coordinates of a point in space, as shown in Fig. 2.2, are given by

$$x = r\cos\phi\ , \qquad y = r\sin\phi\ , \qquad z = z \tag{2.22a,b,c}$$

Table 2.1 Special cases of the general heat conduction equation.

Equation Name	Conditions	Equation
Fourier-Biot	Constant thermo-physical properties	$\nabla^2 T + \frac{\dot{q}}{k} = \frac{1}{\alpha}\frac{\partial T}{\partial t}$
Heat diffusion	Constant thermo-physical properties, no internal heat sources	$\nabla^2 T = \frac{1}{\alpha}\frac{\partial T}{\partial t}$
Poisson	Steady-state, k = constant	$\nabla^2 T + \frac{\dot{q}}{k} = 0$
Laplace	Steady-state, no internal heat sources, k = constant	$\nabla^2 T = 0$

In spherical coordinates $\nabla^2 T$ is given by

$$\nabla^2 T \equiv \frac{1}{r^2}\frac{\partial}{\partial r}\left(r^2 \frac{\partial T}{\partial r}\right) + \frac{1}{r^2 \sin\theta}\frac{\partial}{\partial \theta}\left(\sin\theta \frac{\partial T}{\partial \theta}\right) + \frac{1}{r^2 \sin^2\theta}\frac{\partial^2 T}{\partial \phi^2} \tag{2.23}$$

where the spherical coordinates (r, ϕ, θ) are indicated in Fig. 2.3, and the relations between the rectangular and spherical coordinates of a point in space are given by

$$x = r \sin\theta \cos\phi \, , \qquad y = r \sin\theta \sin\phi \, , \qquad z = r \cos\theta \qquad (2.24a,b,c)$$

A summary of the forms of the Laplacian of the temperature distribution in various coordinate systems is given in Table 2.2.

2.3 INITIAL AND BOUNDARY CONDITIONS

In general, regardless of the mathematical method employed, a solution to the general heat conduction equation will contain seven constants of integration

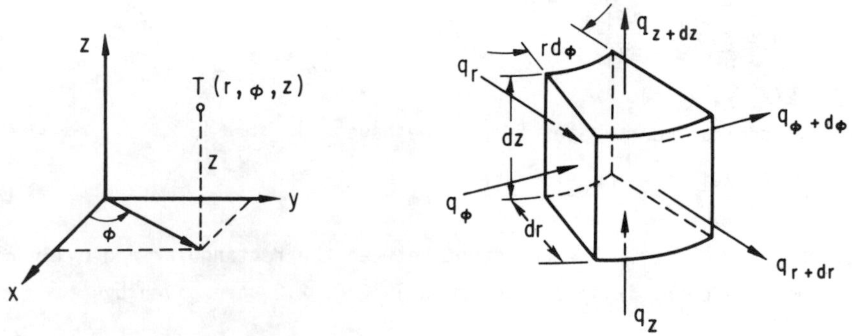

Fig. 2.2 Cylindrical coordinates and the differential volume.

Table 2.2 Laplacian of temperature in various coordinate systems.

Coordinate System	$\nabla^2 T$
Rectangular	$\nabla^2 T = \frac{\partial^2 T}{\partial x^2} + \frac{\partial^2 T}{\partial y^2} + \frac{\partial^2 T}{\partial z^2}$
Cylindrical	$\nabla^2 T = \frac{\partial^2 T}{\partial r^2} + \frac{1}{r}\frac{\partial T}{\partial r} + \frac{1}{r^2}\frac{\partial^2 T}{\partial \phi^2} + \frac{\partial^2 T}{\partial z^2}$
Spherical	$\nabla^2 T = \frac{1}{r^2}\frac{\partial}{\partial r}(r^2\frac{\partial T}{\partial r}) + \frac{1}{r^2 \sin\theta}\frac{\partial}{\partial \theta}(\sin\theta\frac{\partial T}{\partial \theta}) + \frac{1}{r^2\sin^2\theta}\frac{\partial^2 T}{\partial \phi^2}$ or $\nabla^2 T = \frac{1}{r^2}\frac{\partial}{\partial r}(r^2\frac{\partial T}{\partial r}) + \frac{1}{r^2}\frac{\partial}{\partial \mu}\left[(1-\mu^2)\frac{\partial T}{\partial \mu}\right] + \frac{1}{r^2(1-\mu^2)}\frac{\partial^2 T}{\partial \phi^2}$ where $\mu = \cos\theta$

because of the first order derivative with respect to the time variable and second order derivatives with respect to each space variable. To complete the formulation of a problem we need to have conditions to determine the constants of integration. These conditions will be the initial and boundary conditions. That is to say, there is an infinite number of solutions to the differential equation (2.14) or (2.16), but there is only one solution for the prescribed initial and boundary conditions. It can be shown that initial and boundary conditions together with the heat conduction equation uniquely define a problem (see Problem 2.9). The number of initial (for time-dependent problems) and boundary conditions in the direction of each independent variable of a problem is equal to the order of the highest derivative of the governing differential equation in the same

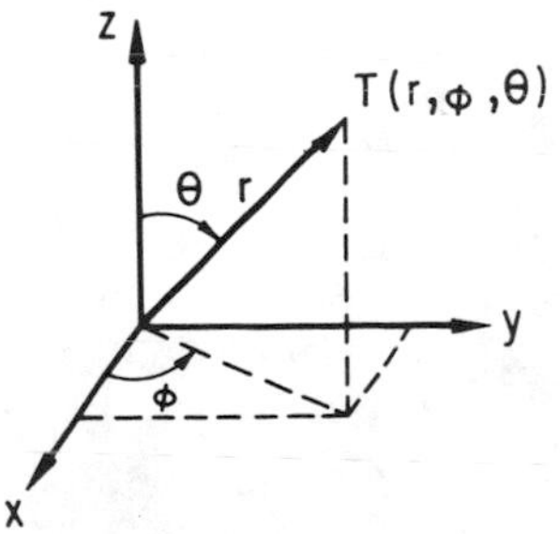

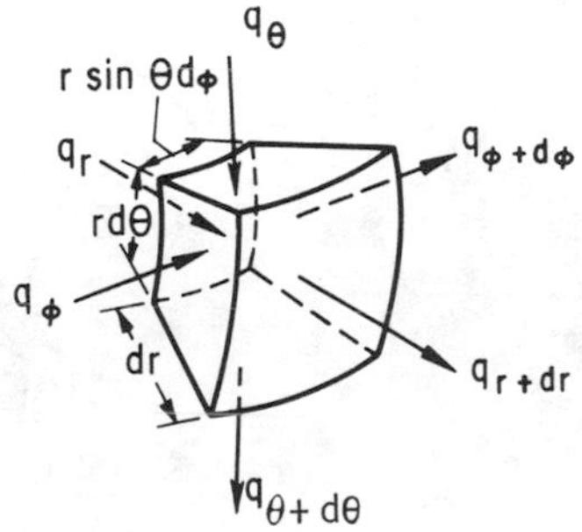

Fig. 2.3 Spherical coordinates and the differential volume.

direction. That is, we need to specify one initial condition (for time-dependent problems) and two boundary conditions in each coordinate direction.

2.3.1 Initial Condition

Initial condition for a time-dependent problem is the temperature distribution in the medium under consideration at some instant of time, usually at the beginning of the heating or cooling process; that is, at t = 0. Mathematically speaking, if the initial condition is given by $T_0(\vec{r})$ then the solution of the problem must be such that at all points of the medium

$$T(\vec{r},t)\Big|_{t\to 0} \to T_0(\vec{r}) \tag{2.25}$$

where $\vec{r}$ is the position vector.

2.3.2 Boundary Conditions

Boundary conditions specify the temperature or the heat flow at the boundary surfaces of a region. For convenience in the analysis we separate the boundary conditions in heat conduction problems into the following groups:

(a) *Prescribed Boundary Temperature*: The distribution or the value of temperature may be prescribed at a boundary surface as shown in Fig. 2.4. This temperature may, in general, be a function of space and time variables; that is,

$$T(\vec{r},t)\Big|_{\vec{r}=\vec{r}_s} = T_s(\vec{r}_s,t) \tag{2.26}$$

A boundary condition of this form is also called the *boundary condition of the first kind*. The temperature distribution at the boundary surface S, in special cases, can be a function only of position or time or it can be a constant. If the temperature on the boundary surface vanishes; that is,

$$T(\vec{r},t)\Big|_{\vec{r}=\vec{r}_s} = 0 \tag{2.27}$$

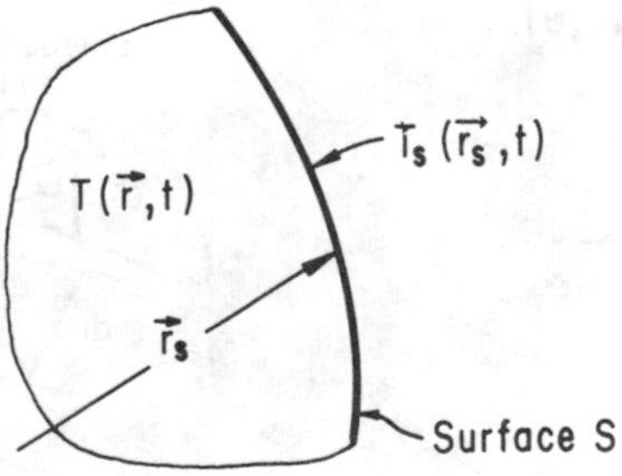

Fig. 2.4 Temperature boundary condition.

then the boundary condition is called the *homogeneous boundary condition of the first kind.*

(b) *Prescribed Heat Flux:* The distribution or the value of the heat flux across a boundary may be specified to be constant or a function of space and/or time. Consider a boundary surface at $\vec{r} = \vec{r}_s$ and let $\hat{n}$ be the outward-drawn normal of this surface as shown in Fig. 2.5a. If the heat flux q_w leaving the surface is specified, then the boundary condition can be written as

$$-k\left(\frac{\partial T}{\partial n}\right)_s = q_w \tag{2.28a}$$

which is a statement of energy balance on the surface.

When the heat flux q_w coming to the surface is prescribed as shown in Fig. 2.5b, the boundary condition becomes

$$k\left(\frac{\partial T}{\partial n}\right)_s = q_w \tag{2.28b}$$

When the heat flux is specified either from the surface or to the surface, it mathematically means that we are specifying the normal derivative of temperature at the boundary. A boundary condition of this form is also called the *boundary condition of the second kind.*

If the derivative of temperature normal to the boundary surface is zero; that is,

$$\left(\frac{\partial T}{\partial n}\right)_s = 0 \tag{2.29}$$

then the boundary condition is called the *homogeneous boundary condition of the second kind*, and such a boundary condition indicates either a *thermally insulated boundary* (*i.e.*, no heat transfer) or a *symmetry* condition at the boundary.

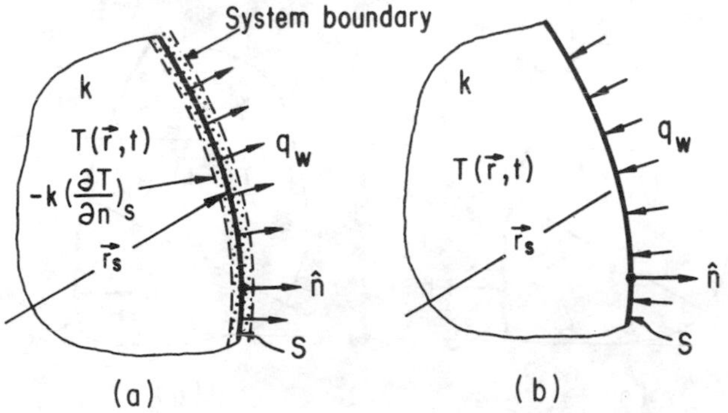

Fig. 2.5 Heat flux boundary condition.

(c) *Heat Transfer by Convection*: When the boundary surface is subjected to convective heat transfer into an ambient fluid at a prescribed temperature T_∞ as shown in Fig. 2.6a, the boundary condition at the boundary surface can be expressed using Newton's law of cooling as

$$-k\left(\frac{\partial T}{\partial n}\right)_s = h[T(\vec{r}_s,t) - T_\infty] \tag{2.30a}$$

or

$$\left[k\frac{\partial T}{\partial n} + h\,T(\vec{r},t)\right]_s = h\,T_\infty \tag{2.30b}$$

where h is the heat transfer coefficient on the surface.

If $\hat{n}$ represents the inward-drawn normal of the boundary surface as shown in Fig. 2.6b the boundary condition can be written as

$$k\left(\frac{\partial T}{\partial n}\right)_s = h[T(\vec{r}_s,t) - T_\infty] \tag{2.31a}$$

or

$$\left[-k\frac{\partial T}{\partial n} + h\,T(\vec{r},t)\right]_s = h\,T_\infty \tag{2.31b}$$

A boundary condition of the form of either (2.30b) or (2.31b) is also called the *boundary condition of the third kind.*

The ambient fluid temperature, T_∞, may be a constant or a function of space and/or time. If the ambient fluid temperature $T_\infty = 0$; that is,

$$\left[\pm k\frac{\partial T}{\partial n} + hT\right]_s = 0 \tag{2.32}$$

where the plus and minus signs correspond to the differentiations along outward and inward normals respectively, then the boundary condition is called the

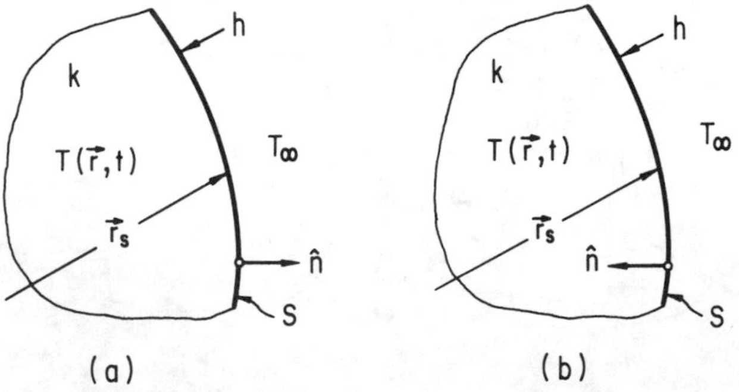

Fig. 2.6 Convection boundary condition.

homogeneous boundary condition of the third kind.

When $h \to \infty$ the boundary condition of the third kind reduces to the boundary condition of the first kind.

As an example, consider the cooling of an electrically heated tube as shown in Fig. 2.7. Boundary conditions can be written as

$$k\left(\frac{\partial T}{\partial r}\right)_{r_1} = h_1[T - T_{f_1}]_{r_1} \tag{2.33a}$$

and

$$-k\left(\frac{\partial T}{\partial r}\right)_{r_2} = h_2[T - T_{f_2}]_{r_2} \tag{2.33b}$$

(d) *Heat Transfer by Radiation*: When the boundary surface is subjected to radiative heat transfer into an environment at an "effective" blackbody temperature T_e as shown in Fig. 2.8, the boundary condition from Eq. (1.56) can be written as

$$-k\left(\frac{\partial T}{\partial n}\right)_s = \sigma\varepsilon[T^4(\vec{r}_s,t) - T_e^4] \tag{2.34}$$

where ε is the emissivity of the surface and σ is the Stefan-Boltzmann constant. Since it involves the fourth power of the surface temperature (dependent variable) it is a non-linear boundary condition.

(e) *Interface Condition*: At an interface between two materials such as shown in Fig. 2.9, the rate of heat flow must be continuous since energy cannot be destroyed or generated there; that is,

$$-k_1\left(\frac{\partial T_1}{\partial n}\right)_s = -k_2\left(\frac{\partial T_2}{\partial n}\right)_s \tag{2.35}$$

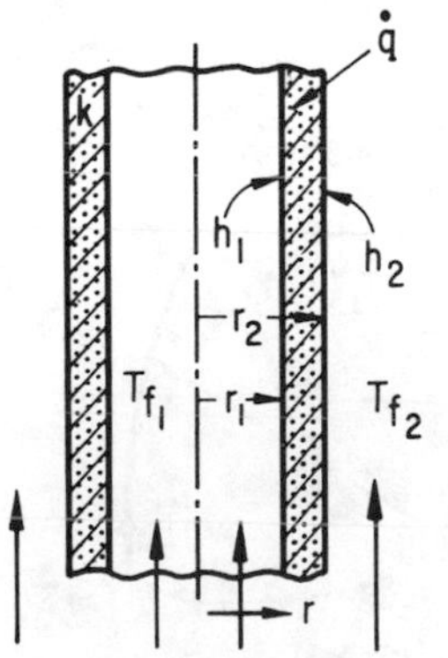

Fig. 2.7 An electrically heated tube.

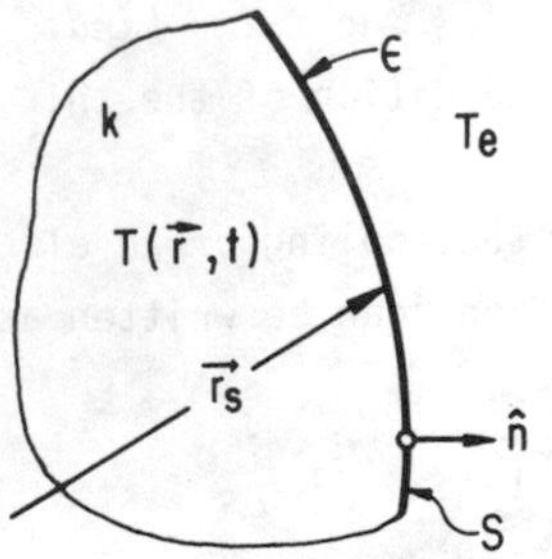

Fig. 2.8 Radiation boundary condition.

If the materials are in perfect thermal contact, the temperature of the surfaces at the interface must be equal to each other; that is,

$$[T_1(\vec{r},t)]_s = [T_2(\vec{r},t)]_s \tag{2.36}$$

If the thermal contact is not perfect, some form of thermal contact resistance or heat transfer coefficient must be introduced at the interface. We shall defer the discussion of thermal contact resistance until Section 3.2.9.

Other types of boundary conditions such as change of phase (moving interface of two media), the interface of two media in relative motion, free convection boundary condition with heat transfer proportional to the 5/4 power of temperature difference, etc., can be written following the same procedure we used in formulating the above boundary conditions.

2.4 TEMPERATURE-DEPENDENT THERMAL CONDUCTIVITY AND KIRCHHOFF TRANSFORMATION

Consider the general heat conduction equation for heterogeneous solids given in the form

$$\vec{\nabla}\cdot[k(T)\vec{\nabla}T] + \dot{q}(\vec{r},t) = \rho c \frac{\partial T}{\partial t} \tag{2.37}$$

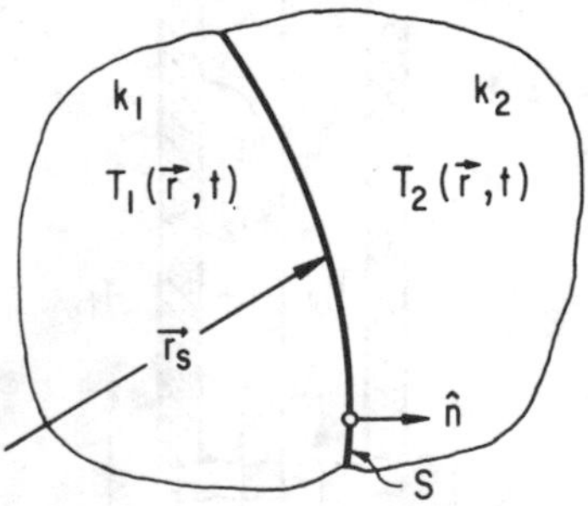

Fig. 2.9 Interface condition.

Because of the dependence of thermal conductivity k on temperature T, Eq. (2.37) is a non-linear differential equation and hence is difficult to work with. Equation (2.37) can be reduced to a linear differential equation by introducing a new temperature function θ by means of the *Kirchhoff transformation* as

$$\theta(\vec{r},t) = \frac{1}{k_0}\int_{T_0}^{T(\vec{r},t)} k(T')\,dT' \tag{2.38}$$

where T_0 is a reference temperature and $k_0 = k(T_0)$. From Eq. (2.38) it follows that

$$\vec{\nabla}\theta = \frac{k(T)}{k_0}\vec{\nabla}T \tag{2.39a}$$

and

$$\frac{\partial\theta}{\partial t} = \frac{k(T)}{k_0}\frac{\partial T}{\partial t} \tag{2.39b}$$

Therefore, Eq. (2.37) can be rearranged as

$$\nabla^2\theta + \frac{\dot{q}(\vec{r},t)}{k_0} = \frac{1}{\alpha}\frac{\partial\theta}{\partial t} \tag{2.40}$$

where $\alpha = k/\rho c$. If α is independent of temperature then Eq. (2.40) is a linear differential equation. If α is temperature dependent then Eq. (2.40) is still non-linear. For many solids, however, the dependence of α on temperature can be neglected compared to that of k. Hence, if α varies little with temperature, it can be assumed constant and Eq. (2.40) becomes linear. For steady-state problems Eq. (2.40) is a linear differential equation regardless of whether α is temperature dependent or not since the right-hand side vanishes identically for steady-state problems.

The transformations of the boundary conditions of the first and second kinds which prescribe T or $\partial T/\partial n$ on a boundary by means of the Kirchhoff transformation pose no difficulty and yield again boundary conditions of the first and second kinds, respectively. The transformation of a boundary condition of the third kind is, in general, not possible; only under certain restrictions on the heat transfer coefficient h the transformation may become possible [3].

REFERENCES

1. Arpacı, V. S., *Conduction Heat Transfer*, Addison-Wesley Pub. Co., Inc., Reading, Massachusetts, 1966.
2. Özışık, M. N., *Boundary Value Problems of Heat Conduction*, International

Textbook Co., Scranton, Pennsylvania, 1968.
3. Özışık, M. N., *Heat Conduction*, John Wiley and Sons, Inc., New York, 1980.
4. Luikov, A. V., *Analytical Heat Diffusion Theory*, Academic Press, New York, 1968.

PROBLEMS

2.1 Consider a homogeneous and isotropic solid in which temperature gradients exist and the temperature distribution is expressed in rectangular coordinates, *i.e.*, $T = T(x,y,z,t)$. By applying the first law of thermdynamics (conservation of energy) to a differential volume element $dx.dy.dz$ at a point (x,y,z) in this solid, derive the diffusion equation (2.17).

2.2 For the analysis of heat conduction in the cylindrical and the spherical coordinates the differential volume elements $dr.rd\phi.dz$ and $dr.r\sin\theta d\phi.rd\theta$, shown in Figs. 2.2 and 2.3, can be used, respectively. Following the same methodology repeat Problem 2.1 to derive the diffusion equation (a) in the cylindrical coordinates, and (b) in the spherical coordinates.

2.3 The general heat conduction equation in the rectangular coordinates with constant thermal conductivity is given by

$$\frac{\partial^2 T}{\partial x^2} + \frac{\partial^2 T}{\partial y^2} + \frac{\partial^2 T}{\partial z^2} + \frac{\dot{q}}{k} = \frac{1}{\alpha}\frac{\partial T}{\partial t}$$

Making coordinate transformations obtain the general heat conduction equation in (a) the cylindrical coordinates, and (b) the spherical coordinates.

2.4 A plane wall having a thermal conductivity k is receiving a radiative heat flux q_s from the sun (all of which is assumed to be absorbed by the wall) on one of its surfaces and at the same time transfers heat to the surrounding air at temperature T_∞ from the same surface with a heat transfer coefficient h. Give an expression for the boundary condition on this surface.

2.5 Consider two solids in perfect contact, one moving relative to the other. The local pressure on the common boundary is p, the coefficient of dry friction is μ, and the relative velocity is V. How can you express the boundary conditions at the interface of the solids?

2.6 Consider a solid spherical ball of radius r_0, and of constant thermal conductivity k. The ball is first heated to a uniform temperature T_i in an oven and then suddenly immersed, at time $t = 0$, in a large oil bath maintained at temperature T_∞. Assuming a constant heat transfer coefficient h on the surface of the sphere, formulate the problem (*i.e.*, give the differential equation, and the initial and boundary conditions) which can be solved to determine the unsteady temperature distribution in the ball for times $t \geq 0$ as a function of position and time.

2.7 The lower surface of a solid cylindrical bar, of radius r_o and height H respectively, is maintained in contact with boiling water at temperature T_f, and the upper surface is insulated. The bar is heated by an electric current and, in the meantime, losses heat from its peripheral surface to an environment at temperature T_∞. Let the heat transfer coefficient h be constant on this surface. Give the differential equation and the boundary conditions which can be solved to determine the steady temperature distribution in the bar as a function of position.

2.8 Consider a long solid semi-cylinder of radius r_o. Assume that the surface at $r = r_o$ is held at an arbitrary temperature distribution $T = f(\phi)$ and the diameter surface is maintained at a uniform and constant temperature T_o. Formulate the problem.

2.9 Prove that the following one-dimensional heat conduction problem has a unique solution:

$$\frac{\partial^2 T}{\partial x^2} = \frac{1}{\alpha}\frac{\partial T}{\partial t}$$

$$T(x,0) = T_i(x)$$

$$T(0,t) = T_1; \qquad T(L,t) = T_2$$

where T_1 and T_2 are two constant temperatures.

2.10 Transform the following one-dimensional non-linear heat conduction problem

$$\frac{d}{dx}\left[k(T)\frac{dT}{dx}\right] + \dot{q} = 0$$

$$\frac{dT(0)}{dx} = 0 \qquad \text{and} \qquad T(L) = T_w$$

into a linear problem in terms of a new temperature function defined as

$$\theta(x) = \frac{1}{k_w}\int_{T_w}^{T(x)} k(T)\,dT$$

where $k_w = k(T_w)$.

2.11 Determine the restriction on the heat transfer coefficient h, so that the boundary condition of the third kind (2.30a) can be transformed by the Kirchoff transformation (2.38).

3

One-Dimensional Steady-State Heat Conduction

3.1 INTRODUCTION

In the preceding chapter we established the general formulation of conduction heat transfer. In this chapter we study a number of simple systems in which the temperature distribution and heat flow are functions of one space variable only. For such problems, the heat conduction equation can be obtained directly from one of the forms of the general heat conduction equation developed in Chapter 2 by neglecting unnecessary terms to suit the given problem. This is possible, but may not always be convenient. The heat conduction equation can also be derived for each problem individually from the basic principles. By doing so we can, in fact, bring the physics of the problem into each phase of the derivation of the equation.

3.2 ONE-DIMENSIONAL STEADY-STATE HEAT CONDUCTION WITHOUT HEAT SOURCES

In this section we discuss one-dimensional steady-state heat conduction problems without heat sources in the rectangular, cylindrical and spherical coordinates, and introduce a number of physical and mathematical facts in terms of representative examples.

3.2.1 Plane Wall

Consider a plane wall or a slab of thickness L in the x-direction as shown in Fig. 3.1. Let the temperatures of the surfaces of the wall be T_1 and T_2. If the depth and height of the wall are very large compared to its thickness, L, then heat transfer will be (away from the ends) in the x-direction and the temperature distribution within the wall will depend on x only.

For steady-state heat conduction through such a wall with constant thermal conductivity Eq. (2.19) reduces to

$$\frac{d^2T}{dx^2} = 0 \tag{3.1}$$

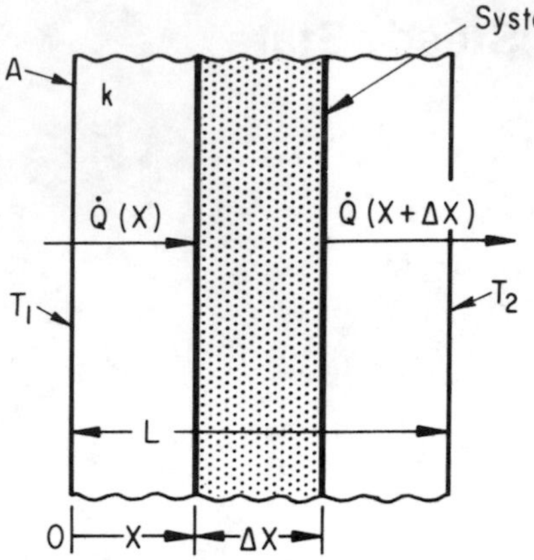

Fig. 3.1 Plane wall.

This equation can also be derived from the basic principles. In order to do this, we first define a system or a control volume as shown in Fig. 3.1. Since the heat transfer process is steady, time rate of change of the internal energy of the system will be zero. Also there is no work done by or on the system. Therefore, the application of the first law of thermodynamics, Eq. (1.4b), to the system shown, in the absence of the internal heat sources, yields

$$\dot{Q}(x) = \dot{Q}(x + \Delta x) \tag{3.2}$$

where $\dot{Q}(x)$ is the rate of heat transfer across the isothermal surface at the location x. As $\Delta x \to 0$, the right-hand side of Eq. (3.2) can be written as

$$\dot{Q}(x + \Delta x) = \dot{Q}(x) + \frac{d\dot{Q}}{dx}\Delta x \tag{3.3}$$

Substitution of Eq. (3.3) into Eq. (3.2) yields

$$\frac{d\dot{Q}}{dx} = 0 \tag{3.4}$$

Hence, the first law gives only $\dot{Q}(x)$ = constant. Substituting the value of $\dot{Q}(x)$ from the Fourier's law of heat conduction (particular law), Eq. (1.30), into Eq. (3.4) we get

$$\frac{d}{dx}\left(-kA\frac{dT}{dx}\right) = 0 \tag{3.5}$$

Since k and A are constants, Eq. (3.5) reduces to Eq.(3.1).

Once the differential equation of the problem is obtained, boundary conditions have to be specified to complete the formulation of the problem.

For the problem under consideration suppose that the wall extends from $x = 0$ to $x = L$. The boundary conditions can now be written as

$$T(0) = T_1 \quad \text{and } T(L) = T_2 \tag{3.6a,b}$$

As seen from this example, in order to formulate a problem from the basic principles, the following *five* steps have to be followed:

(a) Select a coordinate system appropriate to the geometry of the problem.

(b) Select a system or a control volume suited to the one-, two-, or three-dimensional nature of the problem.

(c) State the first law of thermodynamics for the system or the control volume selected in Step (b).

(d) Introduce the Fourier's law of heat conduction into the resulting equation in Step (c), and obtain the governing differential equation for the temperature distribution.

(e) Specify the origin of the coordinate system and state the necessary boundary conditions (and the initial condition for time-dependent problems) on the temperature.

Now, integrating Eq. (3.1) twice we get

$$T(x) = C_1 x + C_2 \tag{3.7}$$

Application of the boundary conditions (3.6) yields

$$C_1 = \frac{T_2 - T_1}{L} \quad \text{and} \quad C_2 = T_1$$

Substituting these values into Eq. (3.7), the temperature distribution in the plane wall is found to be

$$T(x) = T_1 - (T_1 - T_2)\frac{x}{L} \tag{3.8}$$

It follows that, with constant thermal conductivity, the temperature distribution in the wall is a linear function of x.

The rate of heat transfer through the wall is obtained by applying the Fourier's law as

$$\dot{Q} = -kA\frac{dT}{dx} = kA\frac{T_1 - T_2}{L} \tag{3.9}$$

which is consistent with Eq. (1.26) as expected.

If $k = k(x)$, then from Eq. (3.5) or Eq. (2.14) the one-dimensional heat conduction equation becomes

$$\frac{d}{dx}\left[k(x)\frac{dT}{dx}\right] = 0 \tag{3.10}$$

Integrating this equation twice and using the boundary conditions (3.6) we get

$$T(x) = T_1 - \frac{T_1 - T_2}{\int_0^L \frac{dx}{k(x)}} \int_0^x \frac{dx}{k(x)} \tag{3.11}$$

The rate of heat transfer through the wall is then given by

$$\dot{Q} = \frac{A(T_1 - T_2)}{\int_0^L \frac{dx}{k(x)}} \tag{3.12}$$

Hence, all the equations derived for the plane wall with constant thermal conductivity apply with L/k replaced by $\int_0^L dx/k(x)$.

If $k = k(T)$, then Eq. (3.5) becomes

$$\frac{d}{dx}\left[k(T)\,\frac{dT}{dx}\right] = 0 \tag{3.13}$$

which is a non-linear differential equation. Integrating this equation once yields

$$k(T)\,\frac{dT}{dx} = C_1 = -\frac{\dot{Q}}{A} \tag{3.14}$$

One more integration from $x = 0$ ($T = T_1$) to $x = L$ ($T = T_2$) gives

$$\int_{T_1}^{T_2} k(T)\,dT = -\frac{\dot{Q}}{A}\,L \tag{3.15}$$

from which we get

$$\dot{Q} = \frac{A}{L}\int_{T_2}^{T_1} k(T)dT \tag{3.16a}$$

This result for $\dot{Q}$ can also be written in terms of a *mean thermal conductivity*, k_m, as

$$\dot{Q} = k_m A\,\frac{T_1 - T_2}{L} \tag{3.16b}$$

where the mean thermal conductivity is defined as

$$k_m = \frac{1}{T_1 - T_2}\int_{T_2}^{T_1} k(T)dT \tag{3.17}$$

The temperature distribution in the wall can be obtained by integrating

Eq. (3.14) as follows

$$\int_{T_1}^{T(x)} k(T)dT = -\frac{\dot{Q}}{A}\int_0^x dx = -\frac{\dot{Q}}{A}x \tag{3.18}$$

This expression may not be written explicitly for $T(x)$ until the relation $k = k(T)$ is specified.

As we discussed in Section 1.9, with many materials the thermal conductivity k is not constant, but varies in a nearly linear manner with temperature as

$$k(T) = k_0[1 + \gamma(T - T_0)] \tag{3.19}$$

The mean thermal conductivity then becomes

$$k_m = \frac{1}{T_1 - T_2}\int_{T_2}^{T_1} k_0[1 + \gamma(T - T_0)]dT$$

$$= k_0[1 + \gamma\left(\frac{T_1 + T_2}{2} - T_0\right)] \tag{3.20}$$

and Eq. (3.18) yields

$$T^2(x) + \frac{2}{\gamma}[1 - \gamma T_0]T(x) + \frac{2}{\gamma}[T_1(\gamma T_0 - \frac{\gamma}{2}T_1 - 1) + \frac{k_m}{k_0}(T_1 - T_2)\frac{x}{L}] = 0 \tag{3.21}$$

As seen from Eq. (3.21), the temperature distribution is no longer linear. Two cases for positive and negative γ are shown by the dotted lines in Fig. 3.2.

Example 3.1: A plane wall of 50 cm thickness is to be constructed from a material having a thermal conductivity which varies linearly with temperature according to the relation $k(W/m\cdot K) = 1 + 0.0015T(C°)$. Calculate the rate of heat loss per unit area if one side of the wall is maintained at 1000°C and the other side at 0°C.

Solution: Equation (3.20) may be used first to calculate the average thermal conductivity, and then (3.16b) gives heat flow through the wall:

$$k_m = k_0[1 + \gamma\left(\frac{T_1 + T_2}{2} - T_0\right)]$$

where $k_0 = 1.0$ W/m·K, $\gamma = 0.0015$ 1/K, $T_0 = 0°C$. Hence

$$k_m = 1.0x\left[1 + 0.0015x\ \frac{1000}{2}\right] = 1.750 \text{ W/m·K}$$

Therefore, the rate of heat loss per unit area is

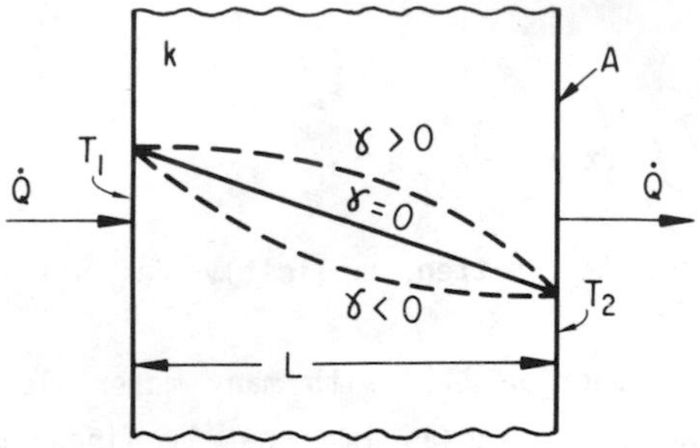

Fig. 3.2 Temperature distribution in a plane wall with $k = k_0[1 + \gamma(T - T_0)]$.

$$\frac{\dot{Q}}{A} = \frac{k_m(T_1 - T_2)}{L} = \frac{1.750(1000 - 0)}{0.50} = 3500 \text{ W/m}^2$$

3.2.2 Conduction Through a Plane Wall From One Fluid To Another

Consider the problem of heat transfer through a homogeneous plane wall separating two moving fluids as shown in Fig. 3.3.

Heat will be convected from the fluid of higher temperature to the wall, conducted through the wall and then convected again from the wall to the fluid of lower temperature. Under steady-state conditions, in the absence of internal heat sources, the rate of heat transfer will be the same on both surfaces and through the wall. In terms of the thickness, L, of the wall, thermal conductivity, k, of the wall material, surrounding fluid temperatures, T_{f_1} and T_{f_2}, and the constant heat transfer coefficients, h_1 and h_2, we can write the following system of equations:

$\dot{Q} = Ah_1(T_{f_1} - T_1)$ for the heat transfer at surface x = 0.

$\dot{Q} = A\frac{k}{L}(T_1 - T_2)$ for conduction through the wall.

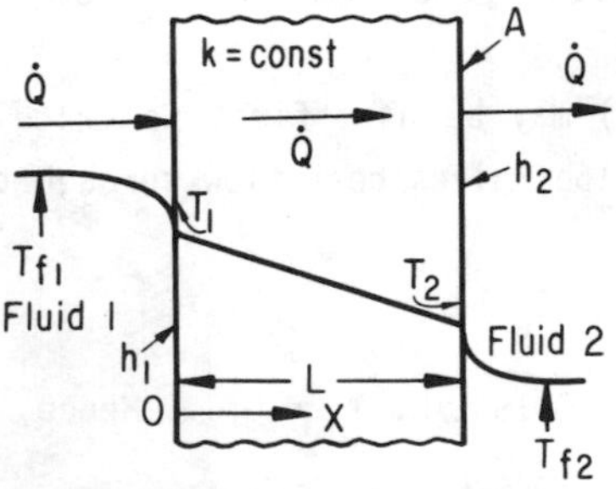

Fig. 3.3 Heat transfer through a homogeneous plane wall.

$$\dot{Q} = Ah_2(T_2 - T_{f_2}) \quad \text{for heat transfer at surface } x = L.$$

After solving these equations for the temperature differences, elimination of T_1 and T_2 gives

$$\dot{Q} = \frac{A(T_{f_1} - T_{f_2})}{\frac{1}{h_1} + \frac{L}{k} + \frac{1}{h_2}} \tag{3.22}$$

Example 3.2: A wall of 25 cm thickness and 10 m^2 surface area is to be constructed from a material which has an average thermal conductivity of 0.5 W/m·K. The average inner and outer fluid temperatures are 20°C and -20°C, and the corresponding heat transfer coefficients are 8 $W/m^2 \cdot K$ and 23 $W/m^2 \cdot K$, respectively. Calculate the rate of heat loss through the wall and the temperatures of the inner and outer surfaces of the wall.

Solution: We can use Eq. (3.22) to compute the rate of heat loss through the wall

$$\dot{Q} = \frac{A(T_{f_1} - T_{f_2})}{\frac{1}{h_1} + \frac{L}{k} + \frac{1}{h_2}} = \frac{10[20 - (-20)]}{\frac{1}{8} + \frac{0.25}{0.5} + \frac{1}{23}} = \frac{400}{0.125 + 0.5 + 0.0435} = 598.4\text{W}$$

The inner and outer surface temperatures are

$$T_1 = T_{f_1} - \frac{(\dot{Q}/A)}{h_1} = 20 - \frac{(598.4/10)}{8} = 12.52°\text{C}$$

$$T_2 = T_{f_2} + \frac{(\dot{Q}/A)}{h_2} = -20 + \frac{(598.4/10)}{23} = -17.40°\text{C}$$

3.2.3 Hollow Cylinder

Consider a hollow circular cylinder of inside and outside radii r_1 and r_2 and length L as shown in cross-section in Fig. 3.4. The inside and outside surfaces are maintained uniformly at temperatures T_1 and T_2, respectively. Let the cylinder material be homogeneous. If the cylinder is sufficiently long so that the end effects may be neglected or if the ends are perfectly insulated then the temperature distribution will be one dimensional; that is, the temperature distribution will be a function of the radial coordinate only. For such a case, the heat conduction equation (2.19) reduces to

$$\frac{d^2T}{dr^2} + \frac{1}{r}\frac{dT}{dr} = 0 \tag{3.23a}$$

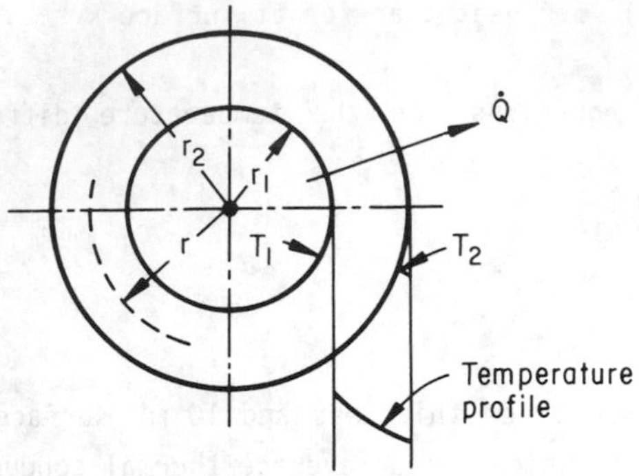

Fig. 3.4 Heat flow through a hollow cylinder.

or

$$\frac{d}{dr}\left(r \frac{dT}{dr} \right) = 0 \tag{3.23b}$$

Integrating Eq. (3.23b) twice yields

$$T(r) = C_1 \ln r + C_2 \tag{3.24}$$

The boundary conditions are

$$T(r_1) = T_1 \quad \text{and} \quad T(r_2) = T_2 \tag{3.25a,b}$$

Application of these boundary conditions gives

$$T_1 = C_1 \ln r_1 + C_2 \tag{3.26a}$$

$$T_2 = C_1 \ln r_2 + C_2 \tag{3.26b}$$

Hence, we obtain

$$C_1 = - \frac{T_1 - T_2}{\ln \frac{r_2}{r_1}}$$

$$C_2 = T_1 + \frac{T_1 - T_2}{\ln \frac{r_2}{r_1}} \ln r_1$$

Finally, substituting the constants C_1 and C_2 into the solution (3.24) we get

$$T(r) = T_1 - \frac{T_1 - T_2}{\ln \frac{r_2}{r_1}} \ln \frac{r}{r_1} \tag{3.27}$$

Thus, the temperature distribution in a hollow circular cylinder is a logarithmic function of the radial coordinate r.

The rate of heat transfer through the cylinder wall is obtained from Fourier's law of heat conduction:

$$\dot{Q} = -k\ 2\pi rL \frac{dT}{dr} = k\ 2\pi rL \frac{T_1 - T_2}{r \ln \frac{r_2}{r_1}}$$

$$\dot{Q} = \frac{2\pi Lk\ (T_1 - T_2)}{\ln \frac{r_2}{r_1}} \tag{3.28}$$

If k is a function of the radial coordinate; that is, if $k = k(r)$ then from Eq. (2.14) we get

$$\frac{d}{dr}\left[r\ k(r)\ \frac{dT}{dr}\right] = 0 \tag{3.29}$$

Integration of this equation over r yields

$$r\ k(r)\ \frac{dT}{dr} = C_1 = -\frac{\dot{Q}}{2\pi L} \tag{3.30}$$

One more integration from r_1 to r_2 gives

$$\dot{Q} = \frac{2\pi L(T_1 - T_2)}{\displaystyle\int_{r_1}^{r_2} \frac{dr}{r\ k(r)}} \tag{3.31}$$

For k = constant, Eq. (3.31) reduces to Eq. (3.28).

3.2.4 Spherical Shells

Consider a hollow sphere of inside and outside radii r_1 and r_2, and at uniform inside and outside surface temperatures T_1 and T_2, respectively. If the material of the sphere is homogeneous, then the temperature distribution will be a function of r only; that is, $T = T(r)$, and the heat conduction equation (2.19) takes the form

$$\frac{d^2T}{dr^2} + \frac{2}{r}\frac{dT}{dr} = 0 \tag{3.32a}$$

or

$$\frac{1}{r^2}\frac{d}{dr}\left(r^2\ \frac{dT}{dr}\right) = 0 \tag{3.32b}$$

The boundary conditions are

$$T(r_1) = T_1 \quad \text{and} \quad T(r_2) = T_2 \tag{3.33a,b}$$

Integrating Eq. (3.32b) twice yields

$$T(r) = -\frac{C_1}{r} + C_2 \tag{3.34}$$

Substituting the boundary conditions (3.33) into Eq. (3.34) we get

$$T_1 = -\frac{C_1}{r_1} + C_2 \tag{3.35a}$$

$$T_2 = -\frac{C_1}{r_2} + C_2 \tag{3.35b}$$

Hence, we obtain

$$C_1 = \frac{T_1 - T_2}{\frac{1}{r_2} - \frac{1}{r_1}} \tag{3.36a}$$

$$C_2 = T_1 + \frac{T_2 - T_1}{r_1\left(\frac{1}{r_2} - \frac{1}{r_1}\right)} \tag{3.36b}$$

Substitution of these constants into Eq. (3.34) yields the temperature distribution in the sphere:

$$T(r) = T_1 + \frac{T_1 - T_2}{\frac{1}{r_2} - \frac{1}{r_1}}\left(\frac{1}{r_1} - \frac{1}{r}\right) \tag{3.37}$$

The rate of heat transfer through the spherical wall is obtained from the Fourier's law of heat conduction:

$$\dot{Q} = -k\ 4\pi r^2 \frac{dT}{dr} = -k\ 4\pi r^2 \frac{T_1 - T_2}{r^2\left(\frac{1}{r_2} - \frac{1}{r_1}\right)}$$

$$= \frac{4\pi k(T_1 - T_2)}{\frac{1}{r_2} - \frac{1}{r_1}} = \frac{k\ A_m(T_1 - T_2)}{r_2 - r_1} \tag{3.38}$$

where

$$A_m = 4\pi r_1 r_2 = 4\sqrt{\pi r_1^2 \times \pi r_2^2} = \sqrt{A_1 A_2}$$

and A_1 and A_2 are the inside and outside surface areas.

3.2.5 Thermal Resistance Concept

Temperature can be thought of as a potential or driving force for heat flow. The flow of heat should then be governed by the potential difference across the heat flow path and the resistance of it. This concept suggests that heat flow is analogous to electric current flow. If we consider steady flows, then

$$I = \frac{\Delta V}{R} \quad \text{is current flow in Amperes}$$

and

$$\dot{Q} = \frac{\Delta T}{R_t} \quad \text{is heat flow in Watts.}$$

where R_t is the so-called *thermal resistance.* Comparison of this definition with Eqs. (1.49), (3.9) and (3.28) results in expressions for the thermal resistance as follows

$$\begin{aligned} &\text{Surface to fluid} : && R_t = \frac{1}{hA} \\ &\text{Plane wall} : && R_t = \frac{L}{kA} \\ &\text{Hollow cylinder} : && R_t = \frac{\ln(r_2/r_1)}{2\pi Lk} \end{aligned} \tag{3.39a,b,c}$$

Figure 3.5 illustrates analogy between the heat transfer through a plane wall and flow of electric current over a resistance R. The thermal resistance (3.39a) is called *convective* or *surface* resistance. The resistances (3.39b,c), on the other hand, are called *conductive* or *internal* resistances of a plane wall of thickness L and of a hollow cylinder of inside and outside radii r_1 and r_2, respectively.

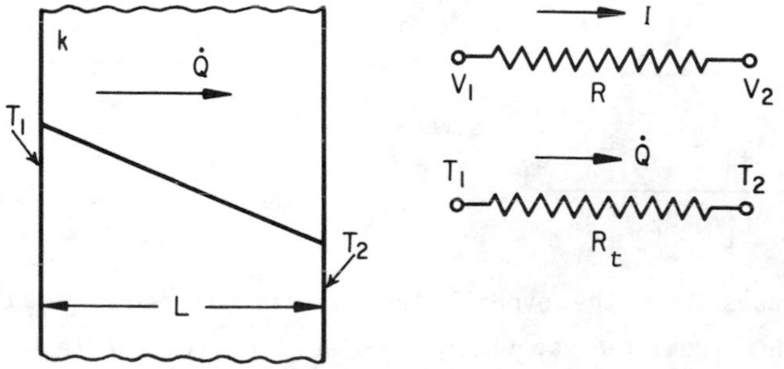

Fig. 3.5 Electrical analogy.

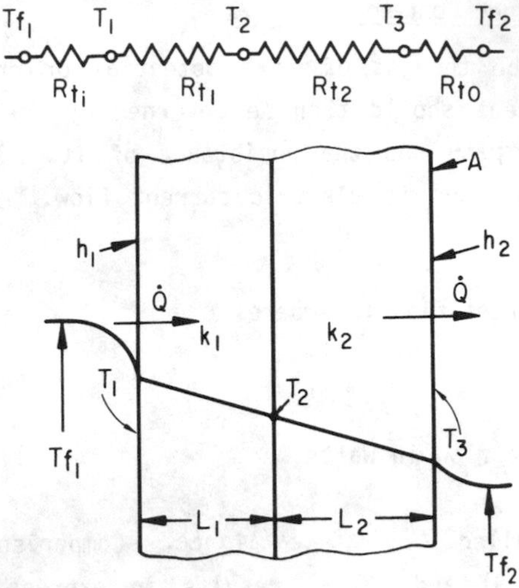

Fig. 3.6 A composite plane wall and electrical analogy.

3.2.6 Composite Plane Walls

By the use of the resistance concept, the following system of equations can be written for the heat transfer through the composite plane wall of Fig. 3.6:

$$
\begin{aligned}
T_{f_1} - T_1 &= \dot{Q}\, R_{t_i} \\
T_1 - T_2 &= \dot{Q}\, R_{t_1} \\
T_2 - T_3 &= \dot{Q}\, R_{t_2} \\
T_3 - T_{f_2} &= \dot{Q}\, R_{t_o}
\end{aligned}
\tag{3.40a,b,c,d}
$$

Adding these four equations we get

$$T_{f_1} - T_{f_2} = \dot{Q}(R_{t_i} + R_{t_1} + R_{t_2} + R_{t_o}) \tag{3.41a}$$

or

$$\dot{Q} = \frac{T_{f_1} - T_{f_2}}{R_{t_i} + R_{t_1} + R_{t_2} + R_{t_o}} \tag{3.42b}$$

This result shows that the overall thermal resistance is equal to the sum of the individual resistances, which is also true for electrical resistances.

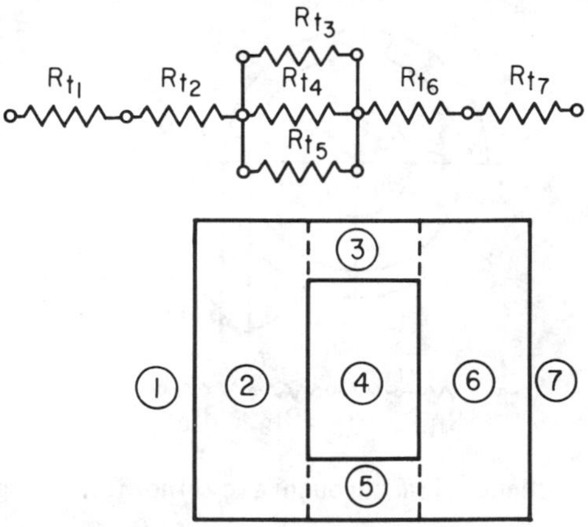

Fig. 3.7 A composite wall with both series and parallel resistances.

For the plane composite wall of Fig. 3.6, the resistances in Eq. (3.42b) are given by Eqs. (3.39a,b) and therefore Eq. (3.42b) can be written as

$$\dot{Q} = \frac{A(T_{f_1} - T_{f_2})}{\frac{1}{h_1} + \frac{L_1}{k_1} + \frac{L_2}{k_2} + \frac{1}{h_2}} \tag{3.43}$$

Since the overall thermal resistance of a composite wall is the sum of the thermal resistances of individual layers, it is clear that the rate of heat transfer through a composite wall of n layers is

$$\dot{Q} = \frac{A(T_{f_1} - T_{f_2})}{\frac{1}{h_1} + \sum_{i=1}^{n} \frac{L_i}{k_i} + \frac{1}{h_2}} \tag{3.44}$$

The thermal resistance concept may also be used to solve more complex problems involving both series and parallel thermal resistances using the rules for combining electrical resistances as illustrated in Fig. 3.7. Some care must be exercised, however, in representing thermal systems with parallel resistances since multi-dimensional effects are likely to be present.

3.2.7 Cylindrical Composite Walls

For the cylindrical composite wall of Fig. 3.8 the reasoning leading to Eq. (3.43) is also applicable. The resistances are given by Eqs. (3.39b,c) and, therefore, the equation for the rate of heat transfer, $\dot{Q}$, becomes

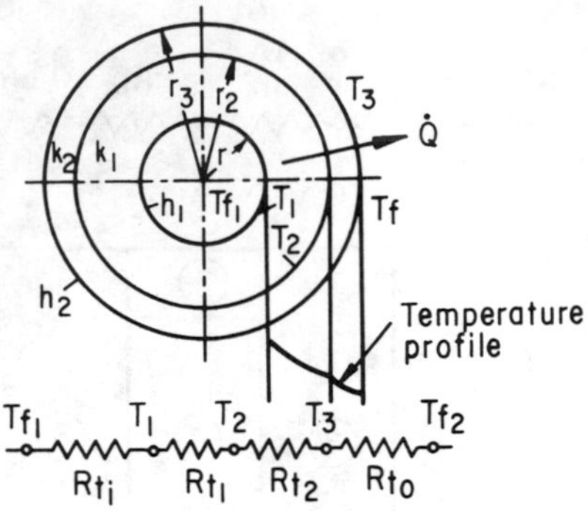

Fig. 3.8 Heat conduction through a cylindrical composite wall.

$$\dot{Q} = \frac{2\pi L(T_{f_1} - T_{f_2})}{\dfrac{1}{h_1 r_1} + \dfrac{\ln(r_2/r_1)}{k_2} + \dfrac{\ln(r_3/r_2)}{k_3} + \dfrac{1}{r_3 h_2}} \tag{3.45}$$

The rate of heat flow through a composite wall of n layers can then be written as

$$\dot{Q} = \frac{2\pi L(T_{f_1} - T_{f_2})}{\dfrac{1}{h_1 r_1} + \displaystyle\sum_{i=1}^{n} \dfrac{1}{k_i} \cdot \ln \dfrac{r_{i+1}}{r_i} + \dfrac{1}{h_2 r_{n+1}}} \tag{3.46}$$

Example 3.3: A steel pipe with an 8-inch standard diameter (OD = 8.625 in., ID = 8.071 in.) is covered with a 4 cm layer of high temperature insulation, k = 0.085 W/m·K, followed by a 6 cm layer of low temperature insulation, k = 0.07 W/m·K. The pipe carries steam at 400°C and the surrounding air temperature is 20°C. The heat transfer coefficients are 6000 $W/m^2 \cdot K$ and 12 $W/m^2 \cdot K$ on the inside and outside surfaces, respectively. The thermal conductivity of the steel pipe is 32 W/m·K. Calculate the rate of heat loss per unit length of the pipe.

Solution: Equation (3.46) will be used for the solution of the problem:

$$\dot{Q}/L = \frac{2\pi(T_{f_1} - T_{f_2})}{\dfrac{1}{h_1 r_1} + \dfrac{1}{k_1} \ln \dfrac{r_2}{r_1} + \dfrac{1}{k_2} \ln \dfrac{r_3}{r_2} + \dfrac{1}{k_3} \ln \dfrac{r_4}{r_3} + \dfrac{1}{h_2 r_4}}$$

where we have

$$T_{f_1} = 400°C, \qquad T_{f_2} = 20°C$$

$$h_1 = 6000\ W/m^2\cdot K, \qquad h_2 = 12\ W/m^2\cdot K$$

$$k_1 = 32\ W/m\cdot K, \qquad k_2 = 0.085\ W/m\cdot K, \qquad k_3 = 0.07\ W/m\cdot K$$

$$r_1 = \frac{8.071}{2} \times 2.54 \times 10^{-2} = 10.25 \times 10^{-2} m$$

$$r_2 = \frac{8.625}{2} \times 2.54 \times 10^{-2} = 10.95 \times 10^{-2} m$$

$$r_3 = 10.95 \times 10^{-2} + 4 \times 10^{-2} = 14.95 \times 10^{-2} m$$

$$r_4 = 14.95 \times 10^{-2} + 6 \times 10^{-2} = 20.95 \times 10^{-2} m$$

Hence

$$\frac{1}{h_1 r_1} = \frac{1}{6000 \times 10.25 \times 10^{-2}} = 16.26 \times 10^{-4} m\cdot K/W$$

$$\frac{1}{h_2 r_4} = \frac{1}{12 \times 20.95 \times 10^{-2}} = 39.78 \times 10^{-2} m\cdot K/W$$

$$\frac{1}{k_1} \ln \frac{r_2}{r_1} = \frac{1}{32} \ln \frac{10.95}{10.25} = 20.64 \times 10^{-4} m\cdot K/W$$

$$\frac{1}{k_2} \ln \frac{r_3}{r_2} = \frac{1}{0.085} \ln \frac{14.95}{10.95} = 3.6632\ m\cdot K/W$$

$$\frac{1}{k_3} \ln \frac{r_4}{r_3} = \frac{1}{0.07} \ln \frac{20.95}{14.95} = 4.8204\ m.K/W$$

Substituting these results into the above equation we obtain

$$\dot{Q}/L = \frac{2\pi(400 - 20)}{(16.26 + 20.64 + 36632 + 48204 + 3978) \times 10^{-4}} = 268.7\ W/m$$

3.2.8 Overall Heat Transfer Coefficient

It is sometimes convenient to simplify the equations for the rate of heat transfer through composite walls by writing them in terms of the so-called *overall coefficient of heat transfer*, U, which is defined by the equation

$$\dot{Q} = AU(T_{f_1} - T_{f_2}) \qquad (3.47)$$

Comparing this equation with Eqs. (3.43) and (3.45) we get the following expressions for U for the cases of composite plane and cylindrical walls:

Plane wall:

$$\frac{1}{U} = \frac{1}{h_1} + \frac{L_1}{k_1} + \frac{L_2}{k_2} + \frac{1}{h_2} \tag{3.48}$$

Cylindrical wall:

$$\frac{1}{U} = A\left[\frac{1}{2\pi L\ r_1 h_1} + \frac{\ln(r_2/r_1)}{2\pi L\ k_1} + \frac{\ln(r_3/r_2)}{2\pi L\ k_2} + \frac{1}{2\pi L\ r_3 h_2}\right] \tag{3.49}$$

Equation (3.47) defines U in terms of a heat transfer area, A. In the cylindrical wall case, A is not constant, but varies from $2\pi r_1 L$ to $2\pi r_3 L$. Therefore, the definition of U in the cylindrical wall case depends on the area selected. There is no accepted practice in this matter. If, for example, $A = A_1 = 2\pi r_1 L$ is taken, then Eq. (3.49) becomes

$$\frac{1}{U_1} = \frac{1}{h_1} + \frac{r_1}{k_1}\ln(r_2/r_1) + \frac{r_1}{k_2}\ln(r_3/r_2) + \frac{r_1}{h_2 r_3} \tag{3.50}$$

This definition of U was arbitrarily based on A_1; it can also be based on either A_2 or A_3 or any area in between these. It is to be noted, however, that in any case

$$A_1U_1 = A_2U_2 = A_3U_3 \tag{3.51}$$

3.2.9 Thermal Contact and Scale Resistances

In addition to the resistances discussed above, two other thermal resistances are often encountered in heat transfer systems; namely *thermal contact resistance* and *scale resistance*.

All machined surfaces which are said to be flat are, in fact, wavy with a regular pitch owing to the periodical nature of machining processes. When two such surfaces are brought into contact, they actually touch only at a limited number of spots, the total of which is usually only a small fraction of the *apparent* contact area as illustrated in Fig. 3.9. The remainder of the space between the surfaces may be filled with air or another fluid, or may even be a vacuum. When heat flows from one solid to the other, heat flow lines converge towards the actual contact spots, since the thermal conductivities of solids are generally greater than those of fluids. This causes a thermal resistance to heat flow, which is usually high compared to the resistances away from the contact spots.

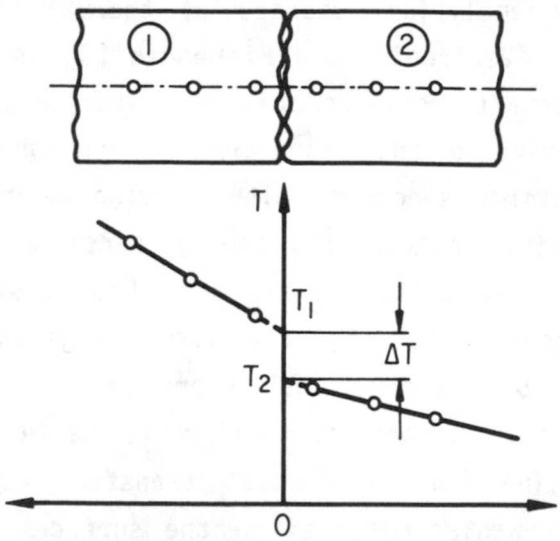

Fig. 3.9 Temperature drop at an interface due to thermal contact resistance.

Thermal contact resistances have been the subject of many investigations, both theoretical and experimental, since such resistances must be accurately estimated for a reliable heat transfer analysis of a given system. In theoretical studies, usually geometrically simple *contact elements* are considered. The simplest of these has a two-dimensional planar configuration. It consists of two solids of rectangular cross-sectional area having an actual strip of contact in the middle of their bases facing each other with an interstitial fluid between the remainder of the contact surfaces. The exact solutions for the thermal resistances of planar and circular contacts have been presented in Reference [2].

Thermal contact resistance depends upon several inter-dependent parameters. The actual contact area is the most important one. In addition to the holes and vacant spaces, traces of poorly conducting material, such as oxide films, may be present between the surfaces. It is usually difficult to estimate accurately the thickness of such films. The actual contact area, on the other hand, strongly depends upon the applied load. At the interface of two solids a *contact heat transfer coefficient* is defined as

$$h_c = \frac{\dot{Q}/A}{\Delta T} \tag{3.52}$$

where ΔT is the temperature drop and $\dot{Q}/A$ is the heat flux across the interface. Thermal contact resistance can, therefore, be defined as

$$R_c = \frac{1}{h_c A} \tag{3.53}$$

where A is the apparent contact area.

One of the first analytical studies of thermal contact resistance was presented by Cetinkale (Veziroglu) and Fishenden [3] in 1951. They obtained an expression for contact resistance. An experimental investigation was also conducted by Veziroglu and Fishenden at atmospheric conditions using steel, brass and aluminium specimens with varying degrees of surface roughnesses. They used air, spindle oil and glycerol as interstitial fluids.

Literature on thermal contact resistance has grown considerably during the last decade. A good review of these works is given in References [4-7].

It is important to recognize that there may be an additional thermal resistance on fluid-solid interfaces, which is called *scale resistance* or *fouling resistance*. The fouling of heat transfer surfaces may be defined as the deposition of unwanted material on the surfaces causing a degradation in performance. The fouling of solid-fluid interfaces can be described in six categories:

Precipitation Fouling - the crystallization from solution of dissolved substances onto the heat transfer surface, sometimes called *scaling*. Normal solubility salts precipitate on subcooled surfaces, while the more troublesome inverse solubility salts precipitate on superheated surfaces.

Particulate Fouling - the accumulation of finely divided solids suspended in the process fluid onto the heat transfer surface. In a minority of instances settling by gravity prevails, and the process may then be referred to as *sedimentation fouling*.

Chemical Reaction Fouling - deposit formation at the heat transfer surface by chemical reactions in which the surface material itself is not a reactant (*e.g.* in petroleum refining, polymer production, food processing).

Corrosion Fouling - the accumulation of indigenous corrosion products on the heat transfer surface.

Biological Fouling - the attachment of macro-organisms (*macro-biofouling*) and/or micro-organisms (*micro-biofouling* or *microbial fouling*) to a heat transfer surface, along with the adherent slimes often generated by the latter.

Solidification Fouling - the *freezing* of a pure liquid or a higher melting point constituent of a multi-component solution onto a subcooled surface.

The functional effect of fouling on a heat transfer surface may be expressed by the thermal fouling resistance

$$R_f = \frac{x_f}{k_f} \tag{3.54}$$

where x_f is the thickness and k_f is the thermal conductivity of the deposit. It is seldom practical to measure the thickness of the fouling. In addition, the thermal conductivity of the deposit may not be known, and may also vary with the thickness.

Over the last decade, increasing efforts have been directed towards a better understanding of fouling and several models were proposed to predict the fouling and fouling resistance in design and operation of heat exchangers [8,9].

If contact and scale resistances are appreciable, additional resistances R_c and R_f must be included in the summation of resistances and then terms like $1/Ah_c$ and $1/Ah_f$ should appear in Eqs. (3.44) and (3.46).

Example 3.4: Calculate the overall heat transfer coefficient for the following cases:

(a) Heat is transferred through 1 m^2 of a 3-cm thick steel plate, k = 36.4 W/m.K. A liquid flows on one side with a heat transfer coefficient h_1 = 4000 $W/m^2 \cdot K$ and the other side is exposed to air with h_2 = 12 $W/m^2 \cdot K$. Assume a scale coefficient h_f = 3000 $W/m^2 \cdot K$ on the liquid side.

(b) Same as part (a), with the air replaced by condensing steam, h_2 = 6000 $W/m^2 \cdot K$.

(c) Same as part (a) with the flowing liquid replaced by a flowing gas, h_1 = 6 $W/m^2 \cdot K$.

Solution: The overall heat transfer coefficient, U, can be calculated as follows:

(a) $$\frac{1}{U} = \frac{1}{h_1} + \frac{L}{k} + \frac{1}{h_2} + \frac{1}{h_f} = \frac{1}{4000} + \frac{0.03}{36.4} + \frac{1}{12} + \frac{1}{3000}$$

$$= 0.00025 + 0.000824 + 0.08333 + 0.000333$$

$$U = 11.8\ W/m^2 \cdot K$$

(b) $$\frac{1}{U} = \frac{1}{4000} + \frac{0.03}{36.4} + \frac{1}{6000} + \frac{1}{3000}$$

$$= 0.00025 + 0.000824 + 0.00016667 + 0.000333$$

$$U = 635.25\ W/m^2 \cdot K$$

(c) $$\frac{1}{U} = \frac{1}{6} + \frac{0.03}{36.4} + \frac{1}{12} + \frac{1}{3000}$$

$$= 0.16667 + 0.000824 + 0.08333 + 0.000333$$

$$U = 4\ W/m^2 \cdot K$$

We now conclude that in part (a) the only significant thermal resistance is on the air side surface, in part (b) none of the thermal resistances is negligible compared to the others, but in part (c) the wall and scale resis-

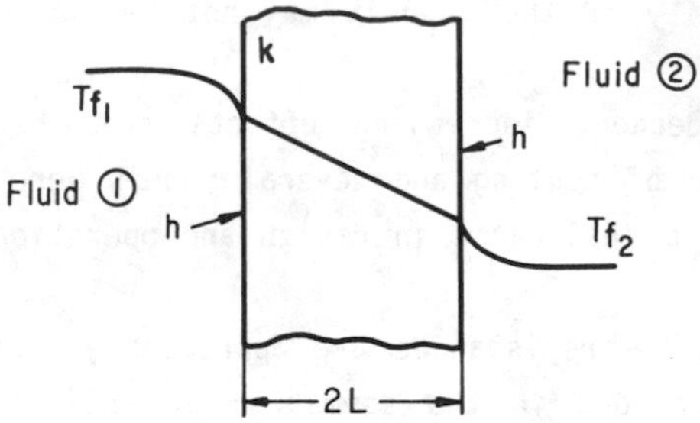

Fig. 3.10 A plane wall with thickness 2L.

tances are negligible compared to the others.

3.2.10 Biot Number

Consider a plane wall which separates two fluids having temperatures T_{f_1} and T_{f_2} as shown in Fig. 3.10. The thickness of the wall is 2L and its thermal conductivity k is constant. Let the heat transfer coefficients h on both surfaces be the same. The rate of heat transfer from fluid 1 to fluid 2 through the wall is given by

$$\dot{Q} = \frac{T_{f_1} - T_{f_2}}{\Sigma R_t} \tag{3.55}$$

where

$$\Sigma R_t = \frac{2}{hA} + \frac{2L}{kA}$$

As seen, there are two kinds of resistances to the heat flow from fluid 1 to fluid 2; namely, the surface resistance 2/hA and internal resistance 2L/kA. The ratio of these two resistances is named as the Biot number.

$$\text{Biot Number} = Bi = \frac{\text{Internal resistance}}{\text{Surface resistance}} = \frac{hL}{k} \tag{3.56}$$

Two limiting cases of Bi number are important:

(a) Bi number may be very large; that is, $Bi \to \infty$: In that case, the total internal resistance is very large compared to the total surface resistance. That is,

$$\Sigma R_t \simeq \frac{2L}{kA}$$

Therefore, there will be no temperature drops on the surfaces and temperatures

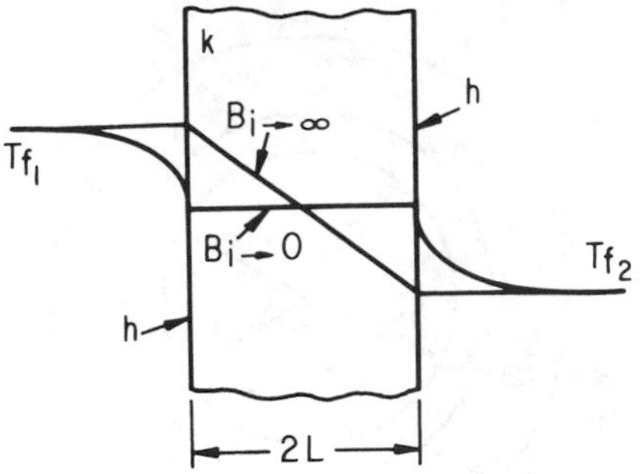

Fig. 3.11 Temperature distribution in a plane wall for two limiting cases of Bi number.

of the surfaces and the fluids will be the same. In this case, the temperature distribution in the wall will appear as shown in Fig. 3.11.

(b) Bi number may be very small; that is, Bi → 0: In this case, the total surface resistance is very large compared to the total internal resistance. That is,

$$\Sigma R_t \simeq \frac{2}{hA}$$

Therefore, there will be no temperature drop in the wall and the temperature distribution will appear as shown in Fig. 3.11.

3.2.11 Critical Thickness of Cylindrical Insulation

Adding insulation to the outside of small pipes does not always reduce heat transfer. As the thickness of the insulation layer around a pipe is increased, thermal resistance of the insulation layer increases logarithmically, but at the same time thermal resistance of the outer surface decreases linearly. Since the total thermal resistance is proportional to the summation of these two resistances heat transfer may increase instead of decreasing.

Let us now consider the influence of the thickness of cylindrical insulation on heat transfer. For simplicity, let us consider a single layer of insulation as shown in Fig. 3.12. The rate of heat flow per unit length of the cylinder can be written as

$$\frac{\dot{Q}}{L} = \frac{2\pi(T_f - T_\infty)}{\dfrac{1}{h_1 r_1} + \dfrac{1}{k_1}\ln\dfrac{r_2}{r_1} + \dfrac{1}{k}\ln\dfrac{r}{r_2} + \dfrac{1}{hr}} \tag{3.57}$$

The total thermal resistance per unit length, therefore, is

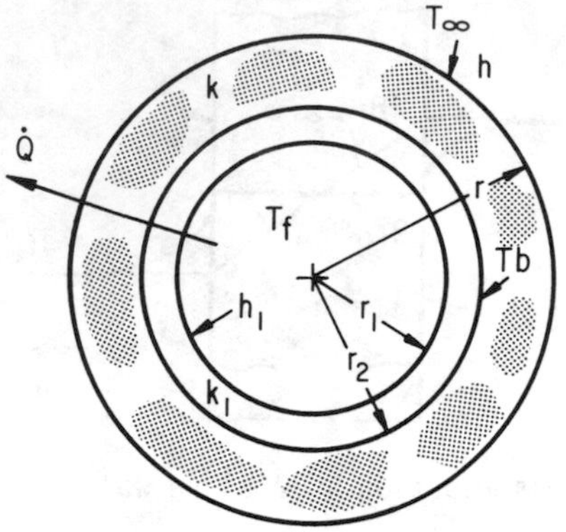

Fig. 3.12 Critical thickness of insulation.

$$R_t = \frac{1}{2\pi}\left(\frac{1}{h_1 r_1} + \frac{1}{k_1}\ln\frac{r_2}{r_1} + \frac{1}{k}\ln\frac{r}{r_2} + \frac{1}{hr}\right) \tag{3.58}$$

Differentiating this expression with respect to r, while holding h constant, we get

$$\frac{dR_t}{dr} = \frac{1}{2\pi r}\left(\frac{1}{k} - \frac{1}{hr}\right)$$

We see that for k = hr the derivative vanishes. The radius at which $dR_t/dr = 0$ is called *critical radius;* that is,

$$(r)_{cr} = \frac{k}{h} \tag{3.59}$$

One more differentiation yields

$$\frac{d^2R_t}{dr^2} = \frac{1}{2\pi r^2}\left(\frac{2}{hr} - \frac{1}{k}\right)$$

At $r = (r)_{cr} = k/h$ the second derivative is

$$\frac{d^2R_t}{dr^2} = \frac{h^2}{k^3} > 0$$

Thus, at $r = (r)_{cr}$, the thermal resistance is a minimum, and therefore $\dot{Q}/L$ is a maximum.

If $r_2 > k/h$, then adding insulation decreases heat loss, but if $r_2 < k/h$ then heat loss is increased by adding insulation until $r = k/h$, where r is the outer radius of the insulation layer. That is, the addition of insulation has a de-insulating effect.

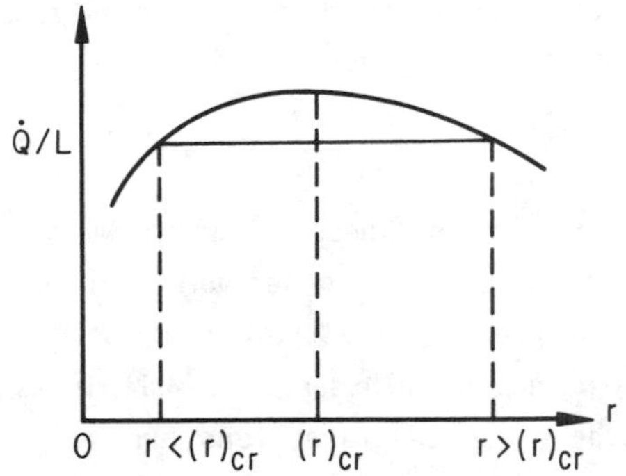

Fig. 3.13 Relationship between heat loss and the thickness of insulation.

Figure 3.13 shows the relationship between the heat loss and the thickness of the cylindrical insulation we have just discussed.

Example 3.5: Will the rate of heat transfer decrease if asbestos insulation, $k = 0.151$ W/m$^2\cdot$K, is added to a 2-inch outer diameter pipe? Assume that the surface heat transfer coefficient is $h = 5$ W/m$^2\cdot$K.

Solution: Since

$$(r)_{cr} = \frac{k}{h} = \frac{0.151}{5} = 0.032 \text{ m} = 3.2 \text{ cm}$$

and the radius of the pipe 2.54 cm < 3.2 cm, the heat loss from the pipe will increase until the insulation thickness is made greater than 0.66 cm.

3.3 ONE-DIMENSIONAL STEADY-STATE HEAT CONDUCTION WITH HEAT SOURCES

Heat conduction problems with heat sources are frequently encountered in several applications. For example, an electric current flowing through a body has the effect of an external energy addition (power input) to the internal portions of the body because of the dissipation due to electrical resistance. This is measured by the quantity $I^2 R_e$ (Current^2 x Electrical Resistance). Since the dissipated energy is to be transferred out of the body by some heat transfer mechanisms the effect of the electric current is said to be a heat source distributed throughout the body. Processes which produce similar effects are chemical reactions distributed throughout a body, nuclear reactions in a fissionable material exposed to a neutron flux, change of phase, and the biological problems of fermentation. Internal energy generation in the latter cases, however, cannot be identified as power input from an external source.

In the following sections we shall consider certain one-dimensional steady-state heat conduction problems which are idealizations of more involved problems frequently encountered in practice.

3.3.1 Plane Wall

Consider a plane wall of thickness 2L as shown in Fig. 3.14. Internal energy is generated at a uniform rate $\dot{q}$ per unit volume throughout this wall. It is exposed to a fluid at temperature T_f on both sides with a constant heat transfer coefficient, h, on both surfaces. It is assumed that the dimensions of the wall in the other two directions are sufficiently large so that the heat flow may be considered as one dimensional. We also assume that the thermal conductivity, k, of the material of the wall is constant. The differential equation which governs the temperature distribution in the wall under steady-state conditions can be obtained from the general heat conduction equation (2.18) as

$$\frac{d^2T}{dx^2} + \frac{\dot{q}}{k} = 0 \tag{3.60}$$

As mentioned in Section 3.2.1, a problem may also be formulated starting from the basic principles. By doing so, one brings the physics of the problem into each phase of the derivation of the differential equation. Proceeding according to the five steps outlined in Section 3.2.1 we now derive Eq. (3.60):

(a) *Coordinate System:* Let the horizontal rightward direction be denoted by the positive x-axis.

(b) *System:* Consider the one-dimensional differential system of thickness, Δx shown in Fig. 3.14.

(c) *First Law:* If internal energy generation is due to an electric current passing through this wall, then application of the first law of thermodynamics (1.4b) to this system gives

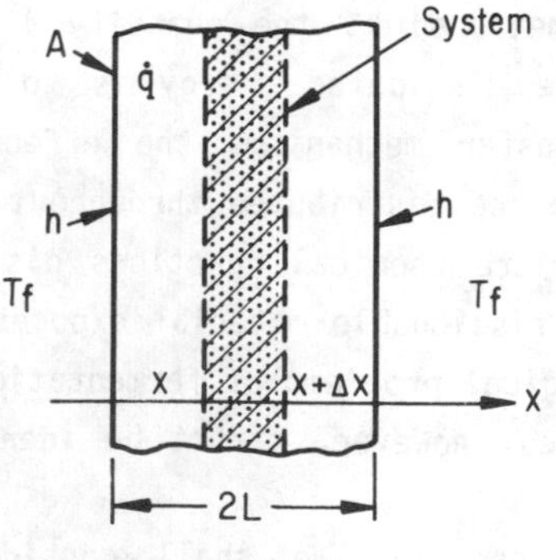

Fig. 3.14 A plane wall with internal energy generation.

$$\dot{Q}_x - \left(\dot{Q}_x + \frac{d\dot{Q}_x}{dx} \Delta x \right) + \dot{q}A\Delta x = 0 \tag{3.61a}$$

since $dE/dt = 0$ and $\dot{W} = \dot{q}A\Delta x$.

If internal energy is generated due to chemical or nuclear reactions distributed throughout the body, application of Eq. (1.4b) yields

$$\dot{Q}_x - \left(\dot{Q}_x + \frac{d\dot{Q}_x}{dx} \Delta x \right) = -\dot{q}A\Delta x \tag{3.61b}$$

since $dE/dt = -\dot{q}A\Delta x$ and $\dot{W} = 0$. In Eqs. (3.61a) and (3.61b) $\dot{Q}_x$ represents the rate of heat transfer across the isothermal surface at location x.

Whatever the source of internal energy generation may be, the first law of thermodynamics gives

$$-\frac{d\dot{Q}_x}{dx} + \dot{q}A = 0 \tag{3.62}$$

(d) *Fourier's Law and Differential Equation:* Introduce Fourier's law of heat conduction:

$$\dot{Q}_x = -kA\frac{dT}{dx} \tag{3.63}$$

Substitution of Eq. (3.63) into Eq. (3.62) yields

$$\frac{d^2T}{dx^2} + \frac{\dot{q}}{k} = 0 \tag{3.64}$$

which is the same as Eq. (3.60).

(e) *Boundary Conditions:* Since the problem is a steady-state problem no initial condition is required. The order of the x-derivative in Eq. (3.64) requires that two boundary conditions be specified in x-direction. Before we specify the boundary conditions, the origin of the coordinate system must be identified. Noting the thermal as well as the geometric symmetries of the problem the origin of the coordinate system is selected as shown in Fig. 3.15. Thus, the boundary conditions can be written in the form

$$\left(\frac{dT}{dx} \right)_{x=0} = 0 \tag{3.65a}$$

and

$$\left[k\frac{dT}{dx} + hT \right]_{x=L} = hT_f \tag{3.65b}$$

Equation (3.64) and the boundary conditions (3.65) complete the formulation of the problem. Integrating Eq. (3.64) twice yields

$$T(x) = -\frac{\dot{q}x^2}{2k} + C_1x + C_2 \tag{3.66}$$

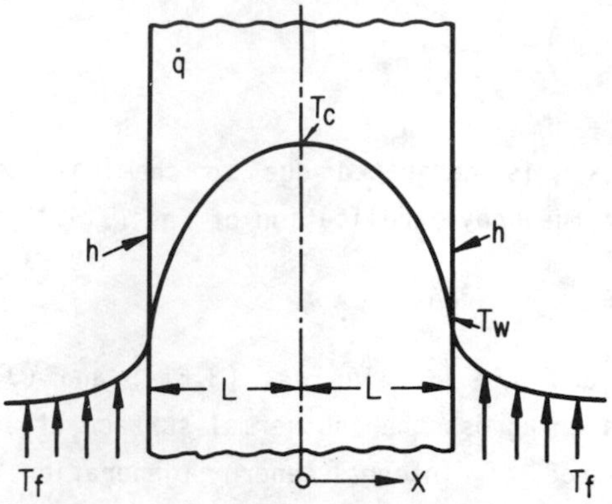

Fig. 3.15 Temperature distribution in a plane wall with uniform internal energy generation.

Applying the boundary condition (3.65a) we get $C_1 = 0$, and the boundary condition (3.65b) gives

$$C_2 = \frac{\dot{q}L^2}{2k} + \frac{\dot{q}L}{h} + T_f$$

Substituting these constants into Eq. (3.66) we obtain the temperature distribution in the wall:

$$T(x) - T_f = \frac{\dot{q}L^2}{2k}\left[1 - \left(\frac{x}{L}\right)^2\right] + \frac{\dot{q}L}{h} \tag{3.67}$$

Temperature of the midplane is then given by

$$T_c - T_f = \frac{\dot{q}L^2}{2k} + \frac{\dot{q}L}{h} \tag{3.68}$$

and the surface temperature is

$$T_w - T_f = \frac{\dot{q}L}{h} \tag{3.69}$$

Hence, the temperature drop from the midplane to the surface is

$$T_c - T_f = \frac{\dot{q}L^2}{2k} \tag{3.70}$$

The temperature distribution (3.67) can be written in dimensionless form as

$$\frac{T - T_f}{\dot{q}L^2/2k} = 1 - \left(\frac{x}{L}\right)^2 + \frac{2}{Bi} \tag{3.71}$$

where $Bi = hL/k$. If $Bi >> 1$, then Eq. (3.71) reduces to

$$\frac{T - T_f}{\dot{q}L^2/2k} = 1 - \left(\frac{x}{L}\right)^2 \tag{3.72}$$

Note that in this case $T_w \to T_f$, and Eq. (3.72) is the solution of the differential equation (3.64) for the following boundary conditions:

$$\left(\frac{dT}{dx}\right)_{x=0} = 0 \tag{3.73a}$$

$$T(L) = T_f \tag{3.73b}$$

The total energy generated in the wall should be equal to the energy lost to the surrounding fluid. Therefore, the energy lost to the surrounding fluid per unit time is

$$\dot{Q}_L = 2LA\dot{q} \tag{3.74}$$

The energy lost to the surrounding fluid per unit time can also be calculated as follows:

$$\dot{Q}_L = 2\left(-kA\frac{dT}{dx}\right)_{x=L} = -2kA\left(-\frac{L\dot{q}}{k}\right) = 2LA\dot{q} \tag{3.75}$$

which, as expected, is the same result as Eq. (3.74).

3.3.2 Solid Cylinder

Consider a long solid cylinder of radius r_1 with uniformly distributed heat sources and constant thermal conductivity as shown in Fig. 3.16. This cylinder is exposed to a fluid at temperature T_f with a constant heat transfer coefficient h on the surface. The governing differential equation from the general heat conduction equation (2.18) is

$$\frac{d^2T}{dr^2} + \frac{1}{r}\frac{dT}{dr} + \frac{\dot{q}}{k} = 0 \tag{3.76}$$

where $\dot{q}$ is the uniform rate of internal energy generation per unit volume. The boundary conditions can be written as

$$\left(\frac{dT}{dr}\right)_{r=0} = 0 \tag{3.77a}$$

$$-k\left(\frac{dT}{dr}\right)_{r=r_1} = h[T(r) - T_f]_{r=r_1} \tag{3.77b}$$

Integrating Eq. (3.76) twice yields

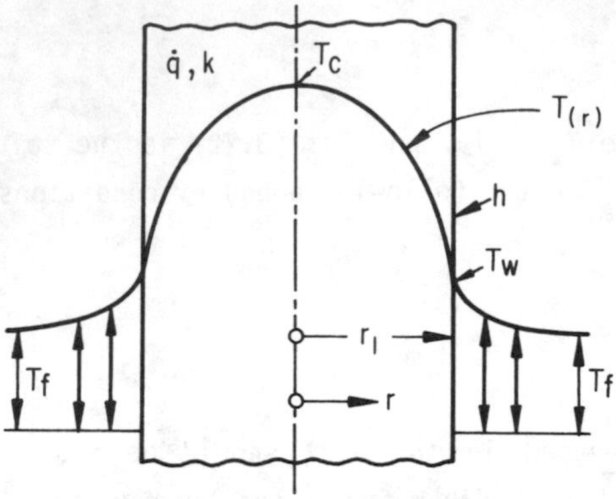

Fig. 3.16. Solid cylinder with uniform internal energy generation.

$$T(r) = \frac{\dot{q}r^2}{4k} + C_1 \ln r + C_2 \tag{3.78}$$

The boundary condition (3.77a), or the fact that the temperature should be finite at $r = 0$, requires $C_1 = 0$. The second constant C_2 follows from the boundary condition (3.77b):

$$C_2 = T_f + \frac{\dot{q}r_1^2}{4k}\left(1 + \frac{2k}{hr_1}\right)$$

Thus, for the temperature distribution we obtain

$$T(r) - T_f = \frac{\dot{q}r_1^2}{4k}\left[1 - \left(\frac{r}{r_1}\right)^2 + \frac{2k}{hr_1}\right] \tag{3.79}$$

The temperature drop from the centerline to the surface of the cylinder is then given by

$$(\Delta T)_{max} = T_c - T_w = \frac{\dot{q}r_1^2}{4k} \tag{3.80}$$

In the limiting case when $h \to \infty$, $T_w \to T_f$ and the temperature distribution (3.79) reduces to

$$T(r) - T_w = \frac{\dot{q}r_1^2}{4k}\left[1 - \left(\frac{r}{r_1}\right)^2\right] \tag{3.81}$$

The rate of heat release per unit length of the cylinder is

$$\dot{Q}_L = \dot{q}\pi r_1^2 \tag{3.82}$$

Hence, Eq. (3.80) can be written as

$$(\Delta T)_{max} = T_c - T_w = \frac{\dot{Q}_L}{4\pi k} \tag{3.83}$$

Thus, for a given heat release rate per unit length, for example, from a fuel rod in a nuclear reactor, the maximum radial temperature difference in the rod is independent of the rod diameter. Conversely, for a fixed surface temperature, T_w, the maximum heat release rate per unit length is determined by the maximum permissible temperature, (*i.e.*, centerline temperature) and is independent of the rod diameter.

Example 3.6: A Cr-Ni steel wire, 2.5 mm in diameter and 30 cm in length, has a voltage of 10 volts applied on it, while its surface is maintained at 90°C. Assuming that the resistivity of the wire is 70 μ ohm-cm and the thermal conductivity is 17.3 W/m·K, calculate the temperature at the centerline of the wire.

Solution: The temperature difference between the centerline and the surface is

$$T_c - T_w = \frac{\dot{q}r_1^2}{4k}$$

where $\dot{q}$ is given by

$$\dot{q} = \frac{I^2R_e}{\pi r_1^2 L} = \left(\frac{V}{R_e}\right)^2 \frac{R_e}{\pi r_1^2 L} = \frac{V^2}{\pi R_e r_1^2 L}$$

The electrical resistance, R_e, of the wire is calculated from

$$R_e = \rho \times \frac{L}{A} = (70 \times 10^{-6}) \times \frac{30}{\pi\left(\frac{0.25}{2}\right)^2} = 4.27 \times 10^{-2} \text{ ohm}$$

Therefore,

$$\dot{q} = \frac{(10)^2}{\pi \times 4.27 \times 10^{-2} \times (1.25 \times 10^{-3})^2 \times 0.3}$$

$$= 1.59 \times 10^9 \text{ W/m}^3$$

The centerline temperature of the wire is

$$T_c = T_w + \frac{\dot{q}r_1^2}{4k} = 90 + \frac{1.59 \times 10^9 \times (1.25 \times 10^{-3})^2}{4 \times 17.3}$$

$$= 90 + 35.9 = 126°C$$

3.3.3 Effect of Cladding

In nuclear reactors, mostly cylindrical fuel elements are used. The simplest arrangement, found in graphite-moderated and gas-cooled reactors fueled with natural uranium, consists of solid uranium rods of approximately 1-inch diameter stacked inside sealed cladding tubes of magnesium alloy (magnox).

Consider now a fuel element as shown in Fig. 3.17. Thickness of the cladding material is $r_2 - r_1$. Internal energy is generated in the fuel rod due to fission reactions only. This is a multi-domain problem. If we denote the properties of the fuel and the cladding by subscripts 1 and 2, respectively, we have

$$\frac{d^2T_1}{dr^2} + \frac{1}{r}\frac{dT_1}{dr} + \frac{\dot{q}}{k_1} = 0, \qquad 0 < r < r_1 \tag{3.84}$$

$$\frac{d^2T_2}{dr^2} + \frac{1}{r}\frac{dT_2}{dr} = 0, \qquad 0 < r < r_2 \tag{3.85}$$

where $\dot{q}$ is the uniform rate of internal energy generation per unit volume in the fuel rod. The boundary conditions can be written as

$$\frac{dT_1(0)}{dr} = 0 \tag{3.86a}$$

$$T_1(r_1) = T_2(r_1) = T_w \tag{3.86b}$$

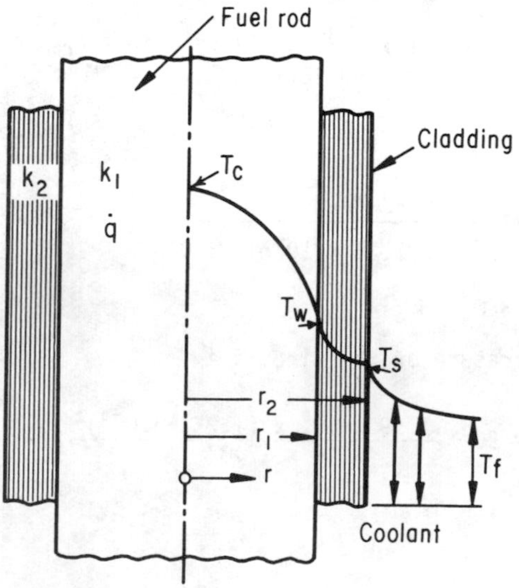

Fig. 3.17 Fuel element with cladding.

$$k_1 \frac{dT_1(r_1)}{dr} = k_2 \frac{dT_2(r_2)}{dr} \tag{3.86c}$$

$$T_2(r_2) = T_s \tag{3.86d}$$

where we have assumed that there is a perfect contact between the fuel rod and the cladding, and the interface temperature is denoted by T_w. Generally we do not expect to achieve perfect thermal contact. In some fuel element designs, in fact, there is usually a space between the fuel rod and the cladding which may, for example, be filled with helium. The surface temperature of the cladding, T_s, is assumed to be given. By solving Eqs. (3.84) and (3.85) with the boundary conditions (3.86) we obtain

$$T_w - T_s = \frac{\dot{Q}_L}{2\pi k_2} \ln \frac{r_2}{r_1} \tag{3.87a}$$

$$T_c - T_s = \frac{\dot{Q}_L}{4\pi k_1} + \frac{\dot{Q}_L}{2\pi k_2} \ln \frac{r_2}{r_1} \tag{3.87b}$$

where $\dot{Q}_L = \dot{q}\pi r_1$.

In general, however, some allowance must be given for thermal contact resistance between the fuel rod and the cladding. The overall temperature drop from the centerline of the fuel rod to the cladding surface may then be expressed in the form.

$$T_c - T_s = \frac{\dot{Q}_L}{4\pi k_1} + \Delta T_c + \frac{\dot{Q}_L}{2\pi k_2} \ln \frac{r_2}{r_1} \tag{3.88}$$

where ΔT_c is the temperature drop at the interface.

If the coolant temperature, T_f, and the surface heat transfer coefficient, h, are known, then the rate of heat flow per unit length can be written as

$$\dot{Q}_L = h\, 2\pi r_2 (T_s - T_f)$$

which yields

$$T_s - T_f = \frac{\dot{Q}_L}{2\pi r_2 h} = \frac{\dot{q} r_1^2}{2 h r_2} \tag{3.89}$$

Combining Eqs. (3.88) and (3.89) we find the temperature difference between the center of the rod and the fluid as

$$T_c - T_f = \frac{\dot{Q}_L}{4\pi k_1} + \Delta T_c + \frac{\dot{Q}_L}{2\pi k_2} \ln \frac{r_2}{r_1} + \frac{\dot{Q}_L}{2\pi r_2 h} \tag{3.90}$$

Similarly, for a long flat plate of a fissionable material with cladding on both sides, the following relation is obtained.

$$T_c - T_f = \frac{\dot{q}L^2}{2k_1} + \frac{\dot{q}L}{k_2} a + \frac{\dot{q}L}{h} + \Delta T_c \tag{3.91}$$

where 2L and a are the thicknesses of the plate and the cladding, respectively.

It should be noted that the following relation exists between the surface heat flux and the heat generation rate:

$$q_s A = \dot{q}V \tag{3.92a}$$

where

V = volume

A = total surface area

q_s = surface heat flux

Therefore,

$$q_s = \frac{\dot{q}V}{A} \tag{3.92b}$$

It follows from Eq. (3.92b) that the surface heat flux depends only on the strength of the internal sources and the ratio of volume to surface area.

In general, it can be shown that

$$(\Delta T)_{max} = T_c - T_w = \frac{\dot{q}R^2}{mk} \tag{3.93}$$

where

Geometry	m	R	V/A
slab	2	L	L
cylinder	4	r_1	$r_1/2$
sphere	6	r_1	$r_1/3$

3.4 TEMPERATURE-DEPENDENT THERMAL CONDUCTIVITY

When the thermal conductivity is temperature dependent, the heat conduction equation under steady-state conditions from Eq. (2.14) is given by

$$\vec{\nabla} \cdot (k\vec{\nabla}T) + \dot{q} = 0 \tag{3.94}$$

Although this is a non-linear partial differential equation, it can be reduced to a linear differential equation by means of the Kirchhoff's transformation we introduced in Section 2.4 by defining a new temperature function, θ, as

$$\theta = \frac{1}{k_0} \int_{T_0}^{T} k(T')dT' \tag{3.95}$$

where T_0 denotes a convenient reference temperature and $k_0 = k(T_0)$.

It follows from Eq. (3.95) that

$$\vec{\nabla}\theta = \frac{k}{k_0} \vec{\nabla}T \tag{3.96}$$

Therefore, Eq. (3.94) can be written as

$$\nabla^2\theta + \frac{\dot{q}}{k_0} = 0 \tag{3.97}$$

which is similar to the heat conduction equation for constant k. Hence, a steady-state problem with temperature-dependent thermal conductivity poses no problem as the transformed equation can be solved with the usual techniques, provided that the boundary conditions can also be transformed. As we have discussed in Section 2.4, if a boundary condition is of either the first or the second kind then it can be transformed. The boundary condition of the third kind, in general, cannot be transformed, and such problems are usually solved using numerical techniques.

We have already discussed in Section 3.2.1 a steady one-dimensional heat conduction problem with temperature-dependent thermal conductivity, but without heat sources. Here we shall consider another one-dimensional problem with temperature-dependent thermal conductivity involving uniform heat sources and employing the Kirchhoff's transformation: Consider a long rod of radius r_1. Assume that the internal energy is generated at a uniform rate $\dot{q}$ per unit volume and the surface is maintained at a uniform temperature T_w. The formulation of the problem is

$$\frac{1}{r}\frac{d}{dr}\left[r\, k(T) \frac{dT}{dr} \right] + \dot{q} = 0 \tag{3.98}$$

$$\left(\frac{dT}{dr}\right)_{r=0} = 0 \tag{3.99a}$$

$$T(r_1) = T_w \tag{3.99b}$$

Defining a new function, θ, as

$$\theta = \frac{1}{k_w} \int_{T_w}^{T(r)} k(T)dT \tag{3.100}$$

where $k_w = k(T_w)$, we may transform Eq. (3.98) and the boundary conditions

(3.99) to

$$\frac{1}{r}\frac{d}{dr}\left(r\frac{d\theta}{dr}\right) + \frac{\dot{q}}{k_w} = 0 \qquad (3.101)$$

$$\left(\frac{d\theta}{dr}\right)_{r=0} = 0 \qquad (3.102a)$$

$$\theta(r_1) = 0 \qquad (3.102b)$$

The solution of this problem is

$$\theta(r) = \frac{\dot{q}r_1^2}{4k_w}\left[1 - \left(\frac{r}{r_1}\right)^2\right] \qquad (3.103)$$

Introducing Eq. (3.103) into Eq. (3.100) we obtain the temperature of the rod T(r) as

$$\int_{T_w}^{T(r)} k(T)dT = \frac{\dot{q}r_1^2}{4}\left[1 - \left(\frac{r}{r_1}\right)^2\right] \qquad (3.104)$$

which at r = 0 reduces to

$$\int_{T_w}^{T_c} k(T)dT = \frac{\dot{q}r_1^2}{4} \qquad (3.105)$$

where T_c is the centerline temperature. For constant k Eq. (3.104) reduces to Eq. (3.81).

Example 3.7: Find the rate of heat generation per unit volume in a rod that will produce a centerline temperature of 2000°C (3632°F) for the following conditions:

$$r_1 = 0.24 \text{ in}$$

$$T_w = 350°C\ (= 662°F)$$

$$k = \frac{3300}{[T + 460]}$$

where k is in Btu/hr·ft·°F and T in °F. What is the heat flux at the surface?

Solution: From Eq. (3.105) we have

$$\dot{q} = \frac{4}{r_1^2}\int_{662}^{3632}\frac{3300}{T + 460}dT = \frac{4x(3300)}{(0.02)^2}\ln\frac{4092}{1122}$$

$$= 4.27 \times 10^7 \text{ Btu/hr·ft}^3$$

$$= 4.41 \times 10^8 \text{ W/m}^3$$

The surface heat flux from Eq. (3.92b) is

$$q_s = \frac{\dot{q}\pi r_1^2 L}{2\pi r_1 L} = \frac{\dot{q} r_1}{2} = \frac{4.41 \times 10^8 \times 0.24 \times 2.54 \times 10^{-2}}{2}$$

$$= 1.345 \times 10^6 \text{ W/m}^2$$

3.5 SPACE-DEPENDENT INTERNAL ENERGY GENERATION

Internal energy generation, or simply heat generation, may be a function of space and/or time. In a cylindrical nuclear fuel element, for example, the rate of heat generation per unit volume will be in the form

$$\dot{q}(r) = \dot{q}_0 \, I_0(\chi r) \tag{3.106}$$

where $\dot{q}_0$ is the heat generation rate per unit volume at the rod axis, $I_0(\chi r)$ is the modified Bessel function of the first kind of zero order (see Appendix B), and χ is the reciprocal of the neutron diffusion length. If we assume that the thermal conductivity is constant, then substituting Eq. (3.106) into Eq. (3.76) we obtain

$$\frac{d^2T}{dr^2} + \frac{1}{r}\frac{dT}{dr} = -\frac{\dot{q}_0 I_0(\chi r)}{k} \tag{3.107}$$

The solution of Eq. (3.107) can be written as

$$T(r) = -\frac{\dot{q}_0 I_0(\chi r)}{\chi^2 k} + A \ln r + B \tag{3.108}$$

Therefore, the temperature distribution T(r) in terms of the surface temperature T_w at $r = r_1$ is given by

$$T(r) - T_w = \frac{\dot{q}_0}{\chi^2 k}[I_0(\chi r_1) - I_0(\chi r)] \tag{3.109}$$

On the other hand, the relation between the surface heat flux, q_s, and the rate of heat generation per unit volume, $\dot{q}$, is

$$2\pi r_1 q_s = \int_0^{r_1} 2\pi r \, \dot{q}(r) dr \tag{3.110}$$

Substituting Eq. (3.106) into Eq. (3.110), and combining the result with Eq. (3.109) we get

$$T(r) - T_w = \frac{q_s}{k}\frac{I_0(\chi r_1) - I_0(\chi r)}{\chi I_1(\chi r_1)} \tag{3.111}$$

If k = k(T), the temperature distribution can be shown to be given by

$$\int_{T_w}^{T(r)} k(T)dT = q_s \frac{I_o(\chi r_1) - I_o(\chi r)}{\chi I_1(\chi r_1)} \tag{3.112}$$

which, of course, reduces to Eq. (3.111) for k = constant.

3.6 EXTENDED SURFACES - FINS AND SPINES

Heat is transferred in conventional heat exchangers from one fluid to another through a metal wall and the rate of heat transfer is directly proportional to the surface area of the wall and the temperature difference between the fluids. In most cases, however, the temperature difference is limited. Therefore, increasing the rate of heat transfer depends on increasing the *effective* heat transfer area. The effective heat transfer area on a wall surface can be increased by attaching thin metal strips, called fins, or spines to the surface. Although addition of metal fins or spines increases the effective heat transfer area, the average surface temperature of the fin or spine will not be the same as the original surface temperature, but will be closer to the surrounding fluid temperature. This causes the rate of heat transfer to be somewhat less than proportional to the extent of total heat transfer area.

Extended surfaces, fins and spines, can be of several types. Some of the commonly used ones are shown in Fig. 3.18.

In this section we shall discuss only the steady-state performances of one-dimensional extended surfaces under the following assumptions:

(a) Heat flow in the extended surface is steady.

(b) Thermal conductivity of the material of the extended surface is constant.

(c) No heat sources within the extended surface.

(d) Temperature of the surrounding fluid is uniform and constant.

(e) Heat transfer coefficient between the extended surface and the surrounding fluid is constant, and it includes the combined effect of convection and radiation.

(f) Temperature of the base of the extended surface is constant.

(g) Thickness of the extended surface is so small compared to its length that the temperature gradients normal to the surface may be neglected; that is, temperature distribution in the extended surface is one-dimensional.

Of these assumptions, (e), (f) and (g) may be questionable. Although the heat transfer coefficient on the surface of a fin or spine varies from point to point, the use of an average value in analytical solutions gives

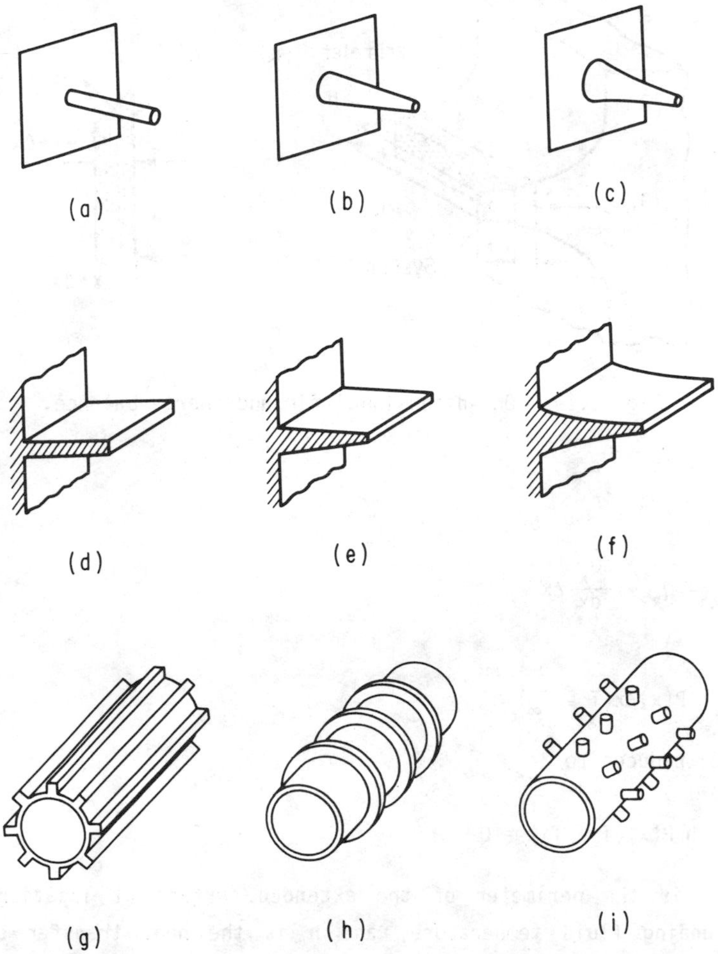

Fig. 3.18 Several types of extended surfaces: (a) cylindrical spine; (b) truncated conical spine; (c) parabolic spine; (d) longitudinal fin of rectangular profile; (e) longitudinal fin of trapezoidal profile; (f) longitudinal fin of parabolic profile; (g) cylindrical tube equipped with straight fins of rectangular profile; (h) cylindrical tube equipped with annular fins of rectangular profile; (i) cylindrical tube equipped with cylindrical spines.

heat transfer results which are in good agreement with experimental measurements. For most fins or spines of practical interest, on the other hand, the error introduced by the assumption (g) is less than 1%.

Let us now consider the diffusion of heat in the extended surface shown in Fig. 3.19. An energy balance (first law of thermodynamics), Eq. (1.4b), applied to the system of Fig. 3.19 gives

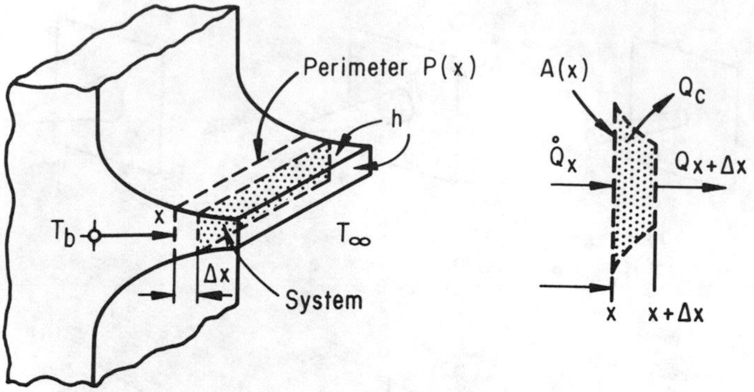

Fig. 3.19 One-dimensional fin and energy balance.

$$\dot{Q}_x = \dot{Q}_{x+\Delta x} + \dot{Q}_c \tag{3.113}$$

Since, as $\Delta x \rightarrow 0$,

$$\dot{Q}_{x+\Delta x} = \dot{Q}_x + \frac{d\dot{Q}_x}{dx}\Delta x \tag{3.114}$$

and

$$\dot{Q}_c = h\,P(x)\Delta x(T - T_\infty) \tag{3.115}$$

Eq. (3.113) reduces to

$$\frac{d\dot{Q}_x}{dx} + h\,P(x)(T - T_\infty) = 0 \tag{3.116}$$

where $P(x)$ is the perimeter of the extended surface at location x, T_∞ is the surrounding fluid temperature, and h is the heat transfer coefficient. From Fourier's law of heat conduction we have

$$\dot{Q}_x = -kA(x)\frac{dT}{dx} \tag{3.117}$$

where $A(x)$ is the cross-sectional area of the extended surface at location x. Substitution of Eq. (3.117) into Eq. (3.116) yields

$$\frac{d}{dx}\left[A(x)\frac{dT}{dx}\right] - \frac{hP(x)}{k}(T - T_\infty) = 0 \tag{3.118}$$

Defining a new temperature function $\theta = T - T_\infty$, and since T_∞ is constant, Eq. (3.118) can be rewritten as

$$\frac{d}{dx}\left[A(x)\frac{d\theta}{dx}\right] - \frac{hP(x)}{k}\theta = 0 \tag{3.119}$$

This is the differential equation which governs the temperature distribution in the extended surface. As this equation is of second order, two boundary conditions are needed in the x-direction; one related to the base and the other to the tip of the extended surface.

3.6.1 Extended Surfaces with Constant Cross-sections

For an extended surface with a constant cross section, Eq. (3.119) reduces to

$$\frac{d^2\theta}{dx^2} - m^2\theta = 0 \tag{3.120}$$

where $m^2 \equiv hP/kA$. The general solution of this differential equation can be written in the form

$$\theta(x) = C_1\, e^{-mx} + C_2\, e^{mx} \tag{3.121a}$$

or

$$\theta(x) = C_3 \sinh mx + C_4 \cosh mx \tag{3.121b}$$

where C_1 and C_2 or C_3 and C_4 are constants of integration to be determined from the boundary conditions. Since the base temperature, T_b, of the extended surface was assumed constant, the boundary condition at x = 0 is

$$T(x)|_{x=0} = T_b \qquad \text{or} \qquad \theta(x)|_{x=0} = T_b - T_\infty = \theta_b \tag{3.122}$$

The second boundary condition at the tip of the extended surface depends upon the nature of the problem. Several cases may be considered:

CASE 1: The extended surface under consideration can be very long as shown in Fig. 3.20. In that case the temperature at the tip is essentially equal to the temperature of the surrounding fluid, and the second boundary condition can therefore be written as

$$T(x)|_{x\to\infty} \to T_\infty \qquad \text{or} \qquad \theta(x)|_{x\to\infty} \to 0 \tag{3.123}$$

The boundary conditions (3.122) and (3.123) give $C_1 = 0$ and $C_2 = \theta_b$ in Eq. (3.121a). Hence the temperature distribution is found to be

$$\theta(x) = \theta_b e^{-mx} \tag{3.124a}$$

or

$$\frac{T - T_\infty}{T_b - T_\infty} = e^{-mx} \tag{3.124b}$$

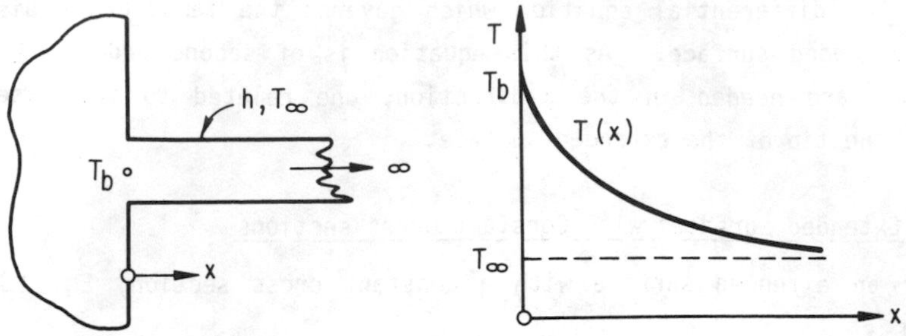

Fig. 3.20 A long extended surface.

The rate of heat transfer from the extended surface can now be determined by integrating the local convective heat transfer over the whole length

$$\dot{Q}_f = \int_0^\infty hPdx(T - T_\infty) = hP\int_0^\infty \theta(x)dx = hP\theta_b\int_0^\infty e^{-mx}dx = \frac{hP\theta_b}{m} = \sqrt{hPkA}\theta_b \tag{3.125}$$

The heat transferred from the extended surface by convection to the surrounding fluid is supplied to the base of the extended surface by conduction. Hence, we may also evaluate the heat transfer rate, $\dot{Q}_f$, from the extended surface by applying Fourier's law at the base:

$$\dot{Q}_f = -kA\left(\frac{dT}{dx}\right)_{x=0} = -kA\left(\frac{d\theta}{dx}\right)_{x=0} = -kA\theta_b\frac{d}{dx}\left[e^{-mx}\right]_{x=0} = kA\theta_b m = \sqrt{hPkA}\theta_b \tag{3.126}$$

Since it involves differentiation, this method is usually easier to apply than the first method which requires integration.

Example 3.8: A very long rod of 2.5 cm diameter is heated at one end. Under steady-state conditions, the temperatures at two different locations which are 7.5 cm apart are measured to be 125°C and 90°C, while the surrounding air temperature is 25°C. Assuming that the heat transfer coefficient is 20 W/m²·K, find the value of the thermal conductivity of the material of the rod.

Solution: Since the rod is very long, the temperature distribution in it will be given by Eq. (3.124b). Therefore, we can write

$$T(x_1) - T_\infty = (T_b - T_\infty)e^{-mx_1}$$

$$T(x_2) - T_\infty = (T_b - T_\infty)e^{-mx_2}$$

from which we obtain

$$\frac{T(x_1) - T_\infty}{T(x_2) - T_\infty} = e^{m(x_2 - x_1)}$$

or substituting the numerical values we get

$$\frac{125 - 25}{90 - 25} = e^{7.5m}$$

which gives

$$m = 5.74 \times 10^{-2} \text{ cm}^{-1}$$

Since

$$m = \sqrt{\frac{hP}{kA}} = \sqrt{\frac{4h\pi D}{k\pi D^2}} = \sqrt{\frac{4h}{kD}}$$

where D is the diameter of the rod, then

$$k = \frac{4h}{m^2 D} = \frac{4 \times 20}{(5.74)^2 \times 2.5 \times 10^{-2}} = 97 \text{ W/m·K}$$

CASE II: The extended surface under consideration may be of finite length L as shown in Fig. 3.21, and heat is lost to the surrounding fluid by convection from its end. In this case, the boundary condition at x = L is

$$-k\left(\frac{dT}{dx}\right)_{x=L} = h_e[T(L) - T_\infty] \tag{3.127}$$

where h_e is the heat transfer coefficient at the end of the extended surface, which may or may not be equal to h. From the boundary condition (3.122) at x = 0 we have $C_4 = \theta_b$ in the general solution (3.121b), and the boundary condition (3.127) at x = L yields

$$C_3 = -\theta_b \frac{mk \sinh mL + h_e \cosh mL}{mk \cosh mL + h_e \sinh mL} \tag{3.128}$$

Substituting these constants into Eq. (3.121b), the temperature distribution

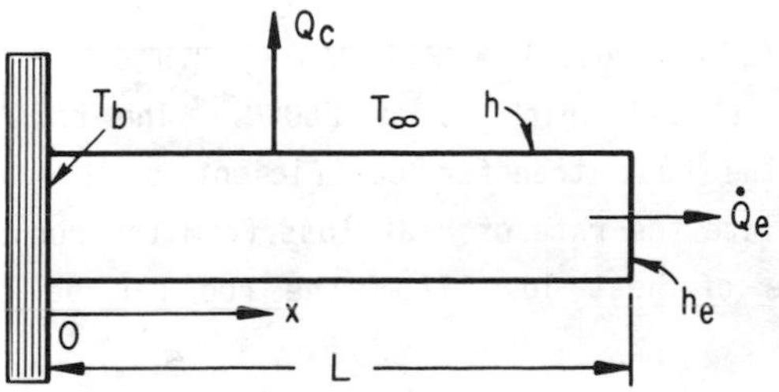

Fig. 3.21 An extended surface of finite length.

is found to be

$$\frac{\theta(x)}{\theta_b} = \frac{T(x) - T_\infty}{T_b - T_\infty} = \frac{\cosh[m(L - x)] + N \sinh[m(L - x)]}{\cosh mL + N \sinh mL} \tag{3.129}$$

where $N = h_e/mk$.

Following the same procedure as in the previous case, one can easily show that the rate of heat transfer from the extended surface to the fluid surrounding it is given by

$$\dot{Q}_f = \sqrt{hPkA}\ \theta_b \frac{\sinh mL + N \cosh mL}{\cosh mL + N \sinh mL} \tag{3.130}$$

If the heat loss from the end of the extended surface under consideration is negligible or if the end is insulated, then the temperature distribution can be obtained from Eq. (3.129) by letting N = 0 (why?) as

$$\frac{\theta(x)}{\theta_b} = \frac{\cosh[m(L - x)]}{\cosh mL} \tag{3.131}$$

and the rate of heat transfer from the extended surface becomes

$$\dot{Q}_f = \sqrt{hPkA}\ \theta_b \tanh mL \tag{3.132}$$

The temperature at the tip of the extended surface (x = L) is found from Eq. (3.129) to be

$$\frac{\theta(L)}{\theta_b} = \frac{1}{\cosh mL + N \sinh mL} \tag{3.133}$$

If the end is insulated or if the heat loss from the end is negligible, then

$$\frac{\theta(L)}{\theta_b} = \frac{1}{\cosh mL} \tag{3.134}$$

The last two equations may be applied to the problem of temperature measurement. That is, they can be used, for example, for determining the tip temperature of a thermocouple immersed in a gas stream at a temperature different than that of the duct wall which supports the thermocouple.

Example 3.9: An aluminum rod, k = 206 W/m·K, 2.5 cm in diameter and 15 cm long protrudes from a wall which is at 260°C. The rod is surrounded by a fluid at 16°C and the heat transfer coefficient on the surface of the rod is 15 W/m²·K. Calculate the rate of heat loss from the rod.

Solution: The rate of heat loss from the rod can be calculated from Eq. (3.130). We have

$$P = \pi D = \pi \times 2.5 \times 10^{-2} = 7.85 \times 10^{-2}\,\text{m}$$

$$A = \frac{\pi D^2}{4} = \frac{\pi \times (2.5 \times 10^{-2})^2}{4} = 4.91 \times 10^{-4} m^2$$

Hence,

$$m = \sqrt{\frac{hP}{kA}} = \left(\frac{15 \times 7.85 \times 10^{-2}}{206 \times 4.91 \times 10^{-4}}\right)^{1/2} = 3.412\ m^{-1}$$

$$mL = 3.412 \times 0.15 = 0.512$$

$$N = \frac{h}{km} = \frac{15}{206 \times 3.412} = 0.0213$$

$$\sinh mL = 0.5346$$

$$\cosh mL = 1.1340$$

$$\sqrt{hPkA} = (15 \times 7.85 \times 10^{-2} \times 206 \times 4.91 \times 10^{-4})^{1/2}$$

$$= 0.345\ W/K$$

Substituting these values into Eq. (3.130) we obtain

$$\dot{Q}_f = 0.345 \times (260 - 16) \times \frac{0.5346 + 0.0213 \times 1.1340}{1.1340 + 0.0213 \times 0.5344} = 41.07\ W$$

Example 3.10: Hot water flows inside a steel tube (k = 41 W/m·K) of 1-inch inside diameter with air flowing on the outside as illustrated in Fig. 3.22. There are 8 longitudinal steel fins of rectangular profile on the outside surface of the tube and each fin is 0.75 inches long. The thickness of both the tube and the fins is 0.25 cm. Heat transfer coefficients on the inside and outside are 5000 W/m²·K and 12 W/m²·K, respectively. Assuming that the heat transfer from the tip of the fins is negligible, calculate the rate of heat transfer per meter length of the tube if the average tube wall and air temperature are 95°C and 15°C, respectively.

Solution: The heat transfer rate from one fin only per meter length of the tube can be calculated from

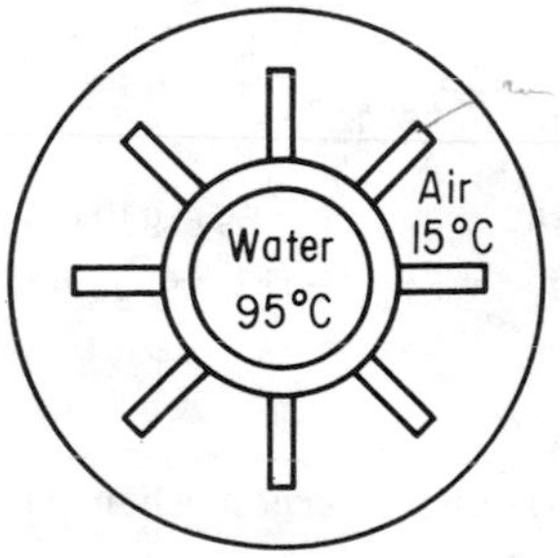

Fig. 3.22 Figure for Example 3.10.

$$\dot{Q} = \sqrt{hPkA}\ \theta_b \tanh mL$$

where

$$P = 2m, \qquad A = 0.25 \times 10^{-2} m^2$$

$$m = \sqrt{\frac{hP}{kA}} = \left(\frac{12 \times 2}{41 \times 0.25 \times 10^{-2}}\right)^{1/2} = 15.3\ m^{-1}$$

$$mL = 15.3 \times 0.75 \times 2.54 \times 10^{-2} = 0.2915$$

$$\tanh mL = 0.2835$$

Hence,

$$\dot{Q} = \left(12 \times 2 \times 41 \times 25 \times 10^{-4}\right)^{\frac{1}{2}} \times (95 - 15) \times 0.2835 = 35.57\ W$$

thus the rate of heat transfer from eight fins is

$$\dot{Q}_f = 8 \times 35.57 = 284.6\ W$$

The rate of heat transfer from the unfinned section of the tube per meter length, on the other hand, can be calculated from

$$\dot{Q}_{unf} = h\ A_{unf}\ \Delta T$$

where

$$A_{unf} = \pi \times OD - 8 \times 0.25 \times 10^{-2}$$

$$= \pi \times (2.54 + 0.5) \times 10^{-2} - 2 \times 10^{-2}$$

$$= 7.55 \times 10^{-2} m^2$$

Hence,

$$\dot{Q}_{unf} = 12 \times 7.55 \times 10^{-2} \times 80 = 72.48\ W$$

Therefore, the rate of heat transfer per meter length of the tube is

$$\dot{Q}_{total} = \dot{Q}_f + \dot{Q}_{unf} = 284.6 + 72.48 = 357.08\ W$$

3.6.2 Rectangular Fin of Least Material

If heat transfer from the end is negligible, the rate of heat transfer from the rectangular fin shown in Fig. 3.23 is given by

$$\dot{Q} = \sqrt{hPk\delta}\ \theta_b \tanh mL \qquad (1.135)$$

Since the width of the fin is much larger than its thickness (*i.e.*, $\delta \ll 1$), $P = 2$. Therefore,

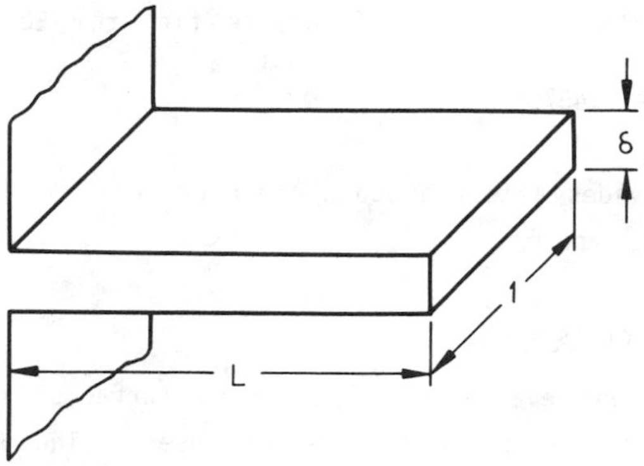

Fig. 3.23. A rectangular fin.

$$\dot{Q} = \sqrt{2hk\delta}\ \theta_b \tanh\left(\sqrt{\frac{2h}{k}}\ \frac{L\delta}{\delta^{3/2}}\right) \tag{3.136}$$

Schmidt [14] has suggested that the criterian for the most favorable fin dimensions might be the maximum value of $\dot{Q}/\theta_b$ for a given amount of fin material (which is proportional to the product $L\delta$). Taking h, k and the product $L\delta$ to be constants, if we differentiate Eq. (3.136) with respect to δ and set the result equal to zero, we obtain

$$M = \frac{1}{6}\sinh 2M \tag{3.137}$$

where

$$M = mL = \sqrt{\frac{2h}{k}}\ \frac{L\delta}{\delta^{3/2}} \tag{3.138}$$

The transcendental equation (3.137) can be solved numerically, or graphically by plotting both sides against M and determining the point of intersection of the two curves. Thus, the value M = mL = 1.4192 that satisfies Eq. (3.137) is obtained. The most favorable thickness-to-length ratio is then given by

$$\frac{\delta}{L} = \frac{2}{(1.4192)^2}\ \frac{hL}{k} \tag{3.139}$$

Combining this result with Eq. (3.136) we obtain

$$L = \frac{0.7978}{h}\ \frac{\dot{Q}}{\theta_b} \tag{3.140}$$

and

$$\delta = \frac{0.6321}{hk}\left(\frac{\dot{Q}}{\theta_b}\right)^2 \tag{3.141}$$

The temperature at the tip of this favorable fin from Eq. (3.134) will be

$$\frac{\theta(L)}{\theta_b} = \frac{1}{\cosh mL} = 0.45706 \tag{3.142}$$

Equation (3.142) provides a test as to whether or not L is the optimum length by measurement of $\theta(L)$ and θ_b.

3.6.3 Performance Factors

For comparison and evaluation of extended surfaces in augmenting heat transfer from the base area two factors are used: The *fin effectiveness*, ϕ, of an extended surface is defined as the ratio of the rate of heat transfer from the extended surface to the rate of heat transfer that would take place from the base area, A_b, without the extended surface, with the base temperature T_b remaining constant; that is

$$\phi = \frac{\dot{Q}_{fin}}{\dot{Q}_{base}} = \frac{\int_{A_f} h\theta(x)\,dA}{h\theta_b A_b} = \frac{\int_{A_f} \theta(x)\,dA}{A_b\theta_b} \tag{3.143}$$

where A_f is the surface area of the extended surface.

The *fin efficiency*, η_f, of an extended surface is defined as the ratio of the rate of heat transfer from the extended surface to the rate of heat transfer if the extended surface were uniformly at the base temperature, T_b, throughout its length; that is,

$$\eta_f = \frac{\dot{Q}_{fin}}{\dot{Q}_{fin}(\text{when } \theta = \theta_b \text{ along the fin})}$$

$$= \frac{\int_{A_f} h\theta(x)\,dA}{h\,A_f\theta_b} = \frac{\int_{A_f} \theta(x)\,dA}{A_f\theta_b} \tag{3.144}$$

A comparison of the expressions for ϕ and η_f reveals that

$$\phi = \frac{A_f}{A_b}\eta_f \tag{3.145}$$

For example, the effectiveness and efficiency of the extended surface shown in Fig. 3.21 are obtained by combining Eqs. (3.130), (3.143) and (3.144) as

$$\phi = \sqrt{\frac{kP}{hA}}\,\frac{\sinh mL + N\cosh mL}{\cosh mL + N\sinh mL} \tag{3.146}$$

$$\eta_f = \frac{\sqrt{hPkA}}{hPL + h_e A} \; \frac{\sinh mL + N \cosh mL}{\cosh mL + N \sinh mL} \tag{3.147}$$

If the heat transfer from the end is negligible or if the end is insulated (*i.e.*, $N \simeq 0$), then Eqs. (3.146) and (3.147) become

$$\phi = \sqrt{\frac{kP}{hA}} \tanh mL \tag{3.148}$$

$$\eta_f = \frac{\tanh mL}{mL} \tag{3.149}$$

Equations (3.148) and (3.149) can also be obtained from Eq. (3.132). In engineering applications the fin efficiency η_f is more widely used. In literature, the efficiencies for various fin configurations have been presented in graphical forms [13, 15].

For a fin of given material and shape, the efficiency will decrease as h increases. For example, a fin which is highly efficient when used with a gas coolant will usually be found hopelessly inefficient when used with water where the value of h is usually much larger.

3.6.4 Heat Transfer From a Finned Wall

The heat transfer rate from the fins on a wall from Eq. (3.144) is

$$\dot{Q}_f = \eta_f h \, A_f (T_b - T_\infty) \tag{3.150}$$

where A_f is the total fin surface area. The rate of heat removed from the surfaces between the fins is

$$\dot{Q}_w = hA_w (T_b - T_\infty) \tag{3.151}$$

where A_w is the total wall surface area between the fins. Therefore, the total rate of heat transfer is

$$\dot{Q}_T = \dot{Q}_f + \dot{Q}_w = h(A_w + \eta_f A_f)(T_b - T_\infty) = h \, A_{eff}(T_b - T_\infty) \tag{3.152}$$

where A_{eff} is called the *effective heat transfer area* of the wall. In other words, $A_w + \eta_f A_f$ is the heat transfer area which is fully effective. Consequently, if h = constant, the rate of heat transfer is increased by a factor of $(A_w + \eta_f A_f)/(A_w + A_b)$, where A_b is the total fin base area.

Example 3.11: Fuel elements in a nuclear reactor consist of 0.25 cm thick fuel plates of fissionable material (alloy of uranium and zirconium) with a 0.050 cm thick protective cladding of zirconium on each side as illustrated in Fig. 3.24a. The coolant flows over the outside surface of the cladding at 260°C and the heat transfer coefficient is 2500 W/m²·K. Thermal conducti-

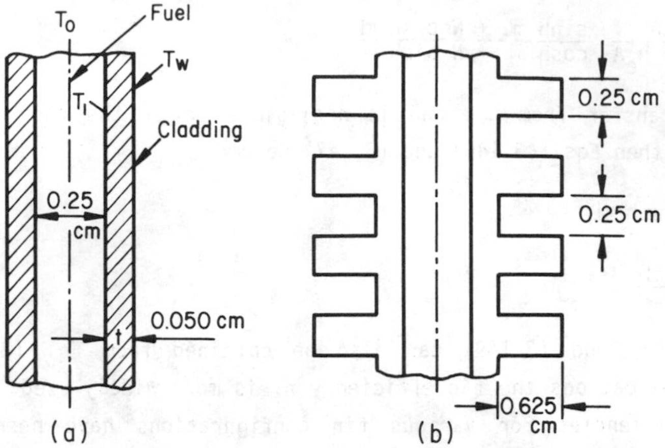

Fig. 3.24 Figure for Example 3.11.

vities are given as

Uranium-zirconium alloy	$k = 24$ W/m·K
Zirconium	$k = 21$ W/m·K

a) Determine the maximum temperature within the fuel elements and the temperature of the outer surface of the cladding when the fuel plates are operating with a uniform internal heat generation of $\dot{q} = 8 \times 10^8$ W/m³.

b) To increase the heat transfer rate to the coolant one engineer proposed that fins be added to the fuel elements as illustrated in 3.24b. Suppose that 0.25 cm thick and 0.625 cm long zirconium fins are added to the cladding surface, spaced so as to provide two fins per cm.

1. By what factor is the heat transfer rate increased?

2. How much is the maximum fuel element temperature reduced?

Assume that the ends of the fins are insulated. This corresponds to the case where the fins between adjacent fuel elements form a continuous finned surface.

Solution: a) The rate of heat transfer through the cladding is given by

$$\dot{Q} = U\,A(T_1 - T_\infty)$$

with

$$\frac{1}{U} = \frac{1}{h} + \frac{t}{k} = \frac{1}{2500} + \frac{0.050 \times 10^{-2}}{21} = 0.424 \times 10^{-3}\ \mathrm{m^2 \cdot K/W}$$

where

T_∞ = coolant temperature

t = thickness of the cladding

Surface heat flux, on the other hand, is

$$q_s = \dot{q} \times L = 8 \times 10^8 \times \frac{0.25 \times 10^{-2}}{2} = 10^6 \text{ W/m}^2$$

Therefore,

$$T_1 - T_\infty = \frac{\dot{Q}}{UA} = \frac{q_s}{U} = 10^6 \times 0.424 \times 10^{-3} = 424°C$$

Temperature on the midplane is given by

$$T_0 - T_1 = \frac{\dot{q}L^2}{2k} = \frac{8 \times 10^8 \times (0.125 \times 10^{-2})^2}{2 \times 24} = 26°C$$

Thus, the maximum temperature within the fuel elements is

$$T_0 = 26 + 424 + 260 = 710°C$$

The outer surface temperature of the cladding is obtained from

$$T_w = \frac{q_s}{h} + T_\infty = \frac{10^6}{2500} + 260 = 400 + 260 = 660°C$$

b) Under steady-state conditions the rate of heat transfer from the fuel elements must be equal to the rate of internal energy generation in the fuel plates. Therefore, the rate of heat transfer is not increased by adding fins. It remains the same as long as $\dot{q}$ and the thickness of the fuel plates are unchanged. Is the maximum fuel element temperature reduced by adding fins?

3.6.5 Limit of Usefulness of Fins

When the surface heat transfer coefficient h is high compared to k/δ for a straight rectangular fin as shown in Fig. 3.25, the addition of such a fin to a solid surface may decrease the heat transfer rate. From the definition of the fin effectiveness, Eq. (3.143), this means that ϕ would be less than unity. To illustrate, consider the fin shown in Fig. 3.25. For this fin

$$N = \frac{h}{mk} = \sqrt{\frac{hA}{kP}} \simeq \sqrt{\frac{h}{k}\,\delta}$$

Substituting this into Eq. (3.146) we get

$$\phi = \frac{1}{N}\,\frac{\sinh\left(N\frac{L}{\delta}\right) + N\cosh\left(N\frac{L}{\delta}\right)}{\cosh\left(N\frac{L}{\delta}\right) + N\sinh\left(N\frac{L}{\delta}\right)}$$

The values of ϕ for various values of L/δ and N are given in Table 3.1.

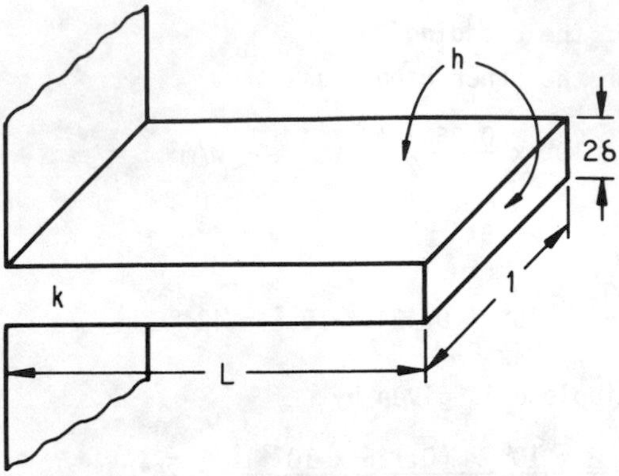

Fig. 3.25 A rectangular fin of thickness 2δ.

It is seen from Table 3.1 that if $N \geq 1$, then $\phi < 1$. Therefore, $N < 1$ is used as the criterian for the useful range of fin performance; that is, if

$$\frac{h\delta}{k} < 1$$

the provision of fins will be worthwhile, and fins will improve the heat transfer from the surface.

3.6.6 Extended Surfaces with Variable Cross-sections

In determining an optimum fin for a given purpose the question may arise as to whether or not weight advantage can be gained by using fins of shapes

Table 3.1 Limit of usefulness of fins.

$\frac{L}{\delta}$	$\phi(N=2)$	$\phi(N=1)$	$\phi(N=0.5)$	$\phi(N=0.25)$
1	0.4153	0.6321	0.8427	0.9518
2	0.4879	0.8646	1.3580	1.7585
5	0.4999	0.9932	1.9909	3.8370
10	0.5000	0.9999	1.9992	3.9597
15	0.5000	0.9999	1.9992	3.9597
20	0.5000	0.9999	1.9994	3.9890

other than the constant cross-section considered thus far. The differential equation which will be satisfied by the temperature distribution in extended surfaces with variable cross-sections has already been given by Eq. (3.119). Since A and P are no longer constants this equation is now a differential equation with variable coefficients. The two boundary conditions in the x-direction, namely one at the base and the other at the tip of the fin, will complete the formulation.

TRIANGULAR FINS: In the discussion that follows, consider the straight fin of triangular profile shown in Fig. 3.26. The mathematical treatment in this case is similar to the case of the fin of rectangular profile except that the area normal to the heat flow is a function of the distance along the fin, decreasing as the fin length increases. From Fig. 3.26 we see that

$$A(x) = \frac{bx}{L}\,\ell \qquad \text{and} \qquad P(x) = 2\left(\frac{bx}{L} + \ell\right)$$

where ℓ is the width of the fin. If we assume that $b \ll \ell$, then $P(x) \simeq 2\ell$. Inserting these values into Eq. (3.119) we get

$$\frac{d}{dx}\left(x\,\frac{d\theta}{dx}\right) - m^2\theta = 0 \tag{3.154a}$$

or

$$x^2\,\frac{d^2\theta}{dx^2} + x\,\frac{d\theta}{dx} - m^2 x\theta = 0 \tag{3.154b}$$

where $m^2 = 2hL/kb$. Defining a new independent variable η as

$$\eta = \sqrt{x} \tag{3.155}$$

Eq. (3.154b) can be written as

$$\eta^2\,\frac{d^2\theta}{d\eta^2} + \eta\,\frac{d\theta}{d\eta} - 4m^2\eta^2\theta = 0 \tag{3.156}$$

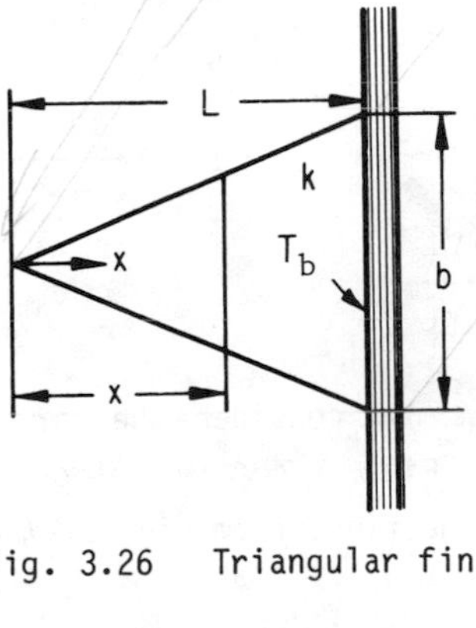

Fig. 3.26 Triangular fin.

Comparing Eq. (3.156) with Eq. (B.13) in Appendix B we get

$$\theta(\eta) = C_1 I_0(2m\eta) + C_2 K_0(2m\eta) \tag{3.157a}$$

or

$$\theta(x) = C_1 I_0(2m\sqrt{x}) + C_2 K_0(2m\sqrt{x}) \tag{3.157b}$$

which could also be obtained directly from Eq. (B.22) in Appendix B. The boundary conditions are

$$T(0) = \text{finite} \quad \rightarrow \quad \theta(0) = \text{finite} \tag{3.158a}$$

$$T(L) = T_b \quad \rightarrow \quad \theta(L) = T_b - T_\infty = \theta_b \tag{3.158b}$$

Since the temperature will be finite at the tip of the fin (x=0) and $K_0(0) \rightarrow \infty$ as can be seen from Fig. B.4 in Appendix B, the application of the boundary condition (3.158a) yields $C_2 = 0$. The second boundary condition (3.158b) gives

$$C_1 = \frac{\theta_b}{I_0(2m\sqrt{L})} \tag{3.159}$$

Thus, the temperature distribution becomes

$$\frac{\theta(x)}{\theta_b} = \frac{I_0(2m\sqrt{x})}{I_0(2m\sqrt{L})} \tag{3.160}$$

The rate of heat transfer from the fin may be obtained by considering the conduction across its base; that is,

$$\dot{Q}_f = kA\left(\frac{dT}{dx}\right)_{x=L} = kA\left(\frac{d\theta}{dx}\right)_{x=L} \tag{3.161}$$

Since, from Appendix B,

$$\frac{d}{dx}[I_0(\alpha x)] = \alpha I_1(\alpha x) \tag{3.162}$$

Eq. (3.161) yields

$$\dot{Q}_f = \ell\sqrt{2hkb}\;\theta_b \frac{I_1(2m\sqrt{L})}{I_0(2m\sqrt{L})} \tag{3.163}$$

CIRCULAR FINS: Let us now consider the circular fins arranged radially around a tube as shown in Fig. 3.27. Equation (3.119) is also applicable for this case. For one of the fins, from Fig. 3.27, we have

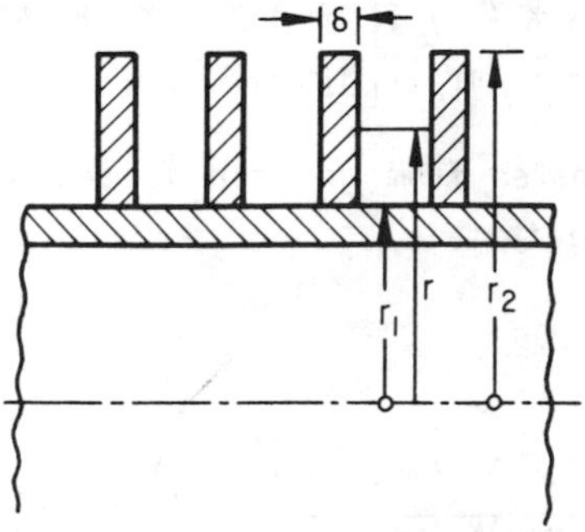

Fig. 3.27 Circular fin.

$$A(r) = 2\pi r\delta \qquad \text{and} \qquad P(r) = 4\pi r$$

Substituting these values into Eq. (3.119) yields the following differential equation:

$$\frac{d}{dr}\left(r\,\frac{d\theta}{dr}\right) - \frac{2hr}{k\delta}\,\theta = 0 \tag{3.164a}$$

or

$$r^2\,\frac{d^2\theta}{dr^2} + r\,\frac{d\theta}{dr} - m^2 r^2\theta = 0 \tag{3.164b}$$

where

$$m^2 = \frac{2h}{k\delta} \tag{3.164c}$$

If the heat loss from the tip of the circular fins is assumed to be negligible then the boundary conditions can be stated as

$$\theta(r_1) = \theta_b = T_b - T_\infty \tag{3.165a}$$

$$\frac{d\theta(r_2)}{dr} = 0 \tag{3.165b}$$

where the base temperature T_b is assumed to be known.

The solution of Eq. (3.164b) can be written (see Appendix B) as

$$\theta(r) = C_1 I_0(mr) + C_2 K_0(mr) \tag{3.166}$$

The constants C_1 and C_2 can be determined by imposing the boundary conditions (3.165). Once these constants are determined, the solution can be shown to be given by

$$\frac{\theta(r)}{\theta_b} = \frac{I_0(mr)\,K_1(mr_2) + K_0(mr)\,I_1(mr_2)}{I_0(mr_1)K_1(mr_2) + K_0(mr_1)I_1(mr_2)} \tag{3.167}$$

The rate of heat transfer from one fin is determined by evaluating the heat conduction at the base; that is,

$$\dot{Q}_f = -k\,2\pi r_1 \delta \left(\frac{d\theta}{dr}\right)_{r_1}$$

$$= 2\pi r_1 \sqrt{2hk\delta}\;\;\theta_b \frac{I_1(mr_2)K_1(mr_1) - I_1(mr_1)K_1(mr_2)}{I_0(mr_1)K_1(mr_2) + I_1(mr_2)K_0(mr_1)} \tag{3.168}$$

Expressions (3.163) and (3.168), and similar ones that are available in literature for various fins of variable cross-sections may not be convenient for practical use. Therefore, in practice mostly the fin efficiencies given in graphical forms are used. Figures 3.28, 3.29 and 3.30 are three such graphs giving efficiencies of several types of fins.

Harper and Brown [15] have shown that if a corrected length, L_c, is defined by increasing the length of a fin by one-half the fin thickness, then all equations which apply to fins with insulated tip can be used and the errors introduced are usually very small, especially when the fin is surrounded by a gas so that h is not too high. They have also shown that the error of such an approximation is less than 8% when

$$\left(\frac{ht}{2k}\right)^{1/2} = 1/2 \tag{3.169}$$

where t is the fin thickness.

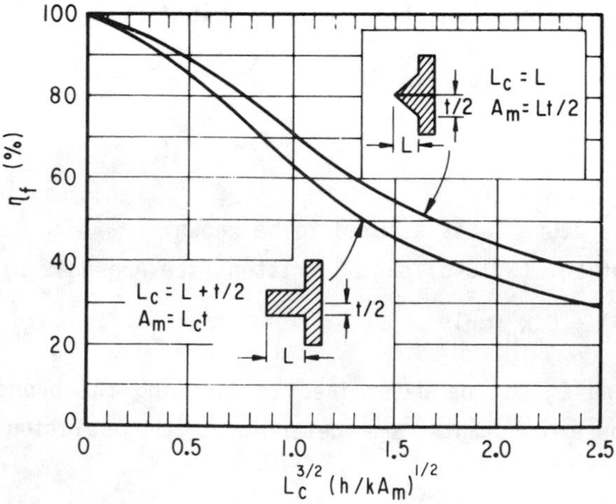

Fig. 3.28 Efficiencies of straight fins of rectangular and triangular profiles [15].

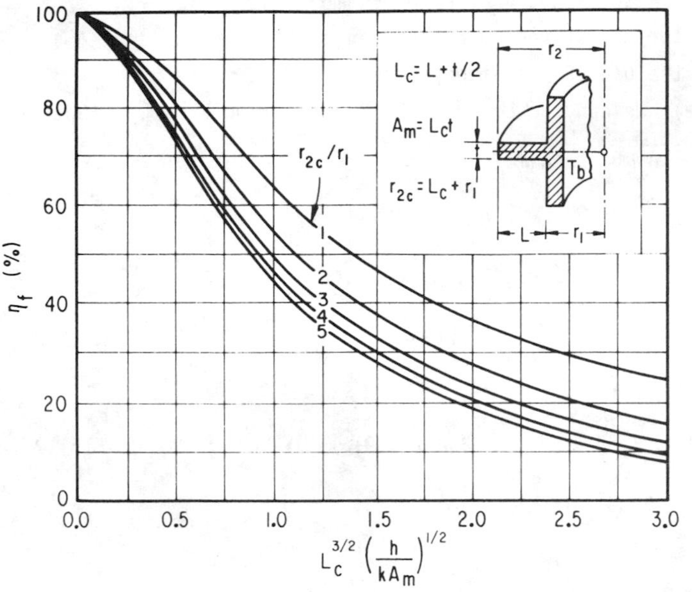

Fig. 3.29 Efficiencies of annular fins of rectangular profile [15].

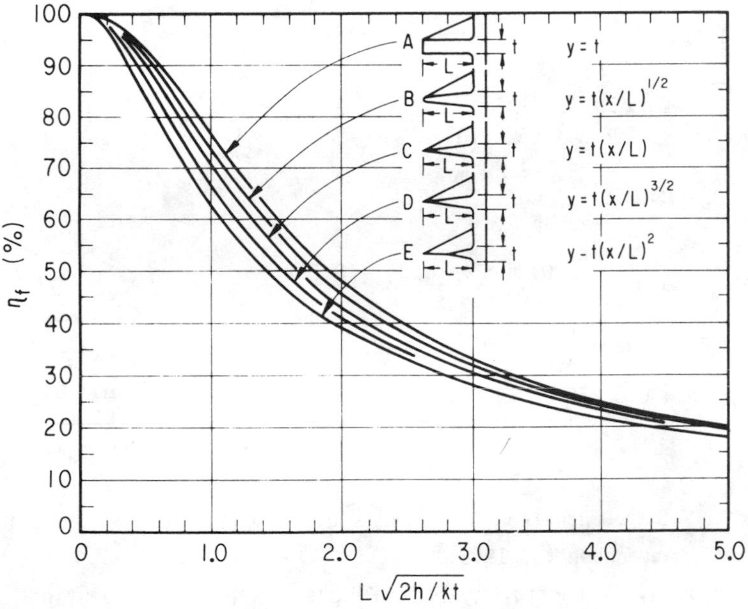

Fig. 3.30 Efficiencies of axial fins. The fin thickness y varies with the distance x from the root of the fin where y = t [13].

Example 3.12: Resolve the problem of Example 3.9

(a) by the use of the corrected length L_c concept, and

(b) by the use of efficiency curves.

Solution: (a) We can use Eq. (3.132) by replacing L by L_c; that is,

$$\dot{Q}_f \simeq \sqrt{hPkA}\ \theta_b \tanh mL_c$$

where

$$L_c = L + \frac{d}{2} = 0.15 + \frac{0.025}{2} = 0.1625 \text{ m}$$

$$m = 3.412 \text{ m}^{-1}, \qquad mL_c = 0.554$$

Thus,

$$\dot{Q}_f \simeq (15 \times 7.85 \times 10^{-2} \times 206 \times 4.91 \times 10^{-4})^{1/2} \times (260 - 16) \times \tanh (0.554)$$

$$\simeq 42.42\text{W}$$

(b) Heat loss can also be found from

$$\dot{Q}_f = h A_f \eta_f (T_b - T_\infty)$$

The parameters needed are

$$A_m = L_c d = 0.1625 \times 0.025 = 0.00406 \text{ m}^2$$

$$A_f = \pi d L_c = \pi \times 0.025 \times 0.1625 = 0.01276 \text{ m}^2$$

$$\left(\frac{h}{kA_m}\right)^{1/2} = \left(\frac{15}{206 \times 4.06 \times 10^{-3}}\right)^{1/2} = 4.234$$

$$(L_c)^{3/2} = 0.0655$$

$$(L_c)^{3/2}\left(\frac{h}{kA_m}\right)^{1/2} = 0.0655 \times 4.234 = 0.272$$

From Fig. 3.28 we have $\eta_f = 0.925$. Thus,

$$\dot{Q}_f \simeq 15 \times 0.01276 \times 0.925 \times 244 \simeq 43.21 \text{ W}$$

REFERENCES

1. Arpacı, V.S., *Conduction Heat Transfer*, Addison-Wesley Pub. Co., Inc., Reading, Massachusetts, 1966.

2. Veziroḡlu, T.N., Huerta, M.A., and Kakac, S., "Exact Solutions for Thermal Conductances of Planar and Circular Contacts", in *Thermal Conductivity 75*, Eds., P.G. Klemens and T.K. Chu, Plenum Press, New York, New York, 1976.

3. Çetinkale (Veziroğlu), T.N., and Fishenden, M., "Thermal Conductance of Metal Surfaces in Contact", *Proc. Int, Heat Transfer Conf.*, Inst. Mech. Eng., London, England, 1951.
4. Moore, C.J., Jr., Blum, H.A., and Atkins, H., "Subject Classification Bibliography for Thermal Contact Resistance Studies", ASME paper 68-WA/HT-18, 1968.
5. Yüncü, H., "Thermal Conductance of Contacts with Interstitial Plates", Ph.D. Thesis, Middle East Technical University, Ankara, Turkey, 1974.
6. Moore, C.J., "Heat Transfer Across Surfaces in Contact: Studies of Transients in One-dimensional Composite Systems", Southern Methodist University, Thermal/Fluid Sci. Ctr. Res. Rep., 67-2, Dallas, Texas, March 1967.
7. Clausing, A.M., "Heat Transfer at the Interface of Dissimilar Metals: The Influence of Thermal Strain", *Int. J. Heat Mass Transfer*, Vol. 9, p. 791, 1966.
8. Kakaç, S., Bergles, A.E., and Mayinger, F., (Eds.), *Heat Exchangers: Thermalhydraulic Fundamentals and Design*, Hemisphere Publishing Corp., New York, New York, pp. 1013-1047, 1981.
9. Kakaç, S., Shah, A.K., and Bergles, A.E., (Eds.), *Low Reynolds Number Flow Heat Exchangers*, Hemisphere Publishing Corp., New York, New York, pp. 951-979, 1983.
10. Holman, J.P., *Heat Transfer*, 5th ed., McGraw-Hill Book Co., New York, New York, 1981.
11. Kakaç, S., *Temperature Distribution and Heat Removal from Nuclear Reactors*, Middle East Technical University, Ankara, Turkey, 1962.
12. Price, P.H., "Notes on Conduction of Heat in Solids", University of Manchester, N.S. 6/20, 1965.
13. Gardner, K.A., "Efficiency of Extended Surfaces", ASME Trans., Vol. 68, No. 8, 1945.
14. Schmidt, E., "Heat Transfer by Fins", *Zeit. Ver. Deutsch. Ing.*, Vol. 70, 1926.
15. Harper, D.P., and Brown, W.M., "Mathematical Equations for Heat Conduction in the Fins of Air Cooled Engines", NACA Report No. 158, 1922.
16. Kern, D.Q., and Kraus, A.D., *Extended Surface Heat Transfer*, McGraw-Hill Book Co., New York, New York, 1972.
17. Eckert, E.R.G., and Drake, R.M., *Analysis of Heat and Mass Transfer*, McGraw-Hill Book Co., New York, New York, 1972.
18. Lottes, P.A., *Nuclear Reactor Heat Transfer*, Argonne National Laboratory, ANL-6469, 1961.

19. Kakaç, S., *Isı Transferine Giriş I: Isı Iletimi*, (in Turkish), Middle East Technical University Publications No. 52, Ankara, Turkey, 1976.

PROBLEMS

3.1 The outside-facing wall of a room is constructed from 25 cm thick brick, 2.5 cm of mortar, 10 cm of limestone (k = 0.186 W/m·K) and 1.25 cm of plaster (k = 0/96 W/m·K). Thermal conductivities of mortar and brick are both 0.52 W/m·K. Assume the heat transfer coefficients on the inside (plaster side) and outside (brick side) surfaces of the wall to be 6 and 12 $W/m^2 \cdot K$, respectively.

(a) Calculate the overall coefficient of heat transfer.

(b) Calculate the overall thermal resistance per m^2.

(c) Calculate the rate of heat transfer per 10 m^2 of the wall surface from the room at 25°C to the outside air at -6°C.

(d) Calculate the temperature at the mortar and limestone interface.

3.2 The walls of a furnace are constructed from 12 cm thick fire brick on the inside and from 25 cm thick red brick on the outside. During steady operation of the furnace the surface temperature on the flame side of the fire brick is 680°C and the outside surface temperature is 120°C. To reduce the heat loss a 5-cm layer of magnesia insulation (k = 0.085 W/m·K) is added onto the outside of the red brick. When steady conditions are reached again the following temperatures are measured: On flame side of the fire brick, 700°C; at the interface between fire brick and red brick, 650°C; at the junction between the red brick and the magnesia, 490°C; on the outer surface of the magnesia insulation, 75°C. Calculate the rate of heat loss from the furnace, expressed as a percentage of the original rate of heat loss.

3.3 A long hollow cylinder is constructed from a material whose thermal conductivity is a strong function of temperature, given by

$$k = 0.137 + 0.002T$$

where T is in °C and k in W/m·K. The inner and outer radii of the cylinder are 12.5 cm and 25 cm, respectively. Under steady-state conditions, the temperature of the outside surface is 80°C.

(a) Taking into account the variation of the thermal conductivity with temperature calculate the rate of heat loss from the cylinder per meter length.

(b) If the heat transfer coefficient on the outside surface is 12 $W/m^2 \cdot K$, calculate the surrounding air temperature.

3.4 The main steam-line of a proposed power plant will carry steam at 113 bar and 400°C. It is agreed that 85% magnesia (k = 0.078 W/m·K) will be used for insulation, but, since magnesia is not satisfactory above 300°C, a layer of an expensive high-temperature insulation (k = 0.2 W/m·K) will be placed next to the pipe. Enough insulation must be used so that the outside surface temperature of the insulation is to be 48°C. The pipe is a 12-in standard diameter carbon steel pipe (OD = 12.75 in; wall thickness = 1.312 in) of k = 40 W/m·K. The film coefficient on the steam side may be taken as 4500 $W/m^2 \cdot K$, and on the air side 12 $W/m^2 \cdot K$. Recommend thicknesses for the high-temperature and magnesia insulations for 30°C environment air temperature.

3.5 Water flows at a rate of 250 kg/hr through a pipe of 2 in OD. The temperature of the water decreases by 0.5 °C per meter length of the pipe. The pipe is covered with a 3-in thick insulation. The temperatures of the inner and outer surfaces of the insulation are 85°C and 65°C, respectively. The film coefficient on the water side is 1000 $W/m^2 \cdot K$, and on the outer surface of the insulation is 125 $W/m^2 \cdot K$. Compute the mean thermal conductivity of the material used for insulation at a temperature of 75°C.

3.6 A furnace wall is constructed from 9-in thick fire brick (k = 1.04 W/m·K), 6-in thick red brick (k = 0.69 W/m·K), 2-in thick insulation (k = 0.70 W/m·K), and 0.125-in thick steel plate (k = 36 W/m·K) on the outside. The heat transfer coefficients on the inside and outside surfaces are h_1 = 25 $W/m^2 \cdot K$ and h_2 = 5 $W/m^2 \cdot K$, respectively, and the contact resistances between the various layers are negligible. The gas temperature inside the furnace is 1000°C, and the outside air temperature is 30°C.
(a) Calculate the rate of heat loss from the furnace per unit area.
(b) Determine the temperatures at all interfaces in the wall.

3.7 Determine the expression for the critical radius of insulation for a sphere.

3.8 At what radius of asbestos insulation (k = 0.151 W/m·K) will the heat transfer rate be the same as for an uninsulated pipe of outer radius 0.5 inch. Assume the heat transfer coefficient to be h = 6 $W/m^2 \cdot K$.

3.9 In a slab extending from x = 0 to x = L internal energy is generated at the rate of $\dot{q}$ per unit volume. The surface at x = 0 is insulated. Find an expression for the temperature difference across the slab under the steady-state conditions.

3.10 Derive an expression for the steady-state temperature distribution $T(x)$ in a plane wall, $0 \leq x \leq L$, having uniformly distributed heat sources and with the surface at $x = 0$ maintained at a constant temperature T_1 while the other at $x = L$ is at T_2. The thermal conductivity of the material of the wall can be assumed constant.

3.11 A rocket motor is in the form of an annulus of inner and outer radii r_i and r_o, respectively, constructed from a radioactive material having an equivalent heat source of strength $\dot{q}$, W/m^3. The outer surface of the annulus is insulated while the inner surface transfers heat to a flowing gas at temperature T_∞ through a film coefficient h. Derive expressions for the temperature distribution in the radioactive material and the rate of heat transfer to the gas.

3.12 A slab of radioactive material may be considered to have a uniform heat source of strength $\dot{q}$ (W/m^3) when it is active. Derive an expression for the steady-state temperature distribution $T(x)$ in a radioactive material of thickness L if the surface at $x = 0$ is insulated and heat is transferred from the other surface at $x = L$ to a fluid medium at temperature T_∞ with a heat transfer coefficient h. Also, determine an expression for the rate of heat transfer to the fluid.

3.13 The fuel elements of a nuclear reactor are in the form of hollow tubes of inner and outer radii r_i and r_o, respectively. The neutron flux results in uniformly distributed heat sources of strength $\dot{q}$ (W/m^3) in the fuel elements. The cooling fluid temperatures on the inside and outside of the tubes are T_{f_1} and T_{f_2} and the respective surface heat transfer coefficients are h_1 and h_2. Obtain expressions for the surface temperatures. What is the maximum temperature within the fuel element.

3.14 A plane wall, 3-in thick, generates heat internally at the rate of 5×10^4 W/m^3. One side of the wall is insulted and the other side is exposed to an environment at 30°C. The heat transfer coefficient between the wall and the environment is 600 $W/m^2 \cdot K$. The thermal conductivity of the wall is 17 $W/m \cdot K$. Calculate the maximum temperature in the wall.

3.15 Consider a thin rod of cross-sectional area A, perimeter P, and thermal conductivity k, supported at its ends by two plates. The ends of the rod are held at constant temperatures T_1 and T_2. The rod is exposed to a fluid at T_∞ and the heat transfer coefficient at the surface is h. Derive expressions,

(a) for the temperature distribution in the rod as a function of distance x along the rod, and

(b) for the total rate of heat loss from the rod.

3.16 A thermometer well mounted through the wall of a steam pipe is a steel tube with a 0.1 in wall thickness, 0.5 OD, 2 in length and k = 26 W/m·K. The steam flow produces an average heat transfer coefficient of 100 $W/m^2 \cdot K$ on the thermometer well. If the thermometer reads 149°C and the temperature of the steam pipe is 65°C, estimate the average steam temperature.

3.17 A thin wire of cross-section area A and perimeter P is extruded at a fixed velocity V through an extrusion nozzle as shown in Fig. 3.31. The temperature of the wire at the extrusion nozzle is T_0, high enough to make the metal extrudable. The wire, after extrusion, passes through air for some distance L and then is rolled onto a large spool, where the temperature reduces to T_L. It is desired to investigate the relationship between the wire velocity V, the spacing distance L, and the air temperature T to obtain a specific value of T_L. Derive a differential equation for determining the wire temperature as a function of the distance x from the extrusion nozzle and state the boundary conditions.

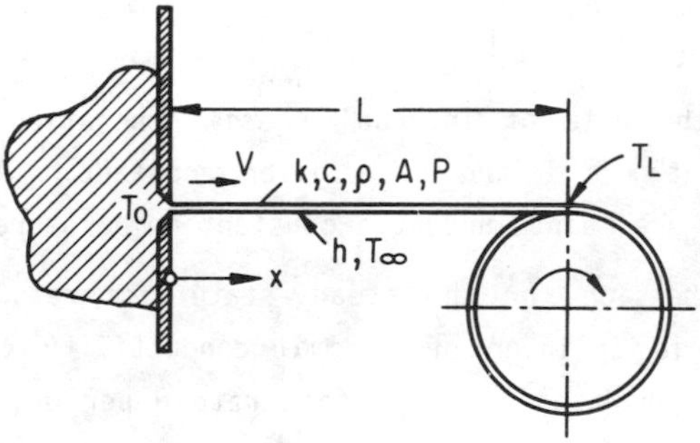

Fig. 3.31 Figure for Problem 3.17.

3.18 Solve the problem formulated in Problem 3.17 to obtain an expression for the wire velocity V in terms of the given parameters of the problem.

3.19 The core of a nuclear reactor is made of cylindrical fuel elements, each composed of a uranium rod of radius r_1 and a stainless steel cladding of thickness $(r_2 - r_1)$. Internal energy is generated in the uranium rods as

$$\dot{q} = \dot{q}_0 \left[1 - \left(\frac{r}{r_1}\right)^2\right], \qquad 0 \le r \le r_1$$

where $\dot{q}_0$ is the rate of internal energy generation per unit volume at the centerline and is a given constant. The temperature of the coolant is held constant at T_f and the heat transfer coefficient h is assumed to be constant. Find the temperature distribution T(r) in the uranium rods and in the claddings under steady-state conditions.

3.20 Starting from the general heat conduction equation (2.16) for homogeneous isotropic solids obtain the governing heat conduction equation (steady-state) for a one-dimensional extended surface with constant cross-section.

3.21 Derive an expression for the steady-state temperature distribution T(r) in a solid cylinder of radius r_0 and constant thermal conductivity k in which internal energy is generated at a rate of

$$\dot{q} = \dot{q}_0\left(1 - \frac{r}{r_0}\right)$$

where $\dot{q}_0$ is the rate of internal energy generation per unit volume at the centerline and is a given constant. The surface of the cylinder at $r = r_0$ is maintained at a constant temperature T_w.

3.22 Derive an expression for the steady-state temperature distribution T(x) in a plane wall of thickness 2L and constant thermal conductivity k in which internal energy is generated at a rate of

$$\dot{q}(x) = \dot{q}_0\left[1 - \left(\frac{x}{L}\right)^2\right]$$

where $\dot{q}_0$ is the rate of internal energy generation per unit volume at the midplane at x = 0 and is a given constant. The surfaces of the wall at $x = \pm L$ are maintained at a constant temperature T_w.

3.23 Derive an expression for the steady-state temperature distribution T(r) in a long solid cylinder of thermal conductivity k and radius r_0, in which heat is generated at a constant rate $\dot{q}$ per unit volume. The outer surface at $r = r_0$ is exposed to a fluid at temperature T with a heat transfer coefficient h.

3.24 A straight fin of triangular profile, 10 cm long and 2 cm wide at its base, is constructed of mild steel (k = 50 W/m·K) and attached to a wall maintained at 40°C. Air at 200 °C flows past the fin and the average heat transfer coefficient is 250 $W/m^2 \cdot K$. Calculate the rate of heat removed by the fin per unit depth.

3.25 A thin-walled disk rotates with angular velocity ω on a stationary disk as illustrated in Fig. 3.32. Both disks have the same dimensions and

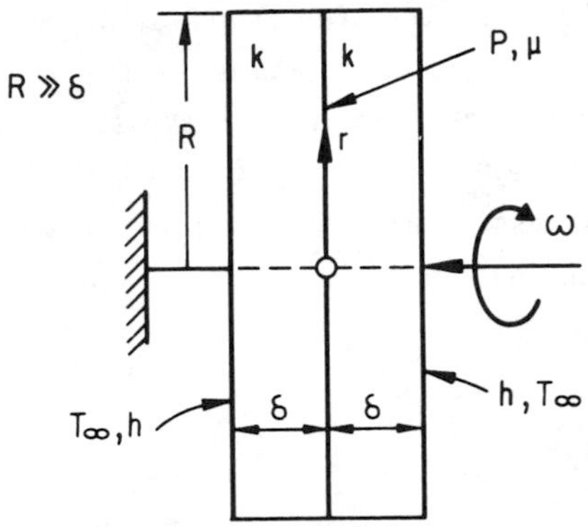

Fig. 3.32 Figure for Problem 3.25.

the same constant thermo-physical properties. The interface pressure is p, the coefficient of dry friction at the interface is constant μ and the ambient temperature is T_∞. The heat transfer coefficient is the same h on both sides and the heat transfer from the end surfaces is negligible. Assuming constant wear at the interface so that pr = C, where C is a given constant, determine the steady-state temperature of the system.

3.26 Solve Problem 3.25 if p = constant.

3.27 A spine attached to a wall at temperature T_w has the shape of a circular cone with base diameter d_0 and height L as illustrated in Fig. 3.33, and is exposed to a fluid at a uniform temperature T_∞.

(a) Assuming that the thermal conductivity k of the material of the

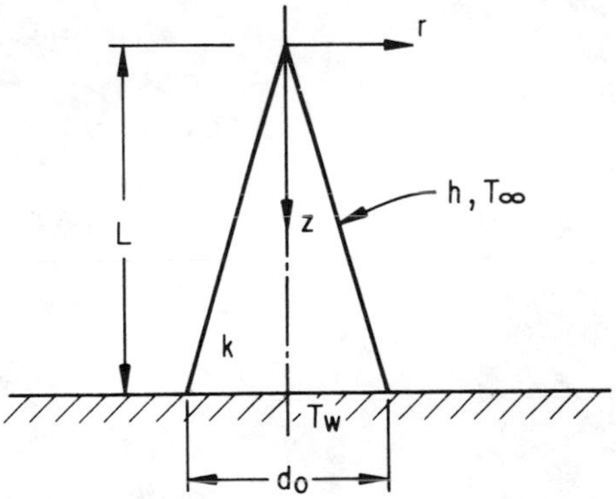

Fig. 3.33 Figure for Problem 3.27.

spine and the heat transfer coefficient h are constants, and that the variation of the temperature in the r-direction is negligible, develop an expression for the steady-state temperature distribution T(z) in the spine.

(b) Obtain an expression for the rate of heat transfer from the spine to the surrounding fluid.

4

Orthogonal Functions, Fourier Expansions, and Finite Fourier Transforms

4.1 INTRODUCTION

We have, so far, discussed one-dimensional steady-state heat conduction problems and seen that formulations of such problems involve ordinary differential equations. Formulations of two- or three-dimensional steady-state and one- or multi-dimensional unsteady-state heat conduction problems, on the other hand, result in partial differential equations. When the boundary surfaces correspond to the coordinate surfaces in a system of coordinates, such as the rectangular, cylindrical or spherical coordinate systems, we may then employ, among several different techniques, the *method of separation of variables, Fourier transforms* or *Laplace transforms* as the method of solution to obtain analytical solutions. In Chapters 5 and 6, we shall first solve some linear heat conduction problems using the method of separation of variables, which was first introduced by d'Alembert, Daniel Bernoulli, and Euler in the middle of the eighteenth century. This method remains to be a method of great value even today and also lies at the heart of Fourier transforms method which we shall discuss in Chapter 7. In this chapter we introduce the basic mathematical concepts related to these two methods. We shall discuss Laplace transforms in Chapter 8 to solve some time-dependent problems.

4.2 CHARACTERISTIC-VALUE PROBLEMS

Let us first review some necessary mathematical concepts: A differential equation or a condition is said to be *linear* if, when rationalized and cleared of fractions, it contains no products of the dependent variable or its derivatives. The heat conduction equation with temperature-dependent thermal conductivity is a good example of a *non-linear* differential equation:

$$\vec{\nabla} \cdot [k(T)\vec{\nabla}T] + \dot{q} = \rho c \frac{\partial T}{\partial t} \qquad (4.1a)$$

An example of a *non-linear* condition would be a radiation boundary condition as illustrated in Fig. 4-1,

$$-k\left(\frac{dT}{dx}\right)_{x=L} = \varepsilon\sigma[T^4(L) - T_e^4] \qquad (4.1b)$$

where T_e is the *effective* blackbody temperature of the environment.

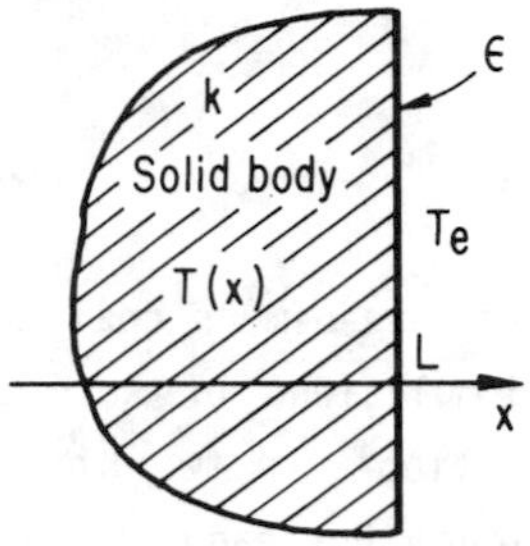

Fig. 4.1 Radiation boundary condition.

A linear differential equation or a linear condition is said to be *homogeneous* if, when satisfied by a particular function y, it is also satisfied by Cy, where C is an arbitrary constant. Thus, for example,

$$\frac{d^2y}{dx^2} + f_1(x)\frac{dy}{dx} + f_2(x)y = f_3(x) \qquad (4.2a)$$

is *non-homogeneous* linear second-order ordinary differential equation. The equation

$$\frac{d^2y}{dx^2} + f_1(x)\frac{dy}{dx} + f_2(x)y = 0 \qquad (4.2b)$$

is a homogeneous linear second-order ordinary differential equation. At the boundary x = a

$$y(a) = C_1, \ \frac{dy(a)}{dx} = C_2 \ \text{ or } \ \alpha y(a) + \beta\frac{dy(a)}{dx} = C_3 \qquad (4.2c)$$

denote non-homogeneous linear boundary conditions, whereas

$$y(a) = 0, \ \frac{dy(a)}{dx} = 0 \ \text{ or } \ \alpha y(a) + \beta\frac{dy(a)}{dx} = 0 \qquad (4.2d)$$

represent homogeneous linear boundary conditions. Here C_1, C_2, C_3, α and β are prescribed constants.

As in the case of many physical problems, in heat conduction problems certain specified conditions must be satisfied by the solution of the differential equation. These conditions often enable the arbitrary constants of integration appearing in the solution of the differential equation to be evaluated to yield a unique solution. If the conditions are specified at two* different values of the independent variable (or variables) at the boundaries, the problem is called a *boundary-value problem*, in distinction with *initial-value problems* wherein all conditions are specified at one value of the independent variable. The heat conduction equation for a one-dimensional steady-state problem together with two boundary conditions, for example, forms a boundary-value problem, whereas the heat conduction equation for an unsteady-state one-dimensional problem together with an initial condition and two boundary conditions forms an *initial-and-boundary-value problem*. Let us now consider the following homogeneous boundary-value problem:

$$\frac{d^2y}{dx^2} + f_1(x)\frac{dy}{dx} + f_2(x)y = 0 \tag{4.3a}$$

$$\alpha_1 y(a) + \beta_1 \frac{dy(a)}{dx} = 0 \tag{4.3b}$$

$$\alpha_2 y(b) + \beta_2 \frac{dy(b)}{dx} = 0 \tag{4.3c}$$

where not both α_1 and β_1 are zero, and also not both α_2 and β_2 are zero. Since the differential equation is linear, the general solution of this differential equation can be written as

$$y(x) = C_1 y_1(x) + C_2 y_2(x) \tag{4.4}$$

where $y_1(x)$ and $y_2(x)$ are two linearly independent solutions of the differential equation and C_1 and C_2 are arbitrary constants. Application of the boundary conditions yields the following set of algebraic equations for the two unknown constants:

$$B_{11}C_1 + B_{12}C_2 = 0 \tag{4.5a}$$

$$B_{21}C_1 + B_{22}C_2 = 0 \tag{4.5b}$$

where we have introduced the following constants:

$$B_{1i} = \alpha_1 y_i(a) + \beta_1 \frac{dy_i(a)}{dx}, \qquad i = 1,2. \tag{4.6a}$$

*Boundary conditions may, of course, be given at more than two values of the independent variable. We shall, however, consider only *two-point boundary-value problems*.

$$B_{2i} = \alpha_2 y_i(b) + \beta_2 \frac{dy_i(b)}{dx} , \quad i = 1,2 \tag{4.6b}$$

One possible solution of Eqs. (4.5a,b) for the unknown constants is $C_1 = C_2 = 0$ leading to the trivial solution $y(x) \equiv 0$. By Cramer's rule, in order to have a non-trivial solution it is necessary that the determinant of coefficients of C_1 and C_2 must vanish; that is,

$$\begin{vmatrix} B_{11} & B_{12} \\ B_{21} & B_{22} \end{vmatrix} = 0 \tag{4.7a}$$

or

$$B_{11}B_{22} - B_{12}B_{21} = 0 \tag{4.7b}$$

If this condition exists, the two equations (4.5a) and (4.5b) become identical and one of the constants can be expressed as a multiple of the other by the use of either equation, second constant then being arbitrary. The use of Eq. (4.5a), for example, yields

$$y(x) = A[B_{12}y_1(x) - B_{11}y_2(x)] \tag{4.8}$$

where A is an arbitrary constant defined as

$$A = \frac{C_1}{B_{12}} = -\frac{C_2}{B_{11}} \tag{4.9}$$

It can easily be shown that Eq. (4.8) satisfies the boundary conditions (4.3b,c) (see Problem 4.1). It should be noticed that Eq. (4.8) is a non-trivial solution only if not both B_{11} and B_{12} are zero. If $B_{11} = B_{12} = 0$, then Eq. (4.5a) is a trivial relation, and in that case we can use Eq. (4.5b) to relate C_1 and C_2. This leads to a non-trivial solution of the form

$$y(x) = B[B_{22}y_1(x) - B_{21}y_2(x)] \tag{4.10}$$

where B is an arbitrary constant defined as

$$B = \frac{C_1}{B_{22}} = -\frac{C_2}{B_{21}} \tag{4.11}$$

Equation (4.10) will be the non-trivial solution only if not both B_{21} and B_{22} are zero. If $B_{ij} = 0$, $i,j = 1,2$, then Eq. (4.4) will be the solution for arbitrary values of C_1 and C_2.

Example 4.1: Find the non-trivial solution of the following boundary-value problem:

$$\frac{d^2y}{dx^2} + y = 0$$

$$y(0) = 0 \quad \text{and} \quad y(\pi) = 0$$

Solution: The general solution of the differential equation is

$$y(x) = C_1 \sin x + C_2 \cos x$$

Substituting into the boundary conditions, we obtain

$$C_1 \cdot 0 + C_2 \cdot 1 = 0 \quad \text{or} \quad C_2 = 0$$

$$C_1 \cdot 0 - C_2 \cdot 1 = 0 \quad \text{or} \quad C_2 = 0$$

Clearly, the solution of these equations gives $C_2 = 0$, C_1 = arbitrary. Hence,

$$y(x) = C_1 \sin x$$

is the solution of the boundary-value problem for any arbitrary constant C_1.

One or both of the coefficients $f_1(x)$ and $f_2(x)$ in Eq. (4.3a) and hence the solutions $y_1(x)$ and $y_2(x)$ may depend upon a parameter λ. In such problems, the determinant (4.7a) may vanish for certain values of λ, say λ_1, λ_2, λ_3,.... These values are called the *characteristic values*, or *eigenvalues*, and the corresponding solutions are called the *characteristic functions*, or *eigenfunctions*, of the problem. Boundary-value problems of this kind are known as *characteristic-value*, or *eigenvalue*, problems. Let us now consider the following example:

Example 4.2: Find the characteristic values and the characteristic functions of the following characteristic-value problem:

$$\frac{d^2y}{dx^2} + \lambda^2 y = 0$$

$$y(0) = 0 \quad \text{and} \quad y(L) = 0$$

Solution: The general solution of the differential equation is

$$y(x) = C_1 \sin\lambda x + C_2 \cos\lambda x$$

The boundary condition at $x = 0$, *i.e.*, $y(0) = 0$, results in $C_2 = 0$, and the boundary condition at $x = L$, *i.e.*, $y(L) = 0$, yields

$$C_1 \sin\lambda L = 0$$

Hence, a non-trivial solution of this problem exists only if λ has a value such that

$$\sin\lambda L = 0$$

and this is possible if λ is equal to one of the values of

$$\lambda_n = \frac{n\pi}{L}, \qquad n = 1,2,3,...$$

which are the characteristic values of the problem. Thus, the foregoing boundary-value problem has no solution other than the trivial solution $y(x) = 0$, unless λ assumes one of the characteristic values λ_n given above. Corresponding to each characteristic value λ_n the solution of the problem can be written as

$$y_n(x) = C_{1n}\phi_n(x)$$

where C_{1n} is an arbitrary non-zero constant and

$$\phi_n(x) = \sin\frac{n\pi}{L} x$$

which is the characteristic function corresponding to the characteristic value λ_n. Here it should be noted that no new solutions are obtained when n assumes negative integer values.

4.3 ORTHOGONAL FUNCTIONS

Two real-valued functions $\phi_m(x)$ and $\phi_n(x)$ are said to be *orthogonal* with respect to a *weight function* (or *density function*) $w(x)$ over an *interval* (a,b) if

$$\int_a^b \phi_m(x)\phi_n(x)w(x)dx = 0, \qquad m \neq n \tag{4.12}$$

Furthermore, a *set* of real-valued functions $\{\phi_n(x);\ n = 1,2,...\}$ is called *orthogonal* with respect to a weight function $w(x)$ over an interval (a,b) if all the pairs of distinct functions in this set satisfy the orthogonality condition (4.12). If there exists some function $f(x)$, different from zero, which is orthogonal to *all* members of the set; that is, if

$$\int_a^b f(x)\phi_n(x)w(x)dx = 0, \qquad n = 1,2,3,...$$

then the set is called *incomplete*, or otherwise it is said to be *complete*.

For example, the set $\{\sin(n\pi/L)x;\ n=1,2,3,...\}$ is an orthogonal set with respect to the weight function unity over the interval (0,L) because

$$\int_0^L \sin\frac{m\pi}{L}x \sin\frac{n\pi}{L}x\, dx = 0, \quad m \neq n$$

which can be verified by direct integration. It can also be shown that this set is a complete orthogonal set, but such a proof is beyond the scope of this text. However, the matter of completeness of certain class of sets of orthogonal functions will be clarified in the following section.

4.4 STURM-LIOUVILLE PROBLEM

Consider the following characteristic-value problem, composed of the linear homogeneous second-order differential equation of the general form

$$\frac{d}{dx}\left[p(x)\frac{dy}{dx}\right] + \left[q(x) + \lambda w(x)\right] y = 0 \tag{4.13a}$$

and the two homogeneous linear boundary conditions

$$\alpha_1 y(a) + \beta_1 \frac{dy(a)}{dx} = 0, \quad \alpha_1^2 + \beta_1^2 \neq 0 \tag{4.13b}$$

$$\alpha_2 y(b) + \beta_2 \frac{dy(b)}{dx} = 0, \quad \alpha_2^2 + \beta_2^2 \neq 0 \tag{4.13c}$$

prescribed at the ends of the finite interval (a,b). The functions $p(x)$, $q(x)$, $w(x)$ and dp/dx are real-valued and continuous, and $p(x)$ and $w(x)$ are positive over the entire interval (a,b). Furthermore α_1, α_2, β_1, and β_2 are given real constants, and λ is an unspecified parameter independent of x. Characteristic-value problems of this type are known as *Sturm-Liouville problems* or *systems*. The first extensive development of the theory of such systems was published by J.C.F. Sturm (1803-1855), Swiss mathematician, and J. Liouville (1809-1882), French mathematician, in the first three volumes of "Journal de Mathematique", 1836-1838.

As we shall see in Chapters 5, 6 and 7, when the method of separation of variables or integral transforms is used to solve a heat conduction problem, the problem is reduced, at one point in the solution procedure, to a Sturm-Liouville system. Depending on the system of coordinates used, the functions $p(x)$, $q(x)$ and $w(x)$ in Eq. (4.13a) will be of certain forms. The homogeneous boundary conditions of the third kind (4.13b,c) result if the original heat conduction problem has boundary conditions of the third kind at $x = a$ and at $x = b$. If the problem has a boundary condition of the second kind, for example, at $x = a$, then α_1 in Eq. (4.13b) will be zero, and if the boundary condition at $x = a$ is of the first kind, then β_1 in Eq. (4.13b) will be zero.

A non-trivial solution of the problem (4.13) exists, in general, for a particular set of values $\lambda_1, \lambda_2, \ldots, \lambda_n \ldots$ of the parameter λ. These are the characteristic values (or the eigenvalues), and the corresponding solutions are the characteristic functions (or the eigenfunctions) of the problem. Let λ_m and λ_n be any two distinct characteristic values and $\phi_m(x)$ and $\phi_n(x)$ be the corresponding characteristic functions, respectively. These functions satisfy the differential equation (4.13a):

$$\frac{d}{dx}\left[p(x)\frac{d\phi_m}{dx}\right] + \left[q(x) + \lambda_m w(x)\right]\phi_m = 0 \tag{4.14a}$$

$$\frac{d}{dx}\left[p(x)\frac{d\phi_n}{dx}\right] + \left[q(x) + \lambda_n w(x)\right]\phi_n = 0 \tag{4.14b}$$

Multiplying Eq. (4.14a) by $\phi_n(x)$ and Eq. (4.14b) by $\phi_m(x)$ and subtracting we obtain

$$\phi_n \frac{d}{dx}\left[p(x)\frac{d\phi_m}{dx}\right] - \phi_m \frac{d}{dx}\left[p(x)\frac{d\phi_n}{dx}\right] + (\lambda_m - \lambda_n)w(x)\phi_m\phi_n = 0$$

or, after simplifying, we get

$$\frac{d}{dx}\left\{p(x)\left[\phi_n\frac{d\phi_m}{dx} - \phi_m\frac{d\phi_n}{dx}\right]\right\} = (\lambda_n - \lambda_m)w(x)\phi_m\phi_n$$

Integration of this result over the interval (a,b) yields

$$(\lambda_n - \lambda_m)\int_a^b \phi_m(x)\phi_n(x)w(x)dx = \left\{p(x)\left[\phi_n\frac{d\phi_m}{dx} - \phi_m\frac{d\phi_n}{dx}\right]\right\}_{x=a}^{x=b} \tag{4.15}$$

Both $\phi_m(x)$ and $\phi_n(x)$ satisfy the conditions given by Eqs. (4.13b) and (4.13c). For example, at x = a

$$\alpha_1\phi_m(a) + \beta_1\frac{d\phi_m(a)}{dx} = 0 \tag{4.16a}$$

$$\alpha_1\phi_n(a) + \beta_1\frac{d\phi_n(a)}{dx} = 0 \tag{4.16b}$$

Multiplying Eq. (4.16a) by $\phi_n(a)$ and Eq. (4.16b) by $\phi_m(a)$ and subtracting we find

$$\beta_1\left[\phi_n(a)\frac{d\phi_m(a)}{dx} - \phi_m(a)\frac{d\phi_n(a)}{dx}\right] = 0 \tag{4.17}$$

If $\beta_1 \neq 0$, *i.e.*, if the homogeneous boundary condition (4.13b) is either of the second or the third kind, then the term in the bracket in Eq. (4.17) will be zero. If $\beta_1 = 0$, *i.e.*, if the homogeneous boundary condition (4.13b) is of the first kind, then $\phi_m(a) = \phi_n(a)$, and hence the term in the bracket in Eq. (4.17) will again be zero. In either case,

$$\phi_n(a)\frac{d\phi_m(a)}{dx} - \phi_m(a)\frac{d\phi_n(a)}{dx} = 0 \tag{4.18}$$

Following the same procedure one can also show that

$$\phi_n(b)\frac{d\phi_m(b)}{dx} - \phi_m(b)\frac{d\phi_n(b)}{dx} = 0 \tag{4.19}$$

In view of Eqs. (4.18) and (4.19), the right side of Eq. (4.15) vanishes; that is,

$$(\lambda_n - \lambda_m)\int_a^b \phi_m(x)\phi_n(x)w(x)dx = 0 \tag{4.20a}$$

or

$$\int_a^b \phi_m(x)\phi_n(x)w(x)dx = 0\ , \qquad m \neq n \tag{4.20b}$$

Thus, we see that characteristic functions of the Sturm-Liouville problem form an orthogonal set with respect to the weight function $w(x)$ over the interval (a,b).

The following extensions of the Sturm-Liouville problem are also important in certain applications.

Singular end points: If it so happens that $p(a) = 0$, then the right side of Eq. (4.15) vanishes at $x = a$ provided that both $y(x)$ and dy/dx are finite at $x = a$. With this condition the characteristic functions corresponding to different characteristic values will be orthogonal, even if Eq. (4.18) does not hold. In this case, the boundary condition (4.13b) can be dropped from the statement of the problem and the requirement that $y(x)$ and dy/dx be finite at $x = a$ when $p(a) = 0$ can be thought of as its replacement.

Similarly, if $p(b) = 0$, then we can replace the second boundary condition (4.13c) by the requirement that $y(x)$ and dy/dx be finite at $x = b$. If both $p(a) = 0$ and $p(b) = 0$, then neither boundary condition as specified by Eqs. (4.13b,c) is needed to ensure the orthogonality of the characteristic functions provided that both $y(x)$ and dy/dx are finite at $x = a$ and $x = b$.

Periodic boundary conditions: If $p(a) = p(b)$, then the right side of Eq. (4.15) vanishes if the boundary conditions are replaced by

$$y(a) = y(b) \tag{4.21a}$$

$$\frac{dy(a)}{dx} = \frac{dy(b)}{dx} \tag{4.21b}$$

These are called *periodic boundary conditions* and they are satisfied, in particular, if the solution $y(x)$ is required to be *periodic*, of period b-a.

It can be shown that the characteristic values of the Sturm-Liouville problem are real and non-negative. Hence λ in Eq. (4.13a) can be replaced by λ^2 with no loss in the generality of the problem. It can also be proved that the characteristic functions form a complete orthogonal set. Proofs of these statements are beyond the scope of this text.

4.5 GENERALIZED FOURIER SERIES AND GENERALIZED FINITE FOURIER TRANSFORM

Consider a complete set of functions $\{\phi_n(x),\ n = 0,1,2,\ldots\}$ orthogonal with respect to a weight function $w(x)$ over the finite interval (a,b). Expand an arbitrary function $f(x)$ into a series of these functions as

$$f(x) = A_0\phi_0(x) + A_1\phi_1(x) + \ldots + A_n\phi_n(x) + \ldots \tag{4.22a}$$

or

$$f(x) = \sum_{n=0}^{\infty} A_n\phi_n(x) \tag{4.22b}$$

The unknown coefficients A_n can be evaluated by multiplying Eq. (4.22a) or Eq. (4.22b) by $\phi_n(x)w(x)$ and then integrating the resulting equation over the interval (a,b):

$$\int_a^b f(x)\phi_n(x)w(x)dx = \int_a^b \phi_n(x)w(x)\left[\sum_{k=0}^{\infty} A_k\phi_k(x)\right]dx$$

Using orthogonality property of the set we get

$$\int_a^b f(x)\phi_n(x)w(x)dx = A_n\int_a^b [\phi_n(x)]^2 w(x)dx$$

which yields

$$A_n = \frac{1}{N_n}\int_a^b f(x)\phi_n(x)w(x)dx \tag{4.23a}$$

where N_n is the *normalization integral* of the function $\phi_n(x)$ defined as

$$N_n = \int_a^b [\phi_n(x)]^2 w(x)dx \tag{4.23b}$$

The series (4.22a) or (4.22b) with coefficients (4.23a) is called the *generalized Fourier series* of $f(x)$ corresponding to the orthogonal set $\{\phi_n(x),\ n = 0,1,2,\ldots\}$. The coefficients A_n are called *Fourier constants* of the function $f(x)$.

The series (4.22a) or (4.22b) is a *formal* series expansion of the function f(x) in the interval (a,b). The problem of determining whether this expansion actually represents the given function is a difficult one and beyond the scope of this text. However, if $\{\phi_n(x),\ n = 0,1,2,\ldots\}$ is the set of the characteristic functions of a Sturm-Liouville problem and f(x) is a *piecewise-differentiable* function, then the series representation of f(x) converges inside (a,b) to f(x) at all points where f(x) is continuous and converges to the mean value $1/2[f(x^+) + f(x^-)]$ at the points where finite jumps occur. This series representation may or may not converge to the given value of f(x) at x = a and at x = b.

Substitution of the Fourier constants (4.23a) into Eq. (4.22b) yields

$$f(x) = \sum_{n=0}^{\infty} \frac{1}{N_n} \left[\int_a^b f(x')\phi_n(x')w(x')dx' \right] \phi_n(x) \tag{4.24}$$

which can be rewritten in two parts as

$$\bar{f}_n = \int_a^b f(x)K_n(x)w(x)dx \tag{4.25a}$$

and

$$f(x) = \sum_{n=0}^{\infty} \bar{f}_n K_n(x) \tag{4.25b}$$

where $K_n(x)$ denotes the *normalized* functions; that is,

$$K_n(x) = \frac{\phi_n(x)}{\sqrt{N_n}} \tag{4.25c}$$

The term $\bar{f}_n$, defined by the integral (4.25a), is called *generalized finite Fourier transform* of the function f(x) in the interval (a,b) and the series (4.25b) is the corresponding *inversion formula*. The function $K_n(x)$ is called the *kernel* of the transform.

4.6 ORDINARY FOURIER SERIES AND FINITE FOURIER TRANSFORMS

The *ordinary Fourier series*, or simply the *Fourier series*, are those which are developed from the characteristic functions of the following characteristic-value problem for different combinations of the boundary conditions:

$$\frac{d^2y}{dx^2} + \lambda^2 y(x) = 0 \tag{4.26a}$$

$$\alpha_1 y(0) + \beta_1 \frac{dy(0)}{dx} = 0 \tag{4.26b}$$

$$\alpha_2 y(L) + \beta_2 \frac{dy(L)}{dx} = 0 \tag{4.26c}$$

This characteristic-value problem is a special case of the Sturm-Liouville system (4.13) with $p(x) = 1$, $q(x) = 0$, $w(x) = 1$ and λ replaced by λ^2. Replacement of λ by λ^2 is merely for convenience and it has no effect on the generality of the problem as the system (4.26) is a Sturm-Liouville problem and, hence, would have non-negative characteristic values. The characteristic functions $\phi_n(x)$, which will be sinusoidal functions, will therefore form a complete orthogonal set in the interval $(0,L)$ with respect to the weight function unity, and an arbitrary function $f(x)$ which is piecewise-differentiable in the interval $(0,L)$ can be expanded in this interval into a series of these functions as

$$f(x) = \sum_{n=0}^{\infty} A_n \phi_n(x) \tag{4.27}$$

where the coefficients A_n are given by

$$A_n = \frac{1}{N_n} \int_0^L f(x)\phi_n(x)dx \tag{4.28a}$$

with

$$N_n = \int_0^L [\phi_n(x)]^2 dx \tag{4.28b}$$

The series (4.27) where the functions $\phi_n(x)$ are the characteristic functions of the characteristic-value problem (4.26) is called *ordinary Fourier series*, or simply *Fourier series*, of $f(x)$ in the interval $(0,L)$.

Substitution of A_n from Eq. (4.28a) into Eq. (4.27) yields

$$f(x) = \sum_{n=0}^{\infty} \frac{1}{N_n}\left[\int_0^L f(x')\phi_n(x')dx' \right]\phi_n(x) \tag{4.29}$$

which can be rewritten in two parts as

$$\bar{f}_n = \int_0^L f(x')K_n(x')dx' \tag{4.30a}$$

and

$$f(x) = \sum_{n=0}^{\infty} \bar{f}_n K_n(x) \tag{4.30b}$$

where

$$K_n(x) = \frac{\phi_n(x)}{\sqrt{N_n}} \tag{4.30c}$$

The term $\bar{f}_n$, given by Eq. (4.30a) is called *ordinary finite Fourier transform*, or simply *finite Fourier transform*, of f(x) in the interval (0,L) and the series (4.30b) is the corresponding *inversion formula*. The function $K_n(x)$ is the *kernel* of the transform.

There are, in fact, nine different combinations of the boundary conditions (4.26b,c). Hence, corresponding to each combination there will be a series expansion of the form given by Eq. (4.27) and also a transform of the form given by Eq. (4.30a). Let us now discuss the two most common cases in detail.

4.6.1 Fourier Sine Series and Finite Fourier Sine Transform

Consider the following characteristic-value problem:

$$\frac{d^2y}{dx^2} + \lambda^2 y = 0 \tag{4.31a}$$

$$y(0) = 0 \quad \text{and} \quad y(L) = 0 \tag{4.31b,c}$$

We have seen in Section 4.2. (see Example 4.2) that this problem has the following characteristic functions and characteristic values:

$$\phi_n(x) = \sin\lambda_n x, \qquad \lambda_n = \frac{n\pi}{L}, \qquad n = 1,2,3,...$$

This characteristic-value problem is a special case of the problem (4.26) with $\beta_1 = \beta_2 = 0$. Therefore, the series (4.27) can now be written as

$$f(x) = \sum_{n=1}^{\infty} A_n \sin\frac{n\pi}{L}x \tag{4.32}$$

where the coefficients A_n are given by

$$A_n = \frac{1}{N_n}\int_0^L f(x)\sin\frac{n\pi}{L}x\,dx \tag{4.33a}$$

with

$$N_n = \int_0^L \sin^2\frac{n\pi}{L}x\,dx = \frac{L}{2} \tag{4.33b}$$

The series (4.32) with coefficients (4.33a) is known as the *Fourier sine series* expansion of f(x) in the interval (0,L).

Substitution of A_n given by Eq. (4.33a) into Eq. (4.32) yields

$$f(x) = \frac{2}{L}\sum_{n=1}^{\infty}\left[\int_0^L f(x')\sin\frac{n\pi}{L}x'dx'\right]\sin\frac{n\pi}{L}x \tag{4.34}$$

Equation (4.34) can be rewritten in two parts as

$$\bar{f}(\lambda_n) = \int_0^L f(x)K(\lambda_n,x)dx \tag{4.35a}$$

and

$$f(x) = \sum_{n=1}^{\infty} \bar{f}(\lambda_n)K(\lambda_n,x) \tag{4.35b}$$

where $K(\lambda_n,x)$ denotes the normalized characteristic functions; that is,

$$K(\lambda_n,x) = \frac{\sin\frac{n\pi}{L}x}{\sqrt{N_n}} = \sqrt{\frac{2}{L}}\sin\frac{n\pi}{L}x \tag{4.35c}$$

Equations (4.35a) and (4.35b) can also be written as

$$\bar{f}(\lambda_n) = \sqrt{\frac{2}{L}}\int_0^L f(x)\sin\frac{n\pi}{L}x\,dx \tag{4.36a}$$

and

$$f(x) = \sqrt{\frac{2}{L}}\sum_{n=1}^{\infty}\bar{f}(\lambda_n)\sin\frac{n\pi}{L}x \tag{4.36b}$$

Equation (4.36a) is called the *finite Fourier sine transform* of the function $f(x)$ over the interval $(0,L)$ and Eq. (4.36b) is the corresponding *inversion* formula.

It should be noted that since the kernel of this tranform, $K(\lambda_n,x)$, is, by definition, the normalized characteristic function, it satisfies the differential equation (4.31a) as well as the boundary conditions (4.31b,c).

We further note that the right side of Eq. (4.32) is a *periodic* function of *period* $2L$, and also an *odd* function of x. Therefore, if the series (4.32) converges to $f(x)$ in $(0,L)$, it will converge to $-f(-x)$ in $(-L,0)$. In other words, if $f(x)$ is an odd function of x, then the series (4.32) will represent $f(x)$ not only in $(0,L)$, but also in $(-L,L)$.

Example 4.3: Expand $f(x) = x$ into a Fourier sine series of period $2L$ over the interval $(0,L)$.

Solution: The expansion

$$x = \sum_{n=1}^{\infty} A_n \sin\frac{n\pi}{L}x$$

is a Fourier sine series expansion of f(x) = x of period 2L over the interval (0,L). The expansion coefficients A_n can be calculated as

$$A_n = \frac{2}{L}\int_0^L x \sin\frac{n\pi}{L}x\,dx = -\frac{2L}{n\pi}\cos n\pi = \frac{2L}{\pi}\frac{(-1)^{n+1}}{n}$$

Hence,

$$x = \frac{2L}{\pi}\sum_{n=1}^{\infty}\frac{(-1)^{n+1}}{n}\sin\frac{n\pi}{L}x$$

Since f(x) = x is an odd function, this expansion is a valid expansion not only in the interval (0,L), but also in the interval (-L,L). Furthermore, we note that the Fourier series does not converge to f(x) = x at x = ±L as the series converges to zero at these points.

4.6.2 Fourier Cosine Series and Finite Fourier Cosine Transform

A similar series expansion involving cosine terms, rather than sine terms, may be obtained considering the following characteristic-value problem:

$$\frac{d^2y}{dx^2} + \lambda^2 y = 0 \tag{4.37a}$$

$$\frac{dy(0)}{dx} = 0 \quad \text{and} \quad \frac{dy(L)}{dx} = 0 \tag{4.37b,c}$$

This is another special case of the problem (4.26) with $\alpha_1 = \alpha_2 = 0$. Therefore, the characteristic functions form a complete orthogonal set with respect to the weight function unity over the interval (0,L). It can easily be shown that (see Problem 4.5) the characteristic functions and characteristic values are

$$\phi_n(x) = \cos\lambda_n x, \qquad \lambda_n = \frac{n\pi}{L}, \qquad n = 0,1,2,\ldots$$

Here it should be noted that $\phi_0(x) = 1$ is a member of the set of characteristic functions corresponding to $\lambda_0 = 0$. Thus, the series (4.27) can now be written as:

$$f(x) = \sum_{n=0}^{\infty} A_n\cos\frac{n\pi}{L}x = A_0 + \sum_{n=1}^{\infty} A_n\cos\frac{n\pi}{L}x \tag{4.38}$$

where the coefficients A_n are given by

$$A_n = \frac{1}{N_n}\int_0^L f(x)\cos\frac{n\pi}{L}x\,dx \tag{4.39a}$$

with

$$N_n = \int_0^L \cos^2\frac{n\pi}{L}x\,dx = \begin{cases} L, & n = 0 \\ \frac{L}{2}, & n = 1,2,3,\ldots \end{cases} \tag{4.39b}$$

Therefore,

$$A_0 = \frac{1}{L}\int_0^L f(x)\,dx \tag{4.40a}$$

$$A_n = \frac{2}{L}\int_0^L f(x)\cos\frac{n\pi}{L}x\,dx, \qquad n = 1,2,3,\ldots \tag{4.40b}$$

The series (4.38) is known as the *Fourier cosine series* expansion of $f(x)$ in the interval $(0,L)$.

Substituting A_n into Eq. (4.38) we get

$$f(x) = \sum_{n=0}^{\infty} \frac{1}{N_n}\left[\int_0^L f(x')\cos\frac{n\pi}{L}x'\,dx'\right]\cos\frac{n\pi}{L}x \tag{4.41}$$

which can be rewritten in two parts as

$$\bar{f}(\lambda_n) = \int_0^L f(x)K(\lambda_n,x)\,dx \tag{4.42a}$$

and

$$f(x) = \sum_{n=0}^{\infty} \bar{f}(\lambda_n)K(\lambda_n,x) \tag{4.42b}$$

where $K(\lambda_n,x)$ represents the normalized characteristic functions; that is,

$$K(\lambda_n,x) = \frac{1}{\sqrt{N_n}}\cos\frac{n\pi}{L}x \tag{4.42c}$$

Equation (4.42a) is called the *finite Fourier cosine transform* of the function $f(x)$ over the interval $(0,L)$ and Eq. (4.42b) is the corresponding *inversion* formula. Here again we should note that the kernel $K(\lambda_n,x)$ satisfies the differential equation (4.37a) and the boundary conditions (4.37b,c).

We also note that the right side of Eq. (4.38) is a *periodic* function of *period* 2L and also an *even* function of x. Therefore, if the series (4.38) converge to f(x) in (0,L), it will converge to f(-x) in (-L,0). In other words, if f(x) is an even function of x, then the series (4.38) will represent f(x) not only in (0,L), but also in (-L,L).

Example 4.4: Expand f(x) = x into a Fourier cosine series of period 2L over the interval (0,L).

Solution: The expansion

$$x = \sum_{n=0}^{\infty} A_n \cos\frac{n\pi}{L}x = A_0 + \sum_{n=1}^{\infty} A_n \cos\frac{n\pi}{L}x$$

is a Fourier cosine series of f(x) = x of period 2L over the interval (0,L). The expansion coefficients A_n can be calculated as

$$A_0 = \frac{1}{L}\int_0^L x\,dx = \frac{L}{2}$$

and

$$A_n = \frac{2}{L}\int_0^L x \cos\frac{n\pi}{L}x\,dx = -[1 - (-1)^n]\,\frac{2L}{n^2\pi^2}, \qquad n = 1,2,3,\ldots$$

Hence, the expansion for x over the interval (0,L) is given by

$$x = \frac{L}{2} - \frac{2L}{\pi^2}\sum_{n=1}^{\infty} \frac{[1 - (-1)^n]}{n^2} \cos\frac{n\pi}{L}x$$

Since x is an odd function this is a valid expansion only in the interval (0,L). In addition, the Fourier series converges to f(x) = x at x = 0 and x = L.

The Fourier sine and cosine series and finite Fourier sine and cosine transforms that we have discussed can be used in solving certain heat conduction problems. These were developed by considering two special cases of the characteristic-value problem (4.26). Several series similar to these can also be developed by considering other combinations of the boundary conditions of this characteristic-value problem. The characteristic values and the characteristic functions for the remaining seven cases can be obtained following the same procedure. The characteristic values, characteristic functions and the normalization integrals of the characteristic functions for all of these nine cases were evaluated and tabulated by Özışık [4,5,6]. They are summarized here in Table 4.1.

We shall discuss several applications of Fourier series expansions in Chapters 5 and 6 in solving heat conduction problems by the method of separation of variables. We shall see the development of other transforms similar

Table 4.1 Fourier series expansions.

$$\begin{pmatrix}\text{Fourier}\\ \text{Expansion}\end{pmatrix} \quad f(x) = \sum_{n=0}^{\infty} A_n\phi_n(x), \qquad 0 < x < L$$

$$\begin{pmatrix}\text{Fourier}\\ \text{Coefficient}\end{pmatrix} \quad A_n = \frac{1}{N_n}\int_0^L f(x)\phi_n(x)dx$$

$$\begin{pmatrix}\text{Characteristic}\\ \text{Value}\\ \text{Problem}\end{pmatrix} \quad \frac{d^2y}{dx^2} + \lambda^2 y = 0, \quad \alpha_1 y(0) + \beta_1 \frac{dy(0)}{dx} = 0, \quad \alpha_2 y(L) + \beta_2 \frac{dy(L)}{dx} = 0$$

B.C. at x = 0	B.C. at x = L	Characteristic Function* $\phi_n(x)$	$N_n = \int_0^L [\phi_n(x)]^2 dx$	Characteristic Values λ_n Are Positive Roots of*
Third Kind ($\alpha_1 \neq 0$, $\beta_1 \neq 0$)	Third Kind ($\alpha_2 \neq 0$, $\beta_2 \neq 0$)	$\lambda_n \cos\lambda_n x - H_1 \sin\lambda_n x$	$\frac{1}{2}\left\{(\lambda_n^2+H_1^2)\left[L+\frac{H_2}{\lambda_n^2+H_2^2}\right]-H_1\right\}$	$\tan\lambda L = -\frac{\lambda(H_1-H_2)}{\lambda^2+H_1H_2}$
Third Kind ($\alpha_1 \neq 0$, $\beta_1 \neq 0$)	Second Kind ($\alpha_2 = 0$, $\beta_2 \neq 0$)	$\cos\lambda_n(L-x)$	$\frac{1}{2\lambda_n}\left[\lambda_n L+\sin\lambda_n L\cos\lambda_n L\right]$	$\lambda\tan\lambda L = -H_1$
Third Kind ($\alpha_1 \neq 0$, $\beta_1 \neq 0$)	First Kind ($\alpha_2 \neq 0$, $\beta_2 = 0$)	$\sin\lambda_n(L-x)$	$\frac{1}{2\lambda_n}\left[\lambda_n L-\sin\lambda_n L\cos\lambda_n L\right]$	$\lambda\cot\lambda L = H_1$
Second Kind ($\alpha_1 = 0$, $\beta_1 \neq 0$)	Third Kind ($\alpha_2 \neq 0$, $\beta_2 \neq 0$)	$\cos\lambda_n x$	$\frac{1}{2\lambda_n}\left[\lambda_n L+\sin\lambda_n L\cos\lambda_n L\right]$	$\lambda\tan\lambda L = H_2$

Table 4.1 (Continued)

B.C. at x = 0	B.C. at x = L	Characteristic Function* $\phi_n(x)$	$N_n = \int_0^L [\phi_n(x)]^2 dx$	Characteristic Values λ_n Are Positive Roots of*
Second Kind ($\alpha_1 = 0,\ \beta_1 \neq 0$)	Second Kind ($\alpha_2 = 0,\ \beta_2 \neq 0$)	$\cos\lambda_n x$	$\frac{L}{2}$**	$\sin\lambda L = 0$ ($\lambda_n = \frac{n\pi}{L},\ n = 0,1,2,\ldots$)
Second Kind ($\alpha_1 = 0,\ \beta_1 \neq 0$)	First Kind ($\alpha_2 \neq 0,\ \beta_2 = 0$)	$\cos\lambda_n x$	$\frac{L}{2}$	$\cos\lambda L = 0$ ($\lambda_n = \frac{2n-1}{L}\frac{\pi}{2},\ n=1,2,3,\ldots$)
First Kind ($\alpha_1 \neq 0,\ \beta_1 = 0$)	Third Kind ($\alpha_2 \neq 0,\ \beta_2 \neq 0$)	$\sin\lambda_n x$	$\frac{1}{2\lambda_n}\left[\lambda_n L - \sin\lambda_n L \cos\lambda_n L\right]$	$\lambda\cot\lambda L = -H_2$
First Kind ($\alpha_1 \neq 0,\ \beta_1 = 0$)	Second Kind ($\alpha_2 = 0,\ \beta_2 \neq 0$)	$\sin\lambda_n x$	$\frac{L}{2}$	$\cos\lambda L = 0$ ($\lambda_n = \frac{2n-1}{L}\frac{\pi}{2},\ n=1,2,3,\ldots$)
First Kind ($\alpha_1 \neq 0,\ \beta_1 = 0$)	First Kind ($\alpha_2 \neq 0,\ \beta_2 = 0$)	$\sin\lambda_n x$	$\frac{L}{2}$	$\sin\lambda L = 0$ ($\lambda_n = \frac{n\pi}{L},\ n = 1,2,3,\ldots$)

* $H_1 = \alpha_1/\beta_1$, $H_2 = \alpha_2/\beta_2$.

** When n = 0 replace $\frac{L}{2}$ by L.

to the finite Fourier sine and cosine transforms, together with the method of solution of heat conduction problems using integral transforms, in Chapter 7.

4.7 COMPLETE FOURIER SERIES

In the previous section we have seen that an arbitrary function f(x) can be expanded in the interval (0,L) into a series of sine functions as

$$f(x) = \sum_{n=1}^{\infty} a_n \sin\frac{n\pi}{L}x \tag{4.43}$$

If f(x) is an *odd* function; that is, if

$$f(-x) = -f(x)$$

then the expansion given by Eq. (4.43) will be valid expansion not only in the interval (0,L) but also in the interval (-L,L).

We have also seen that we can expand another arbitrary function g(x) in the same interval (0,L) into a series of cosine functions as

$$g(x) = \sum_{n=0}^{\infty} b_n \cos\frac{n\pi}{L}x \tag{4.44}$$

If g(x) is an *even* function; that is, if

$$g(-x) = g(x)$$

then the expansion given by Eq. (4.44) will be valid expansion not only in the interval (0,L) but also in the interval (-L,L).

Any function of x, say F(x), can be written as

$$F(x) = \tfrac{1}{2}[F(x) - F(-x)] + \tfrac{1}{2}[F(x) + F(-x)] \tag{4.45}$$

where the first term on the right side is an odd and the second term is an even function of x. Therefore, the function F(x) can be expanded over the interval (-L,L) into a series of sine and cosine functions as

$$F(x) = \sum_{n=1}^{\infty} a_n \sin\frac{n\pi}{L}x + \sum_{n=0}^{\infty} b_n \cos\frac{n\pi}{L}x, \quad -L < x < L \tag{4.46a}$$

or

$$F(x) = b_0 + \sum_{n=1}^{\infty} \left[a_n \sin\frac{n\pi}{L}x + b_n \cos\frac{n\pi}{L}x\right], \quad -L < x < L \tag{4.46b}$$

where the coefficients a_n and b_n are given by

$$a_n = \frac{2}{L}\int_0^L \left\{ \tfrac{1}{2}[F(x) - F(-x)] \right\} \sin\frac{n\pi}{L}x\,dx = \frac{1}{L}\int_{-L}^{L} F(x)\sin\frac{n\pi}{L}x\,dx\,, \quad n = 1,2,3,\ldots \qquad (4,47a)$$

$$b_0 = \frac{1}{L}\int_0^L \left\{ \tfrac{1}{2}[F(x) + F(-x)] \right\} dx = \frac{1}{2L}\int_{-L}^{L} F(x)dx \qquad (4.47b)$$

$$b_n = \frac{2}{L}\int_0^L \left\{ \tfrac{1}{2}[F(x) + F(-x)] \right\} \cos\frac{n\pi}{L}x\,dx = \frac{1}{L}\int_{-L}^{L} F(x)\cos\frac{n\pi}{L}x\,dx, \quad n = 1,2,3,\ldots \qquad (4.47c)$$

The series (4.46a) or (4.46b) is called the *complete Fourier series* of function $F(x)$ over the interval $(-L,L)$. If $F(x)$ is a periodic function with period $2L$, then the coefficients a_n and b_n can be determined equivalently from

$$a_n = \frac{1}{L}\int_c^{c+2L} F(x)\sin\frac{n\pi}{L}x\,dx\,, \quad n = 1,2,3,\ldots \qquad (4.48a)$$

$$b_0 = \frac{1}{2L}\int_c^{c+2L} F(x)dx \qquad (4.48b)$$

$$b_n = \frac{1}{L}\int_c^{c+2L} F(x)\cos\frac{n\pi}{L}x\,dx\,, \quad n = 1,2,3,\ldots \qquad (4.48c)$$

where c is any real constant. Problem 4.11 describes an alternative way to obtain the complete Fourier series.

4.8 FOURIER-BESSEL SERIES AND FINITE HANKEL TRANSFORMS

Series expansions in terms of Bessel functions arise most frequently in connection with the following characteristic-value problem:

$$r^2\frac{d^2R}{dr^2} + r\frac{dR}{dr} + (\lambda^2r^2 - \nu^2)R = 0 \qquad (4.49a)$$

$$\alpha_1R(a) + \beta_1\frac{dR(a)}{dr} = 0 \qquad (4.49b)$$

$$\alpha_2R(b) + \beta_2\frac{dR(b)}{dr} = 0 \qquad (4.49c)$$

This problem is a special case of the Sturm-Liouville system (4.13) with (see Problem 4.4)

$$p(r) = r\,, \quad q(r) = -\frac{\nu^2}{r} \quad \text{and} \quad w(r) = r$$

Hence, the characteristic functions of this problem form a complete orthogonal set with respect to the weight function r in the interval (a,b).

We shall see in the following chapters that solutions of certain heat conduction problems in the cylindrical coordinates, especially problems with homogeneous boundary conditions in the r-direction, can be developed from the solutions of the above characteristic-value problem. In this text, we shall restrict our discussions to solid cylinders and such problems, in general, result in characteristic-value problems of the following form:

$$r^2\frac{d^2R}{dr^2} + r\frac{dR}{dr} + (\lambda^2r^2 - \nu^2)R = 0 \tag{4.50a}$$

$$R(0) = \text{finite} \tag{4.50b}$$

$$\alpha R(r_0) + \beta\frac{dR(r_0)}{dr} = 0 \tag{4.50c}$$

The general solution of Eq. (4.50a) can be written as (see Appendix B)

$$R(r) = A\ J_\nu(\lambda r) + B\ Y_\nu(\lambda r) \tag{4.51}$$

The boundary condition (4.50b) yields $B \equiv 0$. Hence, the characteristic functions of this problem are $J_\nu(\lambda_n r)$ and the characteristic values are the roots of the characteristic-value equation

$$\alpha J_\nu(\lambda_n r_0) + \beta\frac{dJ_\nu(\lambda_n r_0)}{dr} = 0 \tag{4.52}$$

which results from the application of the boundary condition (4.50c). In view of the fact that $p(0) = 0$ and $R(0)$ = finite (and also $\frac{dR(0)}{dr}$ = finite), the characteristic functions of this system become orthogonal with respect to the weight function $w(r) = r$ over the interval $(0,r_0)$; that is,

$$\int_0^{r_0} J_\nu(\lambda_m r)J_\nu(\lambda_n r)r\,dr = 0\ , \qquad \lambda_m \neq \lambda_n \tag{4.53}$$

and the set of functions $\{J_\nu(\lambda_n r);\ n = 1,2,\ldots\}$ is a complete orthogonal set. We can therefore expand an arbitrary function $f(r)$, which is piecewise-differentiable in the interval $(0,r_0)$, into a series of these characteritic functions in the same interval as

$$f(r) = \sum_{n=1}^{\infty} A_n J_\nu(\lambda_n r) \tag{4.54}$$

which is known as *Fourier-Bessel* series of $f(r)$ in the interval $(0,r_0)$. Making use of the orthogonality of the Bessel functions (4.53), the coefficients A_n may readily be obtained as

$$A_n = \frac{1}{N_n}\int_0^{r_o} f(r)J_\nu(\lambda_n r)rdr \tag{4.55a}$$

with

$$N_n = \int_0^{r_o} J_\nu^2(\lambda_n r)rdr \tag{4.55b}$$

Substitution of Eq. (4.55a) into Eq. (4.54) yields

$$f(r) = \sum_{n=1}^{\infty}\left(\int_0^{r_o} f(r')K_\nu(\lambda_n,r')r'dr'\right)K_\nu(\lambda_n,r) \tag{4.56}$$

where $K_\nu(\lambda_n,r)$ represents the normalized characteristic functions; that is,

$$K_\nu(\lambda_n,r) = \frac{J_\nu(\lambda_n r)}{\sqrt{N_n}} \tag{4.57}$$

Equation (4.56) can be rearranged as an integral transform

$$\overline{f}(\lambda_n) = \int_0^{r_o} f(r)K_\nu(\lambda_n,r)rdr \tag{4.58}$$

with the inversion

$$f(r) = \sum_{n=1}^{\infty} \overline{f}(\lambda_n)K_\nu(\lambda_n,r) \tag{4.59}$$

Equation (4.58) is called *finite Hankel transform* of the function f(r) over the interval $(0,r_o)$ and Eq. (4.59) is the corresponding inversion formula. The function $K_\nu(\lambda_n,r)$ given by Eq. (4.57) is the kernel of the transform.

The normalization integral (4.55b) of the characteristic functions can be determined as follows: The function $J_\nu(\lambda_n r)$ satisfies the differential equation (4.50a) when $\lambda = \lambda_n$; that is,

$$r^2\frac{d^2J_\nu(\lambda_n r)}{dr^2} + r\,\frac{dJ_\nu(\lambda_n r)}{dr} + (\lambda_n^2 r^2 - \nu^2)J_\nu(\lambda_n r) = 0 \tag{4.60a}$$

which can be rewritten as

$$r\frac{d}{dr}\left[r\frac{dJ_\nu(\lambda_n r)}{dr}\right] + (\lambda_n^2 r^2 - \nu^2)J_\nu(\lambda_n r) = 0 \tag{4.60b}$$

Multiplying Eq. (4.60b) by $2\frac{dJ_\nu(\lambda_n r)}{dr}$ and rearranging the resulting expression we get

$$\frac{d}{dr}\left[r\frac{dJ_\nu(\lambda_n r)}{dr}\right]^2 = -(\lambda_n^2 r^2 - \nu^2)\frac{dJ_\nu^2(\lambda_n r)}{dr} \tag{4.61}$$

Integrating Eq. (4.61) with respect to r over $(0,r_0)$ and rearranging the right-hand side by integration by parts we obtain

$$N_n = \frac{r_0^2}{2\lambda_n^2}\left\{\left[\lambda_n^2 - \left(\frac{\nu}{r_0}\right)^2\right]J_\nu(\lambda_n r_0) + \left[\frac{dJ_\nu(\lambda_n r_0)}{dr}\right]^2\right\} \tag{4.62}$$

The above result for the three special cases of the boundary condition at $r = r_0$ can be written explicitly as follows:

CASE I: $\alpha \neq 0$, $\beta = 0$. For this special case the characteristic values are the roots of

$$J_\nu(\lambda_n r_0) = 0 \tag{4.63}$$

and Eq. (4.62) reduces to

$$N_n = \frac{r_0^2}{2\lambda_n^2}\left[\frac{dJ_\nu(\lambda_n r_0)}{dr}\right]^2 \tag{4.64}$$

Noting that (see Appendix B)

$$\frac{dJ_\nu(\lambda_n r)}{dr} = -\lambda_n J_{\nu+1}(\lambda_n r) + \frac{\nu}{r}J_\nu(\lambda_n r) \tag{4.65}$$

Eq. (4.64) can also be written as

$$N_n = \frac{r_0^2}{2}J_{\nu+1}^2(\lambda_n r_0) \tag{4.66}$$

Then the kernel $K_\nu(\lambda_n,r)$ to be used in Hankel transform becomes

$$K_\nu(\lambda_n,r) = \frac{\sqrt{2}}{r_0}\,\frac{J_\nu(\lambda_n r)}{J_{\nu+1}(\lambda_n r_0)} \tag{4.67}$$

CASE II: $\alpha = 0$, $\beta \neq 0$. For this special case the characteristic values are the roots of

$$\frac{dJ_\nu(\lambda r_0)}{dr} = 0 \tag{4.68}$$

and in view of this relation Eq. (4.62) reduces to

$$N_n = \frac{r_0^2}{2}\left[1 - \left(\frac{\nu}{\lambda_n r_0}\right)^2\right]J_\nu^2(\lambda_n r_0) \tag{4.69}$$

We should note, however, that if $\nu = 0$ then $\lambda_0 = 0$ is a characteristic value and the corresponding characteristic function is $J_0(\lambda_0 r) = 1$. Thus, when $\nu = 0$, the function $J_0(\lambda_0 r) = 1$ must be included in the set of characteristic functions. The Fourier-Bessel series of f(r) then becomes

$$f(r) = A_0 + \sum_{n=1}^{\infty} A_n J_0(\lambda_n r) \tag{4.70}$$

where

$$A_0 = \frac{2}{r_0^2} \int_0^{r_0} f(r) r \, dr \tag{4.71}$$

The Hankel transform and the corresponding inversion formula can be modified accordingly for this special case.

CASE III: $\alpha \neq 0$, $\beta \neq 0$. For this special case characteristic values are the roots of

$$HJ_\nu(\lambda_n r_0) + \frac{dJ_\nu(\lambda_n r_0)}{dr} = 0 \tag{4.72}$$

where we have defined

$$H = \frac{\alpha}{\beta}$$

and in view of this relation Eq. (4.62) reduces to

$$N_n = \frac{r_0^2}{2}\left[1 + \frac{1}{\lambda_n^2}\left(H^2 - \frac{\nu^2}{r_0^2}\right)\right] J_\nu^2(\lambda_n r_0) \tag{4.73}$$

Since $J_\nu(\lambda r)$ is defined by (see Apendix B)

$$J_\nu(\lambda r) = \sum_{k=0}^{\infty} \frac{(-1)^k (\lambda r/2)^{2k+\nu}}{k!\,\Gamma(k+\nu+1)} \tag{4.74}$$

then

$$J_\nu(-\lambda r) = (-1)^\nu J_\nu(\lambda r) \tag{4.75}$$

So, replacement of λ_n by $-\lambda_n$ in $J_\nu(\lambda_n r)$ either does not change it (when ν is zero or an even integer) or multiplies $J_\nu(\lambda_n r)$ by the numerical constant $(-1)^\nu$ which is -1 if ν is an odd integer. Therefore, the roots of any one of the above three characteristic-value equations (4.63), (4.68) and (4.72) exist in pairs symmetrically located with respect to r = 0. However, we do not need to consider the negative values of λ_n, as both $\pm\lambda_n$ would lead to the same characteristic function $J_\nu(\lambda_n r)$.

We summarize the Fourier-Bessel series expansions obtained in this section in Table 4.2, and Hankel transforms in Table 7.3 in Chapter 7.

Example 4.5: Expand the function $f(r) = 1$ in the interval $(0,r_0)$ into a Fourier-Bessel series of the form

$$1 = \sum_{n=1}^{\infty} A_n J_0(\lambda_n r)$$

where λ_n's are the positive roots of $J_0(\lambda r_0) = 0$.

Solution: From Eqs. (4.55a) and (4.66) or from Table 4.2 we have

$$A_n = \frac{1}{N_n}\int_0^{r_0} J_0(\lambda_n r) r\, dr$$

and

$$N_n = \frac{r_0^2}{2} J_1^2(\lambda_n r_0)$$

Since (see Appendix B)

$$\frac{d}{dr}\left[rJ_1(\lambda_n r) \right] = \lambda_n r J_0(\lambda_n r)$$

then

$$\int_0^{r_0} J_0(\lambda_n r) r\, dr = \frac{1}{\lambda_n}\int_0^{r_0} \frac{d}{dr}\left[rJ_1(\lambda_n r) \right] dr = \frac{r_0}{\lambda_n} J_1(\lambda_n r_0)$$

and therefore we get

$$A_n = \frac{2}{(\lambda_n r_0) J_1(\lambda_n r_0)} , \qquad n = 1,2,3,\ldots$$

Thus, the desired Fourier-Bessel series is given by

$$1 = \frac{2}{r_0}\sum_{n=1}^{\infty} \frac{1}{\lambda_n}\frac{J_0(\lambda_n r)}{J_1(\lambda_n r_0)}$$

$$= \frac{2}{\lambda_1 r_0}\frac{J_0(\lambda_1 r)}{J_1(\lambda_1 r_0)} + \frac{2}{\lambda_2 r_0}\frac{J_0(\lambda_2 r)}{J_1(\lambda_2 r_0)} + \ldots$$

A table of the first 40 values of the zeros α_n of $J_0(\alpha)$ and the corresponding values of $J_1(\alpha_n)$ is given in Appendix B. For the first three zeros, the values are

$\alpha_1 = 2.4048$ $\quad J_1(\alpha_1) = 0.5191$

$\alpha_2 = 5.5201$ $\quad J_2(\alpha_2) = -0.3403$

$\alpha_3 = 8.6537$ $\quad J_3(\alpha_3) = 0.2715$

Table 4.2 Fourier-Bessel series expansions in the finite interval $(0, r_o)$.

$$f(r) = \sum_{n=1}^{\infty} A_n J_\nu(\lambda_n r), \qquad 0 < r < r_o \qquad\qquad r^2\frac{d^2R}{dr^2} + r\frac{dR}{dr} + (\lambda^2 r^2 - \nu^2)R = 0$$

$$R(0) = \text{finite}$$

$$A_n = \frac{1}{N_n}\int_0^{r_o} f(r)J_\nu(\lambda_n r)r\,dr \qquad\qquad \alpha R(r_o) + \beta\frac{dR(r_o)}{dr} = 0$$

Boundary Condition at $r = r_o$	$N_n^* = \int_0^{r_o} J_\nu^2(\lambda_n r)r\,dr$	Characteristic Values λ_n Are the Positive Roots of*
Third Kind ($\alpha \neq 0,\ \beta \neq 0$)	$\frac{r_o^2}{2}\left[1 + \frac{1}{\lambda_n^2}\left(H^2 - \frac{\nu^2}{r_o^2}\right)\right]J_\nu^2(\lambda_n r_o)$	$HJ_\nu(\lambda r_o) + \frac{dJ_\nu(\lambda r_o)}{dr} = 0$
Second Kind ($\alpha = 0,\ \beta \neq 0$)	$\frac{r_o^2}{2}\left[1 - \left(\frac{\nu}{\lambda_n r_o}\right)^2\right]J_\nu^2(\lambda_n r_o)$	$\frac{dJ_\nu(\lambda r_o)}{dr} = 0$**
First Kind ($\alpha \neq 0,\ \beta = 0$)	$\frac{r_o^2}{2} J_{\nu+1}^2(\lambda_n r_o)$	$J_\nu(\lambda r_o) = 0$

* $H = \alpha/\beta$.

** When $\nu = 0$, $\lambda_o = 0$ is also a characteristic value for this case.

So, the first three leading terms become

$$1 = 1.602\, J_0(2.4048 \frac{r}{r_0})$$

$$-1.065\, J_0(5.5201 \frac{r}{r_0}) + 0.8512\, J_0(8.6537 \frac{r}{r_0}) + \ldots$$

Since the Bessel functions $J_0(\alpha_n)$ are even functions, the above series represents 1 not only in the interval $(0, r_0)$ but also in the symmetrical interval $(-r_0, r_0)$. At the end points $r = \pm r_0$, the series does not converge to the given function, because all the terms in the series vanish at $r = \pm r_0$.

As we shall discuss in Chapter 5, two-dimensional steady-state problems in the cylindrical coordinates can be in one of the following forms:

$$T = f_1(r,\phi), \qquad T = f_2(r,z), \qquad \text{and} \qquad T = f_3(\phi,z)$$

$T = f_1(r,\phi)$ represents temperature distribution in a long cylindrical shell or a solid cylinder of an angular section. $T = f_2(r,z)$ corresponds to a short cylinder in which the axial temperature gradient $\partial T/\partial z$ must be considered. $T = f_3(\phi,z)$ has no physical importance, except in thin-walled cylinders. The first class of problems leads to solutions in the form of Fourier expansions while the second class, when the boundary conditions in the r-direction are homogeneous or can be made homogeneous, leads to expansions in the form of Fourier-Bessel series.

If the temperature distribution in a two-dimensional heat conduction problem in the spherical coordinates depends on the polar angle, θ, then the solution leads to an expansion in terms of *Legendre polynomials*. The necessary mathematical background for this will be deferred until Section 5.4.

REFERENCES

1. Hildebrand, F.B., *Advanced Calculus for Applications*, 2nd ed., Prentice-Hall, Inc., Englewood Cliffs, New Jersey, 1976.
2. Arpacı, V.S., *Conduction Heat Transfer*, Addison-Wesley Publishing Co., Inc., Reading, Massachusetts, 1966.
3. Churchill, R.V., *Operational Mathematics*, 3rd ed., McGraw-Hill Book Co., New York, 1972.
4. Özışık, M.N., *Boundary Value Problems of Heat Conduction*, International Textbook, Co., Scranton, Pennsylvania, 1968.
5. Özışık, M.N., *Heat Conduction*, John Wiley and Sons, Inc., New York, 1980.

6. Özışık, M.N., *Basic Heat Transfer*, McGraw-Hill Book Co., New York, 1977.
7. Sneddon, I.N., *The Use of Integral Transforms*, McGraw-Hill Book Co., New York, 1972.

PROBLEMS

4.1 Show that Eqs. (4.8) and (4.10) satisfy the boundary conditions (4.3b,c).

4.2 Show that the characteristic-value problem

$$\frac{d^2y}{dx^2} - \lambda^2 y = 0$$

$$y(0) = 0 \qquad \text{and} \qquad y(L) = 0$$

cannot have non-trivial solutions for real values of λ.

4.3 Consider the following functions

$$A_0, \qquad A_1 + A_2 x, \qquad \text{and} \qquad A_3 + A_4 x + A_5 x^2$$

where $A_0, \ldots, A_5$ are constants. Determine the constants so that these functions are orthogonal in the interval (0,1) with respect to the weight function unity.

4.4 Show that any equation having the form

$$a_0(x)\frac{d^2y}{dx^2} + a_1(x)\frac{dy}{dx} + [a_2(x) + \lambda a_3(x)]y = 0$$

can be written in the form of Eq. (4.13a) with

$$p(x) = \exp[\textstyle\int (a_1/a_0)dx], \qquad q(x) = \frac{a_2}{a_0}p(x), \qquad \text{and} \qquad w(x) = \frac{a_3}{a_0}p(x)$$

4.5 Find the characteristic values and the characteristic functions of the following characteristic-value problem:

$$\frac{d^2y}{dx^2} + \lambda^2 y = 0$$

$$\frac{dy(0)}{dx} = 0 \qquad \text{and} \qquad \frac{dy(L)}{dx} = 0$$

4.6 Find the characteristic values and the characteristic functions of the following characteristic-value problem:

$$\frac{d^2y}{dx^2} + \lambda^2 y = 0$$

$$\frac{dy(0)}{dx} = 0 \qquad \text{and} \qquad \alpha y(L) + \beta \frac{dy(L)}{dx} = 0$$

where α and β are two given constants.

4.7 Expand an arbitrary function f(x) into a series of the characteristic functions of the characteristic-value problem in Problem 4.6 over (0,L) and determine the expansion coefficients. From the expansion obtained also develop a finite Fourier transform of the function f(x) over (0,L).

4.8 Expand the following function into a Fourier sine series of period 2L in the interval (0,L).

$$f(x) = \begin{cases} 1 & \text{when } x < L/2 \\ 0 & \text{when } x > L/2 \end{cases}$$

4.9 Expand the function given in Problem 4.8 into a Fourier cosine series of period 2L in the interval (0,L).

4.10 Expand the function given in Problem 4.8 into a complete Fourier series of period 2L in the interval (-L,L).

4.11 (a) Determine the characteristic values and the characteristic functions of the following Sturm-Liouville problem:

$$\frac{d^2y}{dx^2} + \lambda^2 y = 0$$

$$y(-L) = y(L)$$

$$\frac{dy(-L)}{dx} = \frac{dy(L)}{dx}$$

(b) Show that the expansion of an arbitrary function f(x) in the interval (L,L) in terms of the characteristic functions found in (a) leads to the complete Fourier series expansion of f(x).

4.12 Expand the function f(r) = r in the interval $(0,r_0)$ into a series of the characteristic functions of the characteristic-value problem

$$\frac{d^2R}{dr^2} + \frac{1}{r}\frac{dR}{dr} + \lambda^2 R = 0$$

$$R(0) = \text{finite} \qquad \text{and} \qquad R(r_0) = 0$$

4.13 Show that the characteristic-value problem

$$r^2 \frac{d^2R}{dr^2} + r\frac{dR}{dr} - (\lambda^2 r^2 + \nu^2)R = 0$$

$$R(0) = \text{finite}$$

$$R(r_0) = 0$$

cannot have non-trivial solutions for real values of λ.

4.14 Determine the characteristic values and the characteristic functions of the following Sturm-Liouville problem:

$$\frac{d^2R}{dr^2} + \frac{1}{r}\frac{dR}{dr} + \lambda^2 R = 0$$

$$R(a) = 0$$

$$R(b) = 0$$

4.15 (a) Expand an arbitrary function $f(r)$ over (a,b) into a series of the characteristic functions of the Sturm-Liouville system in Problem 4.14 and determine the expansion coefficients.

(b) From the expansion obtained in (a), develop a transform of the function $f(x)$ over (a,b).

5

Steady Two- and Three-Dimensional Heat Conduction Solutions with Separation of Variables

5.1 INTRODUCTION

In this chapter we discuss solutions of two- and three-dimensional steady-state heat conduction problems by the method of *separation of variables* and introduce the technique in terms of examples. We shall first consider some representative problems in the rectangular coordinate system and investigate the conditions under which the method of separation of variables is applicable. Later on, we shall consider similar problems in the cylindrical and spherical coordinate systems.

5.2 STEADY TWO-DIMENSIONAL PROBLEMS IN THE RECTANGULAR COORDINATE SYSTEM

As an example of steady two-dimensional heat conduction problems in rectangular coordinate system, consider the fin of rectangular cross-section shown in Fig. 5.1. This fin is free of internal heat sources and sinks and has a constant thermal conductivity. If there is no temperature gradient in the

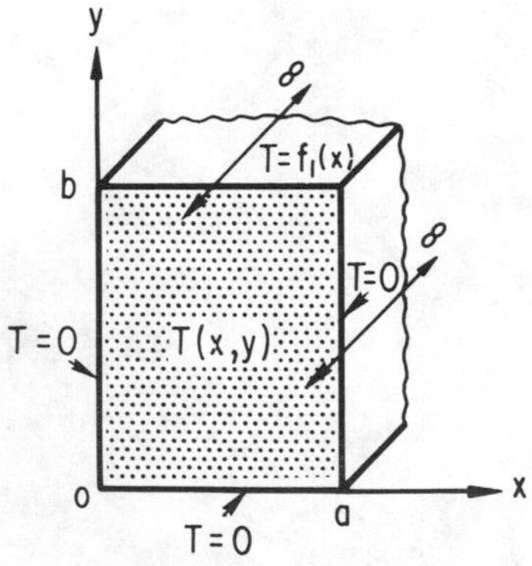

Fig. 5.1 Two-dimensional fin.

z-direction (either it is very long in the z-direction or its surfaces perpendicular to the z-direction at the two ends are perfectly insulated), then under steady-state conditions the temperature field T(x,y) in the fin must satisfy Laplace's equation in two dimensions:

$$\frac{\partial^2 T}{\partial x^2} + \frac{\partial^2 T}{\partial y^2} = 0 \tag{5.1}$$

Assume that the surfaces of the fin at x = 0, x = a and y = 0 are maintained at zero temperature while the temperature of the surface at y = b is given as a function of the x-coordinate; that is, $T(x,b) = f_1(x)$. The boundary conditions can then be written as

$$\begin{aligned} T(0,y) &= 0\,, & 0 < y < b \\ T(a,y) &= 0\,, & 0 < y < b \\ T(x,0) &= 0\,, & 0 < x < a \\ T(x,b) &= f_1(x) & 0 < x < a \end{aligned} \tag{5.2a,b,c,d}$$

The formulation of the problem with the differential equation (5.1) and the above boundary conditions is now complete. In order to find the temperature distribution in the fin, we must determine the solution of Laplace's equation (5.1) which satisfies the prescribed boundary values (5.2). We now seek a solution by the method of *separation of variables*. This method requires the assumption of the existence of a product solution of the form:

$$T(x,y) = X(x)Y(y) \tag{5.3}$$

where X is a function of x alone and Y is a function of y alone. Introducing Eq. (5.3) into Eq. (5.1), there follows

$$Y\frac{d^2X}{dx^2} + X\frac{d^2Y}{dy^2} = 0 \tag{5.4a}$$

or "separating the variables" one gets

$$-\frac{1}{X}\frac{d^2X}{dx^2} = \frac{1}{Y}\frac{d^2Y}{dy^2} \tag{5.4b}$$

Since each term in Eq. (5.4b) is a function of only one of the variables and each variable can be changed independently, the right- and left-hand sides of Eq. (5.4b) will be equal to each other if and only if they are equal to the same constant, say γ. Equation (5.4b) can now be written as two ordinary differential equations as follows:

$$\frac{d^2X}{dx^2} + \gamma\, X(x) = 0 \tag{5.5a}$$

$$\frac{d^2Y}{dy^2} - \gamma\, Y(y) = 0 \tag{5.5b}$$

where so far the separation constant γ is arbitrary. Let us now consider the following possible cases:

Case I: Let γ be zero. In this case, solving Eqs. (5.5) we obtain

$$X(x) = Ax + B \qquad \text{and} \qquad Y(y) = Cy + D$$

Hence,

$$T(x,y) = (Ax + B)(Cy + D) \tag{5.6}$$

$Y(y) = Cy + D$ cannot vanish as it would lead to $T(x,y) = 0$ which does not satisfy the boundary condition (5.2d); imposing the boundary conditions (5.2a,b) we get

$$A = B = 0$$

which in turn leads to $T(x,y) = 0$. Hence, for $\gamma = 0$ there is no solution.

Case II: If $\gamma < 0$, the general solutions of Eqs. (5.5) are of the form

$$X(x) = A \sinh\lambda x + B \cosh\lambda x, \qquad Y(y) = C \sin\lambda y + D \cos\lambda y$$

where we have substituted $\gamma = -\lambda^2$. Hence,

$$T(x,y) = (A \sinh\lambda x + B \cosh\lambda x)(C \sin\lambda y + D \cos\lambda y) \tag{5.7}$$

Application of the boundary conditions (5.2a,b), as in the previous case, yields

$$A = B = 0$$

Hence, for $\gamma < 0$ also there is no solution.

Case III: If $\gamma > 0$, the general solutions of Eqs. (5.5) are of the form

$$X(x) = A \sin\lambda x + B \cos\lambda x, \qquad Y(y) = C \sinh\lambda y + D \cosh\lambda y$$

where we have substituted $\gamma = \lambda^2$. Therefore,

$$T(x,y) = (A \sin\lambda x + B \cos\lambda x)(C \sinh\lambda y + D \cosh\lambda y) \tag{5.8}$$

Application of the boundary condition (5.2a) yields

$$T(0,y) = 0 = B(C \sinh\lambda y + D \cosh\lambda y)$$

so that $B = 0$ as $(C \sinh\lambda y + D \cosh\lambda y) = 0$ would lead to $T(x,y) = 0$. Similarly, the boundary condition (5.2d) gives $D = 0$. Thus,

$$T(x,y) = E \sin\lambda x \sinh\lambda y \tag{5.9}$$

where $E \equiv AC$. Imposing the boundary condition (5.2b) we obtain

$$T(a,y) = 0 = E \sin\lambda a \sinh\lambda y$$

In order not to have $T(x,y) = 0$, it follows that $\sin\lambda a = 0$, which, in turn, implies that λ can have any of the following values

$$\lambda_n = \frac{n\pi}{a}, \qquad n = 1,2,3... \tag{5.10}$$

It is to be noted that $n = 0$ has been excluded because it leads to $T(x,y) = 0$. Furthermore, *negative* integers $n = -1,-2,...$ can be omitted with no loss of generality. For instance, if we put $\lambda = \frac{5\pi}{a}$ and $\lambda = -\frac{5\pi}{a}$ into Eq. (5.9), the results differ only in a minus sign out in front, which can be absorbed by the arbitrary constant E. Thus, a solution of the differential equation (5.1) which satisfies the boundary conditions (5.2a,b,c) can be written as

$$T_n(x,y) = E_n \sin\lambda_n x \sinh\lambda_n y \tag{5.11}$$

Is Eq. (5.11) the solution of the problem for any value of n? That is, does it also satisfy the boundary condition (5.2d)? In general,

$$f_1(x) \neq E_n \sin\lambda_n x \sinh\lambda_n b$$

Therefore Eq. (5.11) does not satisfy the boundary condition at $y = b$. Since the differential equation (5.1) and the boundary conditions (5.2a,b,c) are linear and homogeneous, it follows that a linear combination in the form

$$T(x,y) = \sum_{n=1}^{\infty} E_n \sin\lambda_n x \sinh\lambda_n y \tag{5.12}$$

also satisfies the differential equation (5.1) and the boundary conditions (5.2a,b,c). Finally, imposing the boundary condition (5.2d) on Eq. (5.12) we get

$$f_1(x) = \sum_{n=1}^{\infty} E_n \sin\lambda_n x \sinh\lambda_n b \tag{5.13a}$$

or

$$f_1(x) = \sum_{n=1}^{\infty} a_n \sin\lambda_n x \tag{5.13b}$$

where $a_n = E_n \sinh\lambda_n b$. Is it possible to find a_n's so as to satisfy Eq. (5.13b); if so, how? Observe that

$$T(0,y) = X(0)Y(y) \Rightarrow X(0) = 0$$

$$T(a,y) = X(a)Y(y) \Rightarrow X(a) = 0$$

So, $X(x)$ satisfies

$$\frac{d^2X}{dx^2} + \lambda^2 X(x) = 0 \tag{5.14a}$$

$$X(0) = 0 \tag{5.14b}$$

$$X(a) = 0 \tag{5.14c}$$

Recalling Section 4.6.1, we see that this is a Sturm-Liouville problem with the following characteristic functions and characteristic values:

$$\phi_n(x) = \sin\lambda_n x \qquad \text{and} \qquad \lambda_n = \frac{n\pi}{a}, \qquad n = 1,2,3...$$

Therefore, $f_1(x)$ can be expanded in terms of $\sin \frac{n\pi}{a} x$ and, in fact, Eq. (5.13b) is the Fourier sine expansion of $f_1(x)$ over $(0,a)$, where the expansion coefficients a_n's can be evaluated by using Eqs. (4.33a) and (4.33b) or Table 4.1 as

$$a_n = \frac{2}{a}\int_0^a f_1(x) \sin\lambda_n x \, dx \tag{5.15}$$

Thus,

$$E_n = \frac{2}{a \sinh\lambda_n b}\int_0^a f_1(x) \sin\lambda_n x \, dx \tag{5.16}$$

The final solution for the temperature distribution may then be written as

$$T(x,y) = \frac{2}{a}\sum_{n=1}^{\infty} \frac{\sin \frac{n\pi}{a} x \sinh \frac{n\pi}{a} y}{\sinh \frac{n\pi}{a} b}\int_0^a f_1(x') \sin \frac{n\pi}{a} x' dx' \tag{5.17}$$

If, in particular, $f_1(x) = T_0$ = constant, then

$$E_n = \frac{2}{a \sinh\lambda_n b}\int_0^a T_0 \sin \frac{n\pi}{a} x \, dx = \frac{2T_0}{n\pi} \frac{[1-(-1)^n]}{\sinh\lambda_n b} \tag{5.18}$$

Substituting this result into Eq. (5.12) we find

$$\frac{T(x,y)}{T_0} = \frac{2}{\pi}\sum_{n=1}^{\infty} \frac{[1-(-1)^n]}{n} \frac{\sin \frac{n\pi}{a} x \sinh \frac{n\pi}{a} y}{\sinh \frac{n\pi}{a} b} \tag{5.19}$$

It was not obvious, a priori, that Eq. (5.1) would possess a separable solution of the form (5.11) or that a solution built up from such solutions could be made to satisfy the prescribed boundary conditions (5.2). In view of the above example, we now conclude that the method of separation of variables is applicable to steady two-dimensional problems if and when (a) the differential equation is linear and homogeneous, and (b) the four boundary conditions are linear and three of them are homogeneous, so that one of the directions of the problem is expressed by a homogeneous differential equation subjected to two homogeneous boundary conditions. The sign of the separation constant is chosen so that this boundary-value problem becomes a Sturm-Liouville type characteristic-value problem such as the one given by Eqs. (5.14) in the above problem.

To calculate the temperature at any point numerically, each term of the series (5.19) must be evaluated and summed. Except for very small values of y/a, this series converges rapidly and only the first few terms are sufficient. As a numerical example, let us consider the following problem.

Example 5.1: The temperature is maintained at 0°C along the three surfaces of the two-dimensional fin shown in Fig. 5.1 while the fourth surface at $y = b$ is held at 100°C. If $a = 2b$ calculate the temperature of the center line of the fin under steady-state conditions.

Solution: Equation (5.19) with $x = \frac{a}{2}$ and $y = \frac{b}{2} = \frac{a}{4}$ yields

$$\frac{T(\frac{a}{2},\frac{a}{4})}{100} = \frac{2}{\pi}\sum_{n=1}^{\infty}\frac{[1-(-1)^n]}{n}\,\frac{\sin\frac{n\pi}{2}\,\sinh\frac{n\pi}{4}}{\sinh\frac{n\pi}{2}}$$

$$= 0.48061 - 0.03987 + 0.00502 - \ldots$$

$$= 44.576\ °C$$

As it is seen, this series converges rather rapidly at $x = \frac{a}{2}$ and $y = \frac{a}{4}$.

Let us now consider the two-dimensional fin problem shown in Fig. 5.2. The formulation of the problem is

$$\frac{\partial^2 T}{\partial x^2} + \frac{\partial^2 T}{\partial y^2} = 0 \tag{5.20}$$

$$T(0,y) = 0\ , \qquad 0 < y < b \tag{5.21a}$$

$$T(a,y) = 0\ , \qquad 0 < y < b \tag{5.21b}$$

$$T(x,0) = f_2(x)\ , \qquad 0 < x < a \tag{5.21c}$$

$$T(x,b) = 0\ , \qquad 0 < x < a \tag{5.21d}$$

Assuming the existence of a product solution of the form

$$T(x,y) = X(x)Y(y) \tag{5.22}$$

the solution of the differential equation (5.20) can again be written as

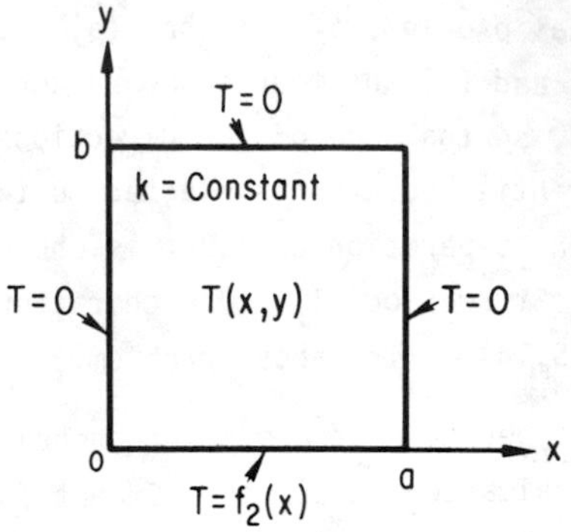

Fig. 5.2 Two-dimensional fin.

$$T(x,y) = (A \sin\lambda x + B \cos\lambda x)(C \sinh\lambda y + D \cosh\lambda y) \tag{5.23}$$

where the sign of the separation constant is chosen such that the homogeneous x-direction results in a Sturm-Liouville type characteristic-value problem. Application of the boundary condition at x = 0 yields B = 0, and the boundary condition at y = b gives

$$A \sin\lambda x\ (C \sinh\lambda b + D \cosh\lambda b) = 0$$

from which we obtain

$$D = -C \frac{\sinh\lambda b}{\cosh\lambda b}$$

Substituting D into Eq. (5.23), together with B = 0, gives

$$T(x,y) = A\ C \sin\lambda x\ (\sinh\lambda y - \frac{\sinh\lambda b}{\cosh\lambda b} \cosh\lambda y) \tag{5.24a}$$

which can also be written as

$$T(x,y) = E \sin\lambda x \sinh\lambda(b - y) \tag{5.24b}$$

where $E = - A\ C/\cosh\lambda b$. Application of the boundary condition at x = a yields

$$E \sin\lambda a \sinh\lambda(b - y) = 0$$

which results in

$$\sin\lambda a = 0$$

or

$$\lambda_n = \frac{n\pi}{a}, \qquad n = 1,2,3...$$

Equation (5.24b) satisfies, for each value of λ_n, the differential equation

(5.20) and the boundary conditions (5.21a,b,d). The solution of Eq. (5.20) which will satisfy all the boundary conditions can now be written as the linear combination of these individual solutions; that is,

$$T(x,y) = \sum_{n=1}^{\infty} E_n \sin\lambda_n x \sinh\lambda_n (b - y) \tag{5.25}$$

Finally, applying the condition $T(x,0) = f_2(x)$, we get

$$f_2(x) = \sum_{n=1}^{\infty} a_n \sin \frac{n\pi}{a} x \tag{5.26}$$

where $a_n = E_n \sinh\lambda_n b$. Equation (5.26) is the Fourier sine series expansion of $f_2(x)$ over (o,a), and the expansion coefficients a_n's may be written from Table 4.1 as

$$a_n = \frac{2}{a}\int_0^a f_2(x)\sin \frac{n\pi}{a} x\, dx$$

Hence, Eq. (5.25) becomes

$$T(x,y) = \frac{2}{a} \sum_{n=1}^{\infty} \frac{\sin \frac{n\pi}{a} x \sinh \frac{n\pi}{a} (b - y)}{\sinh \frac{n\pi}{a} b} \int_0^a f_2(x')\sin \frac{n\pi}{a} x'dx' \tag{5.27}$$

This result could also be obtained directly from the solution (5.17) by substituting b-y for y and $f_2(x)$ for $f_1(x)$.

Similarly, the solution for the problem of Fig. 5.3 can be written as (see Problem 5.2)

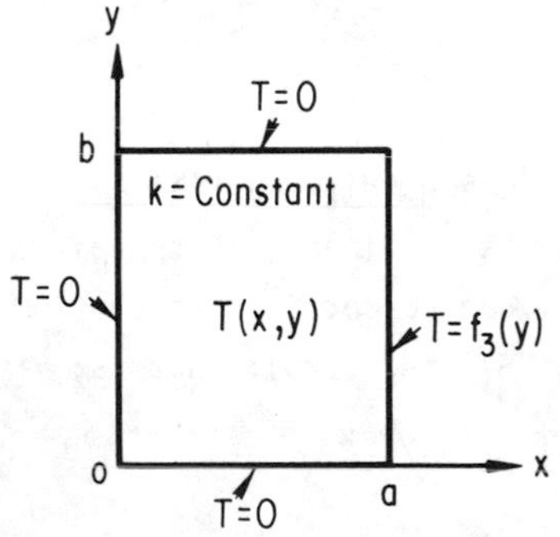

Fig. 5.3 Two-dimensional fin.

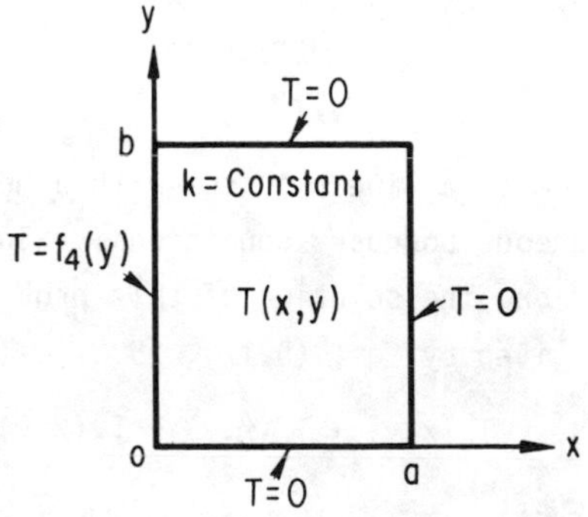

Fig. 5.4 Two-dimensional fin.

$$T(x,y) = \frac{2}{b}\sum_{n=1}^{\infty} \frac{\sinh \frac{n\pi}{b} x \sin \frac{n\pi}{b} y}{\sinh \frac{n\pi}{b} a} \int_0^b f_3(y') \sin \frac{n\pi}{b} y' \, dy' \tag{5.28}$$

and the solution for the problem shown in Fig. 5.4 becomes (see Problem 5.3)

$$T(x,y) = \frac{2}{b}\sum_{n=1}^{\infty} \frac{\sinh \frac{n\pi}{b} (a-x) \sin \frac{n\pi}{b} y}{\sinh \frac{n\pi}{b} a} \int_0^b f_4(y') \sin \frac{n\pi}{b} y' dy' \tag{5.29}$$

5.2.1 Non-Homogeneity in Boundary Conditions

We have seen that the method of separation of variables is applicable to steady two-dimensional linear problems consisting of a homogeneous differential equation subject to three homogeneous and one non-homogeneous boundary conditions. Most steady two-dimensional problems do not satisfy these requirements. Non-homogeneities may result, for example, from more than one non-homogeneous boundary conditions and/or non-homogeneous differential equations. In such cases the problem may be divided into a number of simple problems such that each simple problem satisfies the conditions required for the method of separation of variables. Then the solution to the original problem is obtained by superposing the solutions of the simple problems.

As an example of problems with more than one non-homogeneous boundary conditions, let us consider the following two-dimensional fin problem:

$$\frac{\partial^2 T}{\partial x^2} + \frac{\partial^2 T}{\partial y^2} = 0 \tag{5.30}$$

$$T(0,y) = f_4(y) \,, \qquad 0 < y < b \tag{5.31a}$$

$$T(a,y) = f_3(y) \,, \qquad 0 < y < b \tag{5.31b}$$

$$T(x,0) = f_2(x) \,, \qquad 0 < x < a \tag{5.31c}$$

$$T(x,b) = f_1(x) \,, \qquad 0 < x < a \tag{5.31d}$$

This problem is a linear system with a homogeneous differential equation and four non-homogeneous boundary conditions. Since it is linear, by the *principle of superposition*, the solution of this problem can be obtained as the sum of the four solutions given by Eqs. (5.17), (5.27), (5.28) and (5.29); that is,

$$T(x,y) = T_1(x,y) + T_2(x,y) + T_3(x,y) + T_4(x,y)$$

Figure 5.5 illustrates this superposition.

The solution of linear problems, such as heat conduction problems with

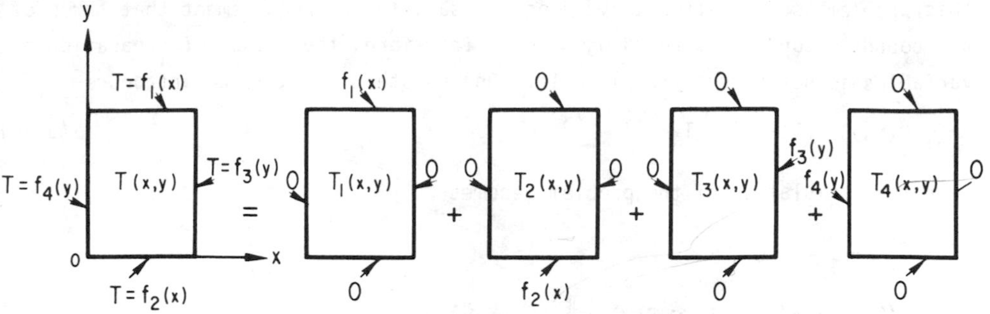

Fig. 5.5 Principle of superposition.

constant properties, may often be reduced to the solution of a number of simpler problems by employing the principle of superposition as explained by the above example. In some cases, however, the problem may be transformed into a form that can be handled by the method of separation of variables by shifting the temperature level. As an example, let us consider the long two-dimensional fin shown in Fig. 5.6, where the surrounding fluid temperature T_∞ and the heat transfer coefficient h are specified. Taking into account the geometric and thermal symmetries with respect to the x-axis, the formulation of the problem under steady-state conditions is given by

$$\frac{\partial^2 T}{\partial x^2} + \frac{\partial^2 T}{\partial y^2} = 0 \tag{5.32}$$

$$T(0,y) = f(y) \; ; \qquad T(\infty,y) = T_\infty \tag{5.33a,b}$$

$$\frac{\partial T(x,0)}{\partial y} = 0 \; ; \qquad -k\,\frac{\partial T(x,b)}{\partial y} = h[T(x,b) - T_\infty] \tag{5.33c,d}$$

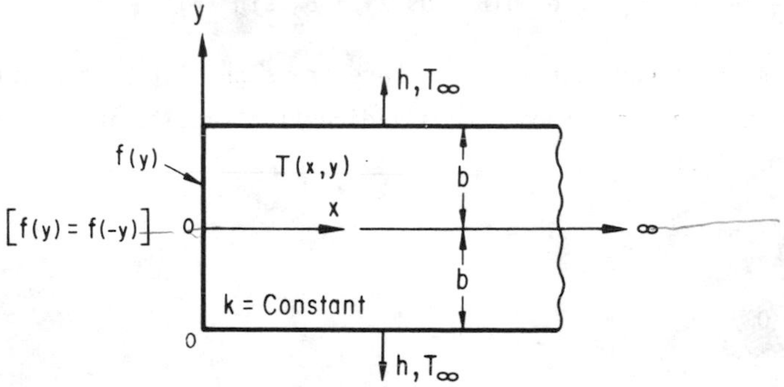

Fig. 5.6 Infinitely long two-dimensional fin with convective boundary conditions.

This problem as formulated above does not satisfy the requirement that three of the boundary conditions be homogeneous. Therefore, the method of separation of variables is not directly applicable. On the other hand, if we introduce

$$\theta(x,y) = T(x,y) - T_\infty \tag{5.34}$$

then the formulation of the problem becomes

$$\frac{\partial^2\theta}{\partial x^2} + \frac{\partial^2\theta}{\partial y^2} = 0 \tag{5.35}$$

$$\theta(0,y) = f(y) - T_\infty \equiv F(y) \; ; \qquad \theta(\infty,y) = 0 \tag{5.36a,b}$$

$$\frac{\partial\theta(x,0)}{\partial y} = 0 \; ; \qquad -k\,\frac{\partial\theta(x,b)}{\partial y} = h\theta(x,b) \tag{5.36c,d}$$

which satisfies the conditions required by the method of separation of variables. Therefore, first assuming the existence of a product solution of the form

$$\theta(x,y) = X(x)Y(y) \tag{5.37}$$

and then introducing this assumption into Eq. (5.35) and dividing each term of th resulting expression by XY, we obtain

$$\frac{1}{X}\frac{d^2X}{dx^2} = -\frac{1}{Y}\frac{d^2Y}{dy^2} = \lambda^2 \tag{5.38}$$

or

$$\frac{d^2Y}{dy^2} + \lambda^2 Y = 0 \tag{5.39a}$$

$$\frac{d^2X}{dx^2} - \lambda^2 X = 0 \tag{5.39b}$$

which lead to the following general solution

$$\theta(x,y) = (A_1\, e^{-\lambda x} + A_2\, e^{\lambda x})(B_1 \cos \lambda y + B_2 \sin \lambda y) \tag{5.40}$$

for the differential equation (5.35). Here the sign of the separation constant λ^2 is chosen such that the homogeneous y-direction results in a characteristic-value problem. That is,

$$\frac{d^2Y}{dy^2} + \lambda^2 Y = 0 \tag{5.41a}$$

$$\frac{dY(0)}{dy} = 0 \tag{5.41b}$$

$$k\,\frac{dY(b)}{dy} + h\,Y(b) = 0 \tag{5.41c}$$

This is a Sturm-Liouville problem with the following characteristic functions (see Table 4.1 or Problem 4.6):

$$\phi_n(x) = \cos \lambda_n y \tag{5.42}$$

where the characteristic values are the positive roots of

$$\lambda_n \tan \lambda_n b = h/k , \qquad n = 1,2,3... \tag{5.43}$$

which are obtained from the application of the boundary condition (5.41c).

Finally, application of the boundary conditions (5.36b,c,d) yields

$$\theta(x,y) = \sum_{n=1}^{\infty} a_n e^{-\lambda_n x} \cos\lambda_n y \tag{5.44}$$

where λ_n's are the positive roots of the transcendental equation (5.43). The non-homogeneous boundary condition (5.36a) requires that

$$F(y) = f(y) - T_\infty = \sum_{n=1}^{\infty} a_n \cos\lambda_n y \tag{5.45}$$

where the expansion coefficients a_n can be obtained from Table 4.1 as

$$a_n = \frac{2\lambda_n}{\lambda_n b + \sin\lambda_n b \cos\lambda_n b} \int_0^b [f(y) - T_\infty]\cos \lambda_n y dy \tag{5.46}$$

Hence, the temperature distribution becomes

$$\theta(x,y) = T(x,y) - T_\infty$$

$$= 2 \sum_{n=1}^{\infty} \frac{\lambda_n e^{-\lambda_n x} \cos\lambda_n y}{\lambda_n b + \sin\lambda_n b \cos\lambda_n b} \int_0^b [f(y') - T_\infty]\cos\lambda_n y' dy' \tag{5.47}$$

If, in particular, $f(y) = T_0$ = constant, then the temperature distribution is given by

$$\frac{T(x,y) - T_\infty}{T_0 - T_\infty} = 2 \sum_{n=1}^{\infty} \frac{\sin\lambda_n b}{\lambda_n b + \sin\lambda_n b \cos\lambda_n b} e^{-\lambda_n x} \cos\lambda_n y \tag{5.48}$$

The characteristic values λ_n in Eqs. (5.47) and (5.48) are to be obtained from the transcendental equation (5.43), which may be rearranged as

$$\tan\lambda_n b = \frac{Bi}{\lambda_n b} \tag{5.49a}$$

or

$$\cot\lambda_n b = \frac{\lambda_n b}{Bi} \tag{5.49b}$$

where Bi = hb/k. It is to be noted that the number of possible solutions for the roots of Eq. (5.49a) or Eq. (5.49b) are infinite. Although they cannot be obtained by ordinary algebraic methods, they can be found numerically or determined graphically as illustrated in Fig. 5.7. A tabulation of the characteristic values, that is, the roots of Eq. (5.49b), is given in References [4,5].

5.2.2 Non-Homogeneity in Differential Equations

So far, we have discussed non-homogeneities in boundary conditions. Let us now discuss non-homogeneities in differential equations.

As an example, consider the rectangular fin shown in Fig. 5.8. Heat is generated in this fin at a constant rate $\dot{q}$ per unit volume (W/m³). Assuming that there is no temperature gradient in the z-direction and the thermal conductivity of the material of the fin is constant, the differential equation for the temperature distribution T(x,y) can be written as

$$\frac{\partial^2 T}{\partial x^2} + \frac{\partial^2 T}{\partial y^2} + \frac{\dot{q}}{k} = 0 \tag{5.50}$$

which is linear but not homogeneous, and is, therefore, not separable. The boundary conditions are

$$\frac{\partial T(0,y)}{\partial x} = 0 \ ; \qquad T(a,y) = 0 \tag{5.51a,b}$$

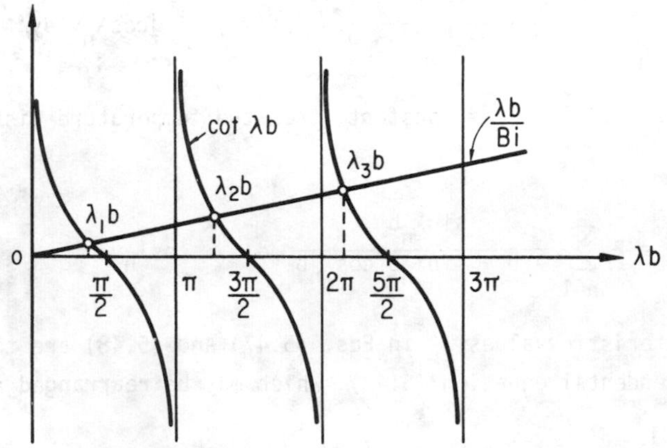

Fig. 5.7 Graphical determination of the roots of the transcendental equation (5.49b).

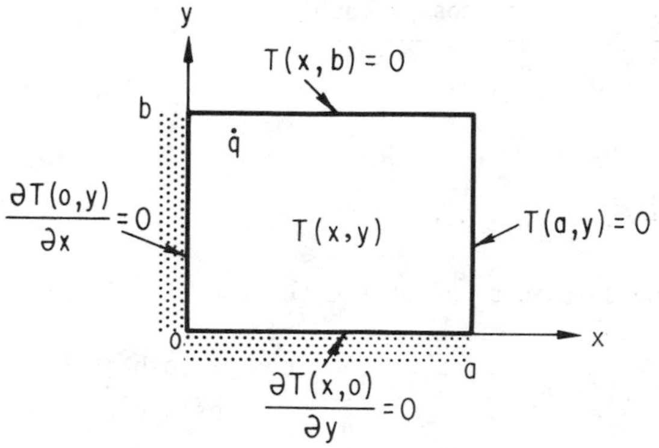

Fig. 5.8 Two-dimensional fin.

$$\frac{\partial T(x,0)}{\partial y} = 0 \ ; \qquad T(x,b) = 0 \tag{5.51c,d}$$

which are linear and homogeneous. If the solution is assumed to be in the form

$$T(x,y) = \psi(x,y) + \phi(x) \tag{5.52}$$

then the problem can be written as a superposition of the following two simple problems:

$$\frac{d^2\phi}{dx^2} + \frac{\dot{q}}{k} = 0 \tag{5.53}$$

$$\frac{d\phi(0)}{dx} = 0 \ ; \qquad \phi(a) = 0 \tag{5.54}$$

and

$$\frac{\partial^2\psi}{\partial x^2} + \frac{\partial^2\psi}{\partial y^2} = 0 \tag{5.55}$$

$$\frac{\partial\psi(0,y)}{\partial x} = 0 \ ; \qquad \psi(a,y) = 0 \tag{5.56a,b}$$

$$\frac{\partial\psi(x,0)}{\partial y} = 0 \ ; \qquad \psi(x,b) = -\phi(x) \tag{5.56c,d}$$

The solution to the problem for $\phi(x)$ can easily be found to be

$$\phi(x) = \frac{\dot{q}a^2}{2k}\left[1 - \left(\frac{x}{a}\right)^2\right] \tag{5.57}$$

The problem for $\psi(x,y)$ can be solved using the method of separation of variables and the solution is

$$\psi(x,y) = -\frac{2\dot{q}}{ak}\sum_{n=0}^{\infty}\frac{(-1)^n}{\lambda_n^3}\frac{\cos\lambda_n x\cosh\lambda_n y}{\cosh\lambda_n b} \tag{5.58}$$

where

$$\lambda_n = \frac{(2n+1)\pi}{2a}, \qquad n = 0,1,2...$$

Combining these two solutions we get the solution for T(x,y) as

$$\frac{T(x,y)}{\dot{q}\ a^2/k} = \frac{1}{2}\left[1 - \left(\frac{x}{a}\right)^2\right] - \frac{2}{a^3}\sum_{n=0}^{\infty}\frac{(-1)^n}{\lambda_n^3}\frac{\cos\lambda_n x\cosh\lambda_n y}{\cosh\lambda_n b} \tag{5.59}$$

It is to be noted that this problem can also be solved by assuming

$$T(x,y) = \psi(x,y) + \phi(y) \tag{5.60}$$

However, if $\dot{q} = \dot{q}(x)$ then only the first assumption (5.52) will be applicable. Similarly, if $\dot{q} = \dot{q}(y)$ then only the second assumption (5.60) can be used. On the other hand, if $\dot{q} = \dot{q}(x,y)$ then, in general, the problem cannot be separated into simpler problems. In such a case, the solution can be obtained, for example, by following the procedure outlined in Problem 5.12, or by employing integral transforms that we shall discuss in Chapter 7. The superposition technique introduced in this section is general in the sense that it can equally be applied to cylindrical and spherical geometries and to unsteady problems as well.

5.3 STEADY TWO-DIMENSIONAL PROBLEMS IN THE CYLINDRICAL COORDINATE SYSTEM

Under steady-state conditions and without heat sources or sinks, the heat conduction equation in the cylindrical coordinates with constant thermal conductivity is obtained from Eqs. (2.19) and (2.21) as:

$$\frac{\partial^2 T}{\partial r^2} + \frac{1}{r}\frac{\partial T}{\partial r} + \frac{1}{r^2}\frac{\partial^2 T}{\partial \phi^2} + \frac{\partial^2 T}{\partial z^2} = 0 \tag{5.61}$$

where $T = T(r,\phi,z)$. From this differential equation it is easily seen that there can be three types of two-dimensional problems in cylindrical systems in the form

$$T = T(r,\phi), \qquad T = T(r,z), \qquad \text{and} \qquad T = T(\phi,z)$$

Problems of the form $T(\phi,z)$ have no physical significance, except in thin-walled cylinders. In the following sections we shall therefore discuss problems of the other two forms, namely $T(r,\phi)$ and $T(r,z)$, in terms of representative examples.

5.3.1 Steady Two-Dimensional Problems in (r,ϕ) Variables

As an example to the problems of the form $T = T(r,\phi)$, consider a long solid semi-cylinder of radius r_0 of cross section shown in Fig. 5.9. Assume that the surface at $r = r_0$ is held at an arbitrary temperature distribution $f(\phi)$ and the diameter surface is maintained at a constant and uniform temperature T_0. The differential equation which governs the steady-state temperature distribution $T(r,\phi)$ from Eq. (5.61) is

$$\frac{\partial^2 T}{\partial r^2} + \frac{1}{r}\frac{\partial T}{\partial r} + \frac{1}{r^2}\frac{\partial^2 T}{\partial \phi^2} = 0 \tag{5.62}$$

and the boundary conditions are

$$T(0,\phi) = T_0 \ ; \qquad T(r_0,\phi) = f(\phi) \tag{5.63a,b}$$

$$T(r,0) = T_0 \ ; \qquad T(r,\pi) = T_0 \tag{5.63c,d}$$

Defining a new temperature function as $\theta = T - T_0$ and assuming a product solution in the form

$$\theta(r,\phi) = R(r)\psi(\phi) \tag{5.64}$$

from the differential equation (5.62) we get

$$r^2 \frac{d^2R}{dr^2} + r\frac{dR}{dr} - \lambda^2 R = 0 \tag{5.65a}$$

$$\frac{d^2\psi}{d\phi^2} + \lambda^2\psi = 0 \tag{5.65b}$$

where the sign of the separation constant is consistent with the fact that the ϕ-direction is the homogeneous direction for $\theta(r,\phi)$. Hence, noting that Eq. (5.65a) is a Cauchy-Euler equation (see Appendix B), we obtain*

$$\theta(r,\phi) = (A_1 r^{\lambda} + A_2 r^{-\lambda})(B_1 \sin\lambda\phi + B_2 \cos\lambda\phi) \tag{5.66}$$

Since $T(0,\phi) = T_0$ and therefore $\theta(0,\phi) = 0$, $r^{-\lambda}$ cannot exist in the solution as this term is not bounded at $r = 0$. Setting $A_2 = 0$ so that $\theta(r,\phi)$ is bounded as $r \to 0$ and also employing the other boundary conditions, we get

$$\theta(r,\phi) = \sum_{n=1}^{\infty} a_n r^{\lambda_n} \sin\lambda_n\phi, \qquad 0 < \phi < \pi \tag{5.67a}$$

with

* Equation (5.66) will be the solution if $\lambda \neq 0$. But in this problem λ is, in fact, not equal to zero as it will be shown; see Eq. (5.67b).

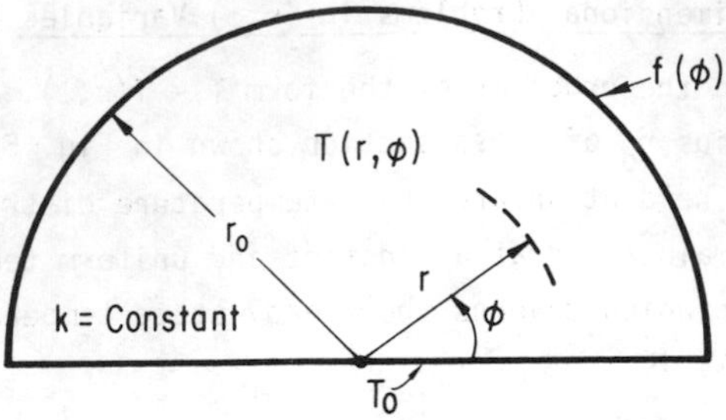

Fig. 5.9 Long solid semi-cylinder.

$$\lambda_n = \frac{n\pi}{\pi} = n, \qquad n = 1,2,3... \tag{5.67b}$$

Finally, imposing the non-homogeneous boundary condition (5.63b) yields

$$F(\phi) = \sum_{n=1}^{\infty} a_n r_0^n \sin n\phi, \qquad 0 < \phi < \pi \tag{5.68}$$

where we have defined

$$F(\phi) = f(\phi) - T_0 \tag{5.69}$$

Equation (5.68) is the Fourier sine expansion of $F(\phi)$ over the interval $(0,\pi)$, and hence from Eqs. (4.33a) and (4.33b) or Table 4.1 we obtain

$$a_n = \frac{2}{\pi r_0^n}\int_0^{\pi} F(\phi')\sin n\phi'\, d\phi' \tag{5.70}$$

Then, the final solution for the temperature distribution $T(r,\phi)$ is given by

$$T(r,\phi) = T_0 + \frac{2}{\pi}\sum_{n=1}^{\infty}\left(\frac{r}{r_0}\right)^n \sin n\phi \int_0^{\pi}[f(\phi') - T_0]\sin n\phi'\, d\phi' \tag{5.71}$$

If, in particular, $f(\phi) = T_1 =$ Constant, then the temperature distribution becomes

$$\frac{T(r,\phi) - T_0}{T_1 - T_0} = \frac{2}{\pi}\sum_{n=1}^{\infty}\frac{[1-(-1)^n]}{n}\left(\frac{r}{r_0}\right)^n \sin n\phi \tag{5.72}$$

As a second example to the problems of the form $T(r,\phi)$, consider a long solid cylinder of radius r_0 of cross section shown in Fig. 5.10. Assume that the surface of the cylinder is held at an arbitrary temperature distribution

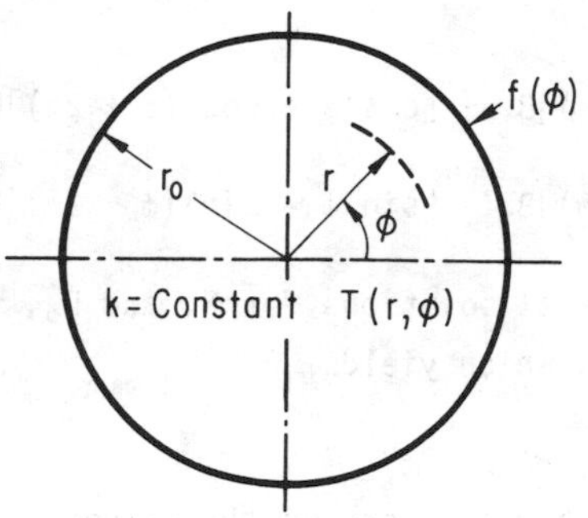

Fig. 5.10 Long solid cylinder.

$f(\phi)$. Under steady-state conditions, this problem can be formulated as follows:

$$\frac{\partial^2 T}{\partial r^2} + \frac{1}{r}\frac{\partial T}{\partial r} + \frac{1}{r^2}\frac{\partial^2 T}{\partial \phi^2} = 0 \tag{5.73}$$

$$T(0,\phi) = \text{finite} \tag{5.74a}$$

$$T(r_0,\phi) = f(\phi) \tag{5.74b}$$

$$T(r,\phi) = T(r,\phi + 2\pi) \tag{5.74c}$$

$$\frac{\partial T(r,\phi)}{\partial \phi} = \frac{\partial T(r,\phi + 2\pi)}{\partial \phi} \tag{5.74d}$$

Seeking a product solution of the form

$$T(r,\phi) = R(r)\psi(\phi) \tag{5.75}$$

yields the following general solution

$$T(r,\phi) = (A_1 r^{\lambda} + A_2 r^{-\lambda})(B_1 \sin\lambda\phi + B_2 \cos\lambda\phi) \tag{5.76}$$

for the differential equation (5.73), where we have chosen the sign of the separation constant such that the homogeneous ϕ-direction results in a Sturm-Liouville type characteristic-value problem; that is,

$$\frac{d^2\psi}{d\phi^2} + \lambda^2\psi = 0 \tag{5.77a}$$

$$\psi(\phi) = \psi(\phi + 2\pi) \tag{5.77b}$$

$$\frac{d\psi(\phi)}{d\phi} = \frac{d\psi(\phi + 2\pi)}{d\phi} \tag{5.77c}$$

This is a special kind of the Sturm-Liouville problem with periodic boundary conditions of period 2π.

If we impose the boundary conditions (5.74c) and (5.74d) on the solution

(5.76) we obtain

$$[\sin\lambda\phi - \sin\lambda(\phi + 2\pi)]B_1 + [\cos\lambda\phi - \cos\lambda(\phi + 2\pi)]B_2 = 0 \tag{5.78a}$$

$$[\cos\lambda\phi - \cos\lambda(\phi + 2\pi)]B_1 - [\sin\lambda\phi - \sin\lambda(\phi + 2\pi)]B_2 = 0 \tag{5.78b}$$

In order to have non-trivial solutions for B_1 and B_2, the determinant of the coefficients must vanish, which yields

$$\cos 2\lambda\pi = 1 \tag{5.79}$$

This is possible if λ is equal to one of the values of

$$\lambda_n = n, \qquad n = 0,1,2... \tag{5.80}$$

which are, in fact, the characteristic values of the problem (5.77). This result means that λ in Eq. (5.76) can only take on zero or integer values if that solution is to satisfy the conditions (5.74c) and (5.74d). Here we have discarded negative integers in Eq. (5.80) as they would lead to the same solutions as the positive ones. Setting $A_2 = 0$ so that the solution will satisfy the condition (5.74a) and employing superposition we can write

$$T(r,\phi) = \sum_{n=0}^{\infty} r^n(a_n \sin n\phi + b_n \cos n\phi) \tag{5.81a}$$

or

$$T(r,\phi) = b_0 + \sum_{n=1}^{\infty} r^n(a_n \sin n\phi + b_n \cos n\phi) \tag{5.81b}$$

This result satisfies the differential equation (5.73) and the conditions (5.74a), (5.74c) and (5.74d). It also satisfies the condition (5.74b) because imposing this condition yields

$$f(\phi) = b_0 + \sum_{n=1}^{\infty} r_0^n(a_n \sin n\phi + b_n \cos n\phi) \tag{5.82}$$

which is the complete Fourier series representation of $f(\phi)$ over the interval $(0,2\pi)$. Since $f(\phi)$ is a periodic function with period 2π, the coefficients in Eq. (5.82) can be determined from Eqs. (4.48) as

$$b_0 = \frac{1}{2\pi}\int_0^{2\pi} f(\phi)d\phi \tag{5.83a}$$

$$a_n = \frac{1}{\pi r_o^n}\int_0^{2\pi} f(\phi)\ \sin n\phi d\phi, \qquad n = 1,2,3... \tag{5.83b}$$

$$b_n = \frac{1}{\pi r_o^n}\int_0^{2\pi} f(\phi)\ \cos n\phi d\phi, \qquad n = 1,2,3... \tag{5.83c}$$

where we have taken c = 0 in Eqs. (4.48). Thus, the final solution can then be written as

$$T(r,\phi) = \frac{1}{2\pi}\int_0^{2\pi} f(\phi)d\phi + \frac{1}{\pi}\sum_{n=1}^{\infty} \left(\frac{r}{r_o}\right)^n [\sin n\phi \int_0^{2\pi} f(\phi')\sin n\phi' d\phi'$$

$$+ \cos n\phi \int_0^{2\pi} f(\phi')\cos n\phi' d\phi'] \tag{5.84a}$$

or

$$T(r,\phi) = \frac{1}{\pi}\int_0^{2\pi}\left[\frac{1}{2} + \sum_{n=1}^{\infty}\left(\frac{r}{r_o}\right)^n \cos n(\phi - \phi')\right] f(\phi')d\phi' \tag{5.84b}$$

Since it can be shown that (see Reference [6])

$$\sum_{n=1}^{\infty}\left(\frac{r}{r_o}\right)^n \cos n(\phi - \phi') = \frac{r_o^2 - rr_o \cos(\phi - \phi')}{r_o^2 - 2rr_o \cos(\phi - \phi') + r^2} - 1 \tag{5.85}$$

the solution (5.84b) can also be written as

$$T(r,\phi) = \frac{1}{2\pi}\int_0^{2\pi} \frac{r_o^2 - r^2}{r_o^2 - 2rr_o \cos(\phi - \phi') + r^2}\ f(\phi')d\phi' \tag{5.86}$$

This result is also known as the *Poisson integral formula*.

Here it is to be observed that for $\lambda = 0$ the product solution (5.75), in fact, yields

$$T_o(r,\pi) = (A_{10} + A_{20} \ln r)(B_{10} + B_{20}\phi) \tag{5.87}$$

But boundedness, *i.e.*, condition (5.74a), implies that $A_{20} = 0$, and 2π-periodicity, *i.e.*, conditions (5.74c) and (5.74d), implies that $B_{20} = 0$; therefore, the omission of the ϕ and $\ln r$ terms in Eq. (5.81a) caused no problem as b_o corresponds to $A_{10}B_{10}$.

At r = 0, Eq. (5.86) reduces to

$$T(0,\phi) = \frac{1}{2\pi}\int_0^{2\pi} f(\phi')d\phi' \tag{5.88}$$

Hence, the centerline temperature is the *average* of the surface temperature distribution. This striking result must apply *locally* as well; that is, the temperature at any point P within $r < r_0$ is equal to the average temperature around the edge of each circle that is centered at P and that lies within $r < r_0$.

5.3.2 Steady Two-Dimensional Problems in (r,z) Variables

As an example to the problems of the form $T = T(r,z)$, consider a fin of circular cross-section with radius r_0 and length L protruding from a hot wall as shown in Fig. 5.11. The base temperature is specified as f(r). The fin is coole by a fluid stream at temperature T_∞. Assuming constant thermal conductivity and an infinite heat transfer coefficient around the fin, this problem under steady-state conditions can be formulated in terms of $\theta = T - T_\infty$ as

$$\frac{\partial^2\theta}{\partial r^2} + \frac{1}{r}\frac{\partial\theta}{\partial r} + \frac{\partial^2\theta}{\partial z^2} = 0 \tag{5.89}$$

$$\theta(0,z) = \text{finite}; \qquad \theta(r_0,z) = 0 \tag{5.90a,b}$$

$$\theta(r,0) = f(r) - T_\infty = F(r); \qquad \theta(r,L) = 0 \tag{5.90c,d}$$

Assume a product solution in the form

$$\theta(r,z) = R(r)Z(z) \tag{5.91}$$

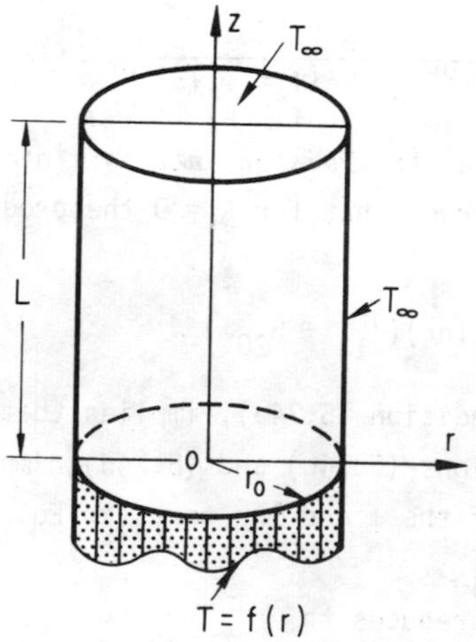

Fig. 5.11 Finite solid cylinder.

Substituting Eq. (5.91) into Eq. (5.89) and observing that the r-direction is the homogeneous direction we get the following two ordinary differential equations:

$$\frac{d}{dr}\left(r\frac{dR}{dr}\right) + \lambda^2 rR = 0 \tag{5.92a}$$

$$\frac{d^2Z}{dz^2} - \lambda^2 Z = 0 \tag{5.92b}$$

Thus, Bessel functions of order zero in r (see Appendix B) and hyperbolic functions in z are obtained, and the solution to Eq. (5.89) can be written as

$$\theta(r,z) = [A_1 J_0(\lambda r) + A_2 Y_0(\lambda r)][B_1 \sinh\lambda z + B_2 \cosh\lambda z] \tag{5.93}$$

Imposing the boundary conditions yields

$$\theta(r,z) = \sum_{n=1}^{\infty} a_n J_0(\lambda_n r)\sinh\lambda_n(L - z) \tag{5.94}$$

where λ_n's are the positive roots of the following equation

$$J_0(\lambda_n r_0) = 0, \qquad n = 1,2,3\ldots \tag{5.95}$$

Now we have to determine the coefficients a_n in Eq. (5.94) in such a way that $\theta(r,0) = F(r)$; that is,

$$F(r) = \sum_{n=1}^{\infty} a_n \sinh\lambda_n L \; J_0(\lambda_n r) \tag{5.96}$$

which is a Fourier-Bessel expansion of F(r) over the interval $(0,r_0)$. Using Eq. (4.55a), together with the help of the relation (4.66), or from Table 4.2, we get

$$a_n \sinh \lambda_n L = \frac{2}{r_0^2 J_1^2(\lambda_n r_0)} \int_0^{r_0} F(r)J_0(\lambda_n r)rdr \tag{5.97}$$

The temperature distribution is then given in the form

$$\theta(r,z) = \frac{2}{r_0^2} \sum_{n=1}^{\infty} \frac{J_0(\lambda_n r)}{J_1^2(\lambda_n r_0)} \frac{\sinh\lambda_n(L - z)}{\sinh\lambda_n L} \int_0^{r_0} F(r')J_0(\lambda_n r')r'dr' \tag{5.98}$$

If, in particular, $f(r) = T_0$ = Constant, the integral in Eq. (5.98) can easily be evaluated [see Example (4.5)] and the temperature distribution becomes

$$\frac{\theta(r,z)}{\theta_0} = \frac{2}{r_0} \sum_{n=1}^{\infty} \frac{1}{\lambda_n} \frac{J_0(\lambda_n r)}{J_1(\lambda_n r_0)} \frac{\sinh \lambda_n (L - z)}{\sinh \lambda_n L} \tag{5.99}$$

where we have introduced $\theta_0 = T_0 - T_\infty$.

Example 5.2: For the solid cylinder shown in Fig. 5.11 calculate the value of $\theta(r,z)/\theta_0$ at $z/L = 0.75$, $r/r_0 = 0.50$. Take $L/r_0 = 2.0$.

Solution: From Eq. (5.99) we have

$$\frac{\theta\left(\frac{r}{r_0}, \frac{z}{L}\right)}{\theta_0} = 2 \sum_{n=1}^{\infty} \frac{1}{\alpha_n} \frac{J_0(\alpha_n \frac{r}{r_0})}{J_1(\alpha_n)} \frac{\sinh[\alpha_n \frac{L}{r_0} (1 - \frac{z}{L})]}{\sinh(\alpha_n \frac{L}{r_0})}$$

which, at $z/L = 0.75$ and $r/r_0 = 0.50$ with $L/r_0 = 2.0$, becomes

$$\frac{\theta(0.5,\ 0.75)}{\theta_0} = 2 \sum_{n=1}^{\infty} \frac{1}{\alpha_n} \frac{J_0(0.5\alpha_n)}{J_1(\alpha_n)} \frac{\sinh(0.5\alpha_n)}{\sinh(2\alpha_n)}$$

where $\alpha_n = \lambda_n r_0$. The first four zeros of $J_0(\alpha)$ and the corresponding values of $J_1(\alpha)$ and $J_0(0.5\alpha)$ from Appendix B are

$\alpha_1 = 2.4048$ $\quad J_1(\alpha_1) = 0.5191$ $\quad J_0(0.5\alpha_1) = 0.6711$

$\alpha_2 = 5.5201$ $\quad J_1(\alpha_2) = -0.3403$ $\quad J_0(0.5\alpha_2) = -0.1680$

$\alpha_3 = 8.6537$ $\quad J_1(\alpha_3) = -0.2715$ $\quad J_0(0.5\alpha_3) = -0.3560$

$\alpha_4 = 11.7915$ $\quad J_1(\alpha_4) = -0.2325$ $\quad J_0(0.5\alpha_4) = -0.1207$

Substituting these numerical values we get

$$\frac{\theta(0.50,\ 0.75)}{\theta_0} = 0.02654 + 4.527 \times 10^{-5} - 2.629 \times 10^{-10}$$

$$-1.809 \times 10^{-9} \ldots$$

$$\simeq 0.02654$$

As it is seen, the series (5.99) converges rather rapidly for $z/L = 0.75$ and $r/r_0 = 0.5$ when $L/r_0 = 2.0$.

Let us now consider the solution of the differential equation (5.89) for the following boundary conditions:

$$\theta(0,z) = \text{finite}; \qquad \theta(r_0,z) = F(z) \tag{5.100a,b}$$

$$\theta(r,0) = 0; \qquad \theta(r,L) = 0 \tag{5.100c,d}$$

Since the boundary conditions in the z-direction are homogeneous, by assuming a product solution of the form

$$\theta(r,z) = R(r)Z(z) \tag{5.101}$$

we obtain from the differential equation (5.89) the following two differential equations:

$$\frac{d}{dr}\left(r\frac{dR}{dr}\right) - \lambda^2 rR = 0 \tag{5.102a}$$

$$\frac{d^2Z}{dz^2} + \lambda^2 Z = 0 \tag{5.102b}$$

Thus, the solution to Eq. (5.89) can be written as (see Appendix B)

$$\theta(r,z) = [A_1 I_0(\lambda r) + B_1 K_0(\lambda r)][A_2 \sin\lambda z + B_2 \cos\lambda z] \tag{5.103}$$

where I_0 and K_0 are the zeroth order modified Bessel functions of the first and second kinds, respectively. Since $K_0(0) = \infty$ and $\theta \neq \infty$ at $r = 0$, $B_1 = 0$. On the other hand, $\theta(r,0) = 0$ requires that $B_2 = 0$. Hence, Eq. (5.103) reduces to

$$\theta(r,z) = a\, I_0(\lambda r)\sin\lambda z \tag{5.104}$$

where $a = A_1 A_2$. Imposing the condition (5.100d) yields

$$\sin\lambda_n L = 0$$

or

$$\lambda_n = \frac{n\pi}{L}, \qquad n = 1,2,3...$$

Therefore, the solution of the problem becomes

$$\theta(r,z) = \sum_{n=1}^{\infty} a_n I_0\left(\frac{n\pi r}{L}\right)\sin\frac{n\pi z}{L} \tag{5.105}$$

Application of the boundary condition at $r = r_0$ yields

$$F(z) = \sum_{n=1}^{\infty} a_n I_0\left(\frac{n\pi r_0}{L}\right)\sin\frac{n\pi z}{L} \tag{5.106}$$

which is the Fourier-sine series expansion of F(z) over the interval (0,L). Hence, we have

$$a_n = \frac{2}{L}\frac{1}{I_0\left(\frac{n\pi r_0}{L}\right)}\int_0^L F(z)\sin\frac{n\pi z}{L}\,dz \tag{5.107}$$

Substitution of this result into Eq. (5.105) gives

$$\theta(r,z) = \frac{2}{L}\sum_{n=1}^{\infty}\frac{I_0\left(\frac{n\pi r}{L}\right)}{I_0\left(\frac{n\pi r_0}{L}\right)}\sin\frac{n\pi}{L}z\int_0^L F(z')\sin\frac{n\pi z'}{L}\,dz' \tag{5.108}$$

In the problems discussed in this section, we were able to apply the method of separation of variables directly because the differential equations were linear and homogeneous. Furthermore, the boundary conditions were also linear and three of them were either homogeneous or could be made homogeneous by defining a new temperature function. If these problems involved internal heat generation and/or more than one non-homogeneous boundary conditions then the principle of superposition would be applied as discussed in Section 5.2 (for an example, see Problem 5.16).

5.4 STEADY TWO-DIMENSIONAL PROBLEMS IN THE SPHERICAL COORDINATE SYSTEM

When the temperature distribution in a heat conduction problem in spherical coordinates is a function of r and θ variables only (Fig. 5.12), the solution for the temperature distribution can be expressed in terms of *Legendre polynomials*. For this reason we will first look at the solutions of Legendre's differential equation and then study briefly the expansion of an arbitrary function into a series of Legendre polynomials.

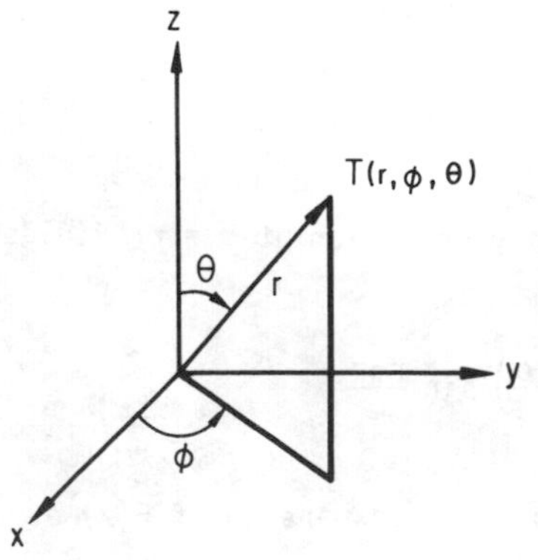

Fig. 5.12 Spherical coordinates.

5.4.1 Legendre Polynomials

Consider the following linear second-order ordinary differential equation with variable coefficients

$$(1 - x^2)\frac{d^2y}{dx^2} - 2x\frac{dy}{dx} + \alpha(\alpha + 1)y = 0 \tag{5.109a}$$

or, equivalently,

$$\frac{d}{dx}\left[(1 - x^2)\frac{dy}{dx}\right] + \alpha(\alpha + 1)y = 0 \tag{5.109b}$$

where α is real and non-negative. This equation is known as *Legendre's differential equation* and its two linearly independent solutions are called *Legendre functions* of the *first* and *second kind*, respectively. When $\alpha = n$, where n is a positive integer or zero, which are the cases commonly arising in practice, the Legendre function of the first kind becomes a polynomial of degree n, whereas the other can be expressed as an infinite series. The general solution of Legendre's equation for n = 0,1,2,3... can be written as

$$y(x) = C_1P_n(x) + C_2Q_n(x) \tag{5.110}$$

where $P_n(x)$ are polynomials called *Legendre polynomials* of *degree* n, and $Q_n(x)$ are the Legendre functions of the second kind.

The Legendre polynomials are given by

$$P_0(x) = 1, \qquad P_n(x) = (-1)^{n/2}\frac{1.3.5...(n - 1)}{2.4.6... \quad n}U_n(x), \quad n=2,4,6... \tag{5.111a,b}$$

and

$$P_1(x) = x, \qquad P_n(x) = (-1)^{(n-1)/2}\frac{1.3.5... \quad n}{2.4.6...(n-1)}V_n(x), \quad n=3,5,7... \tag{5.111c,d}$$

Here we have used

$$U_n(x) = 1 - \frac{n(n+1)}{2!}x^2 + \frac{n(n-2)(n+1)(n+3)}{4!}x^4 - ... \tag{5.112a}$$

$$V_n(x) = x - \frac{(n-1)(n+2)}{3!}x^3 + \frac{(n-1)(n-3)(n+2)(n+4)}{5!}x^5 - ... \tag{5.112b}$$

Note that if n is an even positive integer, the series (5.112a) terminates with the term involving x^n, and hence is a polynomial of degree n. Similarly, if n is an odd positive integer, the series (5.112b) terminates with the term involving x^n. Otherwise, these expressions are infinite series convergent only when $|x| < 1$.

The first six Legendre polynomials are readily found to be

$$P_0(x) = 1 \qquad P_3(x) = \tfrac{1}{2}(5x^3 - 3x)$$

$$P_1(x) = x \qquad P_4(x) = \tfrac{1}{8}(35x^4 - 30x^2 + 3)$$

$$P_2(x) = \tfrac{1}{2}(3x^2 - 1) \qquad P_5(x) = \tfrac{1}{8}(63x^5 - 70x^3 + 15x) \tag{5.113}$$

In all cases $P_n(1) = 1$ and $P_n(-1) = (-1)^n$.

The Legendre polynomials can also be expressed by *Rodrigues' Formula*:

$$P_n(x) = \frac{1}{2^n n!} \frac{d^n}{dx^n} (x^2 - 1)^n \tag{5.114}$$

A useful *recurrence* formula can be established at once by observing that each of these polynomials is a function of the polynomial preceding and succeeding it. Thus,

$$P_0(x) = \frac{1}{x} P_1(x)$$

$$P_1(x) = \frac{1}{3x} [P_0(x) + 2P_2(x)]$$

$$P_2(x) = \frac{1}{5x} [2P_1(x) + 3P_3(x)]$$

or, in general,

$$P_n(x) = \frac{n\, P_{n-1}(x) + (n+1)P_{n+1}(x)}{(2n+1)x} \tag{5.115}$$

The following property of Legendre polynomials is also used frequently:

$$\frac{dP_{n+1}}{dx} - \frac{dP_{n-1}}{dx} = (2n + 1)P_n(x), \qquad n = 1,2,3... \tag{5.116}$$

If $|x| < 1$, the Legendre functions of the second kind are given by

$$Q_0(x) = V_0(x), \quad Q_n(x) = (-1)^{n/2} \frac{2.4.6...\ n}{1.3.5...(n-1)} V_n(x), \quad n=2,4,6... \tag{5.117a,b}$$

and

$$Q_1(x) = -U_1(x), \quad Q_n(x) = (-1)^{(n+1)/2} \frac{2.4.6...(n-1)}{1.3.5...\ n} U_n(x), \quad n=3,5,7... \tag{5.117c,d}$$

It is to be noted that the Legendre functions of the second kind are infinite series, which are convergent when $|x| < 1$ but diverge as $x \to \pm 1$.

It can be shown that the Legendre functions of the second kind $Q_n(x)$ also satisfy the recurrence formulas (5.115) and (5.116). It can further be verified that when $|x| < 1$

$$Q_0(x) = \frac{1}{2} \ln \frac{1 + x}{1 - x} = \tanh^{-1} x \tag{5.118}$$

The use of the recurrence formula (5.115) permits the determination of $Q_n(x)$ for any positive integer value of n:

$$Q_1(x) = P_1(x)Q_0(x) - 1$$

$$Q_2(x) = P_2(x)Q_0(x) - \frac{3}{2}x$$

$$Q_3(x) = P_3(x)Q_0(x) - \frac{5}{2}x^2 + \frac{2}{3}$$

$$Q_4(x) = P_4(x)Q_0(x) - \frac{35}{8}x^3 + \frac{55}{24}x$$

$$Q_5(x) = P_5(x)Q_0(x) - \frac{63}{8}x^4 + \frac{49}{8}x^2 - \frac{8}{15}$$

or, in general,

$$Q_n(x) = P_n(x)Q_0(x) - \frac{(2n-1)}{1-n} P_{n-1}(x) - \frac{(2n-5)}{3(n-1)} P_{n-3}(x). . . \tag{5.119}$$

If $|x| < 1$, the substitution $x = \cos\theta$ transforms Legendre's differential equation from the form (5.109b) into the form

$$\frac{1}{\sin\theta} \frac{d}{d\theta}\left(\sin\theta \frac{dy}{d\theta}\right) + n(n + 1)y = 0 \tag{5.120a}$$

or, equivalently,

$$\frac{d^2y}{d\theta^2} + \frac{dy}{d\theta} \cot\theta + n(n + 1)y = 0 \tag{5.120b}$$

when $\alpha = n$. Hence, the general solution of Eq. (5.120a) is

$$y(\theta) = C_1 P_n(\cos\theta) + C_2 Q_n(\cos\theta) \tag{5.121}$$

Equations of such a form arise in connection with solutions of various heat conduction problems in spherical coordinates. It should be noted here that $Q_n(\cos\theta)$ is not finite when $\cos\theta = \pm 1$; that is, when $\theta = k\pi$, $k = 0, \pm 1, \pm 2...$, whereas $P_n(\cos\theta)$ is merely a polynomial of degree n in $\cos\theta$. In particular we have

$$P_0(\cos\theta) = 1$$

$$P_1(\cos\theta) = \cos\theta$$

$$P_2(\cos\theta) = \frac{1}{2}(3\cos^2\theta - 1) = \frac{1}{4}(3\cos 2\theta + 1)$$

$$P_3(\cos\theta) = \frac{1}{2}(5\cos^3\theta - 3\cos\theta) = \frac{1}{8}(5\cos 3\theta + 3\cos\theta)$$

$$\vdots$$

When $|x| < 1$, the functions

$$P_n^m(x) = (1 - x^2)^{m/2}\,\frac{d^m P_n(x)}{dx^m}, \qquad P_n^0(x) = P_n(x) \tag{5.122a}$$

$$Q_n^m(x) = (1 - x^2)^{m/2}\,\frac{d^m Q_n(x)}{dx^m}, \qquad Q_n^0(x) = Q_n(x) \tag{5.122b}$$

are called the *associated Legendre functions* of *degree* n and *order* m of the *first* and *second kinds*, respectively. They can be shown to satisfy the differential equation

$$(1 - x^2)\frac{d^2y}{dx^2} - 2x\frac{dy}{dx} + \left[n(n+1) - \frac{m^2}{1 - x^2}\right]y = 0 \tag{5.123}$$

For m = 0, Eq. (5.123) reduces to Eq. (5.109a).

5.4.2 Fourier-Legendre Series

Equation (5.109b), when compared with Eq. (4.13a), gives

$$p(x) = 1 - x^2, \qquad q(x) = 0, \qquad w(x) = 1, \qquad \lambda = \alpha(\alpha + 1)$$

Since the function p(x) vanishes at the points $x = \pm 1$, from the results of Section 4.4 we conclude that any two distinct solutions of Eq. (5.109a) or Eq. (5.109b) which are finite and have finite derivatives at $x = \pm 1$ will be orthogonal with respect to the weight function $w(x) = 1$ in the interval (-1,1). Therefore, no specific boundary conditions are needed for Eq. (5.109a) to form a Sturm-Liouville type characteristic-value problem over the interval (-1,1), provided the characteristic functions and their first derivatives are finite at the ends of the interval.

Equation (5.109a) possesses solutions which are finite at $x = \pm 1$ only if α is a positive integer or zero. This condition of finiteness determines the permissible values of α in the form

$$\alpha = n, \qquad n = 0,1,2,3\ldots \tag{5.124}$$

The corresponding solutions which are finite when $x = \pm 1$ are proportional to the Legendre polynomials; that is,

$$\phi_n(x) = P_n(x) \tag{5.125}$$

Thus, we conclude that

$$\int_{-1}^{1} P_m(x)P_n(x)dx = 0, \qquad m \neq n \tag{5.126}$$

Hence, the set $\{P_n(x), n = 0,1,2...\}$ is a complete orthogonal set with respect to the weight function unity over the interval (-1,1). Therefore, any piecewise-differentiable function f(x) can be represented in the interval (-1,1) by a series of the Legendre polynomials as

$$f(x) = A_0P_0(x) + A_1P_1(x) + A_2P_2(x) + \ldots \tag{5.127a}$$

$$= \sum_{n=0}^{\infty} A_nP_n(x), \qquad -1 < x < 1 \tag{5.127b}$$

which is called the *Fourier-Legendre series expansion* of f(x) over the interval (-1,1). The expansion coefficients A_n are given by

$$A_n = \frac{\int_{-1}^{1} f(x)P_n(x)dx}{\int_{-1}^{1} [P_n(x)]^2\, dx} \tag{5.128}$$

In order to evaluate the integrals appearing in Eq. (5.128), it is useful to express $P_n(x)$ by Rodrigues' formula (5.114); that is,

$$P_n(x) = \frac{1}{2^n\, n!} \frac{d^n}{dx^n} (x^2 - 1)^n \tag{5.129}$$

Hence

$$\int_{-1}^{1} f(x)P_n(x)dx = \frac{1}{2^n\, n!} \int_{-1}^{1} f(x) \frac{d^n}{dx^n} (x^2 - 1)^n dx \tag{5.130}$$

If the right-hand member of Eq. (5.130) is integrated by parts N times, where $N \leq n$, assuming that the first N derivatives of f(x) are continuous in (-1,1), and the first (n - 1) derivatives of $(x^2 - 1)^n$ vanish at $x = \pm 1$, then Eq. (5.130) becomes

$$\int_{-1}^{1} f(x)P_n(x)dx = \frac{(-1)^N}{2^n\, n!} \int_{-1}^{1} \left[\frac{d^N}{dx^N} f(x)\right] \left[\frac{d^{n-N}}{dx^{n-N}} (x^2-1)^n\right] dx \tag{5.131}$$

In particular, when N = n there follows

$$\int_{-1}^{1} f(x)P_n(x)dx = \frac{(-1)^n}{2^n\ n!} \int_{-1}^{1} (x^2 - 1)^n \frac{d^n f(x)}{dx^n} dx \tag{5.132}$$

provided that f(x) and its first n derivatives are continuous throughout the interval (-1,1).

Replacing f(x) by $P_n(x)$ in Eq. (5.132) and noticing from Eq. (5.129) that

$$\frac{d^n P_n(x)}{dx^n} = \frac{1}{2^n\ n!} \frac{d^{2n}}{dx^{2n}} (x-1)^n = \frac{1}{2^n\ n!} \frac{d^{2n}}{dx^{2n}} (x^{2n} - x^{2n-2} + \ldots) = \frac{(2n)!}{2^n\ n!} \tag{5.133}$$

we obtain from Eq. (5.132)

$$\int_{-1}^{1} [P_n(x)]^2 dx = - \frac{(2n)!}{2^{2n}(n!)^2} \int_{-1}^{1} (1 - x^2)^n dx \tag{5.134}$$

The integral on the right-hand side can be evaluated by successive reductions (involving integrations by parts) in the form

$$\int_{-1}^{1} (1 - x^2)^n dx = 2 \frac{(2n)(2n - 2)\ldots\ldots.4.2}{(2n + 1)(2n - 1)\ldots 5.3} = \frac{2^{2n+1}\ (n!)^2}{(2n + 1)!} \tag{5.135}$$

Thus, Eq. (5.134) takes the form

$$\int_{-1}^{1} [P_n(x)]^2 dx = \frac{2}{2n + 1} \tag{5.136}$$

Equation (5.128) can now be written as

$$A_n = \begin{cases} \dfrac{2n + 1}{2} \displaystyle\int_{-1}^{1} f(x)P_n(x)dx \\[2ex] \dfrac{2n + 1}{2^{n+1} n!} \displaystyle\int_{-1}^{1} (1 - x^2)^n \frac{d^n f(x)}{dx^n} dx \end{cases} \tag{5.137}$$

The second form is correct only if f(x) and its first n derivatives are continuous in (-1,1). Since $P_n(x)$ is an even function of x when n is even, and an odd function when n is odd, it follows that if f(x) is an even function of x, the coefficients A_n will vanish when n is odd; whereas if f(x) is an odd function of x, the coefficients A_n will vanish when n is even. Thus, for an *even* function f(x), there follows

$$A_n = \begin{cases} 0 & (n \text{ odd}) \\ (2n + 1)\displaystyle\int_0^1 f(x)P_n(x)dx & (n \text{ even}) \end{cases} \tag{5.138}$$

while for an *odd* function f(x)

$$A_n = \begin{cases} (2n + 1)\displaystyle\int_0^1 f(x)P_n(x)dx & (n \text{ odd}) \\ 0 & (n \text{ even}) \end{cases} \tag{5.139}$$

If we are interested in representing a given function f(x) merely over the interval (0,1), it is clear that a series containing only even polynomials can be obtained by using Eq. (5.138), or if preferred, a series containing only odd polynomials can be obtained by using Eq. (5.139). Either of these series will then represent f(x) inside the interval (0,1) and the first series will represent f(-x) and the second -f(x) in the interval (-1,0).

Expansions valid in the more general interval (-a,a) are readily obtained by replacing x by x/a in the preceding development. This procedure leads to the expansion

$$f(x) = \sum_{n=0}^{\infty} A_n P_n(x/a), \qquad -a < x < a \tag{5.140}$$

where

$$A_n = \begin{cases} \dfrac{2n + 1}{2a} \displaystyle\int_{-a}^{a} f(x)P_n\left(\frac{x}{a}\right)dx \\ \dfrac{2n + 1}{2^{n+1} n!\ a^{n+1}} \displaystyle\int_{-a}^{a} (a - x^2)^n \frac{d^n f(x)}{dx^n}\, dx \end{cases} \tag{5.141}$$

which are the general forms of Eq. (5.137).

If f(x) is a polynomial of degree k, all derivatives of f(x) of order n vanish identically when n > k. Hence, it follows from Eq. (5.137) or Eq. (5.141) that any polynomial of degree k can be expressed as a linear combination of the first (k + 1) Legendre polynomials.

5.4.3 Solid Sphere

For the solution of Laplace's equation in spherical coordinates by the application of Fourier-Legendre series, we consider the steady-state temperature distribution in a solid sphere of radius r_0. Assume that surface of the sphere at $r = r_0$ is maintained at some arbitrary temperature distribution $f(\theta)$ and the thermal conductivity of the material of the sphere is constant.

The differential equation from Eqs. (2.19) and (2.23) and the boundary conditions are given by

$$\frac{\partial}{\partial r}\left(r^2 \frac{\partial T}{\partial r}\right) + \frac{1}{\sin\theta}\frac{\partial}{\partial\theta}\left(\sin\theta \frac{\partial T}{\partial\theta}\right) = 0 \tag{5.142}$$

$$T(0,\theta) \neq \infty \tag{5.143a}$$

$$T(r_0,\theta) = f(\theta) \tag{5.143b}$$

Assuming a product solution in the form

$$T(r,\theta) = R(r)\phi(\theta) \tag{5.144}$$

Eq. (5.142) can be separated as follows:

$$\frac{d^2R}{dr^2} + \frac{2}{r}\frac{dR}{dr} - \frac{\lambda}{r^2} R = 0 \tag{5.145}$$

and

$$\frac{1}{\sin\theta}\frac{d}{d\theta}\left(\sin\theta \frac{d\phi}{d\theta}\right) + \lambda\phi = 0 \tag{5.146}$$

The general solution of the Cauchy-Euler equation (5.145) can be written as

$$R(r) = C\, r^{s_1} + B\, r^{s_2} \tag{5.147}$$

where

$$s_{1,2} = -\tfrac{1}{2} \pm\left(\tfrac{1}{4} + \lambda\right)^{\frac{1}{2}} \tag{5.148}$$

The differential equation (5.146) can be transformed into the form of Legendre's equation by redefining the independent variable as $x = \cos\theta$. This gives

$$\frac{d}{dx}\left[(1 - x^2)\frac{d\phi}{dx}\right] + \lambda\phi = 0 \tag{5.149}$$

which can, in turn, be put into the form of Eq. (5.109b) if we let $\lambda = \alpha(\alpha + 1)$. If $\alpha = n$, where n is zero or a positive integer, we know that the solutions of Legendre's equation, which are finite at $\theta = 0$ and $\theta = \pi$ $(x = \pm 1)$, are the Legendre polynomials $P_n(\cos\theta)$. Thus,

$$\phi_n(\theta) = A_n P_n(\cos\theta), \qquad n = 1,2,3... \tag{5.150}$$

and the general solution of Laplace's equation (5.142) for the boundary conditions (5.143a,b) can then be written as

$$T(r,\theta) = \sum_{n=0}^{\infty} A_n[C_n r^n + B_n r^{-(n+1)}]P_n(\cos\theta) \tag{5.151}$$

From the boundary condition at $r = 0$ we get $B_n = 0$. Therefore

$$T(r,\theta) = \sum_{n=0}^{\infty} K_n\, r^n\, P_n(\cos\theta) \tag{5.152}$$

where $K_n \equiv A_n\, C_n$. By the use of the second boundary condition (5.143b), Eq. (5.152) reduces to

$$f(\theta) = \sum_{n=0}^{\infty} K_n r_o^n\, P_n(\cos\theta) \tag{5.153}$$

It should be noted here that the finiteness of the solution defines the characteristic functions and the characteristic values, which corresponds to stating two boundary conditions in the θ-direction as

$$T(r,0) \neq \infty \qquad \text{and} \qquad T(r,\pi) \neq \infty$$

Equation (5.153) is the Fourier-Legendre series expansion of $f(\theta)$. The expansion coefficients $K_n\, r_o^n$ can be determined by employing Eq. (5.137) as

$$K_n r_o^n = \frac{2n+1}{2}\int_0^{\pi} f(\theta)P_n(\cos\theta)\,\sin\theta\, d\theta \tag{5.154}$$

As an application, assume that the surface temperature is specified as

$$f(\theta) = \begin{cases} T_o & 0 < \theta < \pi/2 \\ 0 & \pi/2 < \theta < \pi \end{cases} \tag{5.155}$$

Since $f(\theta)$ is discontinuous in $(0,\pi)$, the second expression in Eq. (5.137) cannot be used. The basic relation in this case is the first expression in Eq. (5.137):

$$K_n r_o^n = T_o\,\frac{2n+1}{2}\int_0^1 P_n(x)dx \tag{5.156}$$

Hence, there follows

$$K_0 = \frac{1}{2}T_0 \int_0^1 dx = \frac{1}{2}T_0$$

$$K_1 r_0 = \frac{3}{2}T_0 \int_0^1 x dx = \frac{3}{4}T_0$$

$$K_2 r_0^2 = (\frac{5}{2})(\frac{1}{2})T_0 \int_0^1 (3x^2 - 1)dx = 0$$

$$K_3 r_0^3 = (\frac{7}{2})(\frac{1}{2})T_0 \int_0^1 (5x^2 - 3x)dx = \frac{7}{16}T_0 \quad (5.157)$$

. . .

. . .

It is seen that $K_0 = \frac{1}{2}T_0$ is the only non-vanishing even coefficient. The first four terms of the desired series (5.153) are therefore in the form

$$f(x) = \frac{1}{2}T_0 P_0(x) + \frac{3}{4}T_0 P_1(x) - \frac{7}{16}T_0 P_3(x) + \frac{11}{32}T_0 P_5(x) + \ldots , \quad 0 < x < \pi \quad (5.158)$$

Hence, the temperature distribution becomes

$$\frac{T(r,\theta)}{T_0} = \frac{1}{2} + \frac{3}{4}(\frac{r}{r_0})P_1(\cos\theta) - \frac{7}{16}(\frac{r}{r_0})^3 P_3(\cos\theta) + \frac{11}{32}(\frac{r}{r_0})^5 P_5(\cos\theta) + \ldots \quad (5.159)$$

This solution can also be extended for a hollow sphere assuming, for example, $T(r_1,\theta) = T_0$ and $T(r_2,\theta) = f(\theta)$, where r_1 and r_2 are the inner and outer radii of the sphere, respectively.

5.5 STEADY THREE-DIMENSIONAL SYSTEMS

In order to illustrate the solution of Laplace's equation in three-dimensional steady-state problems, let us consider the rectangular parallelepiped shown in Fig. 5.13. Assume that the boundary conditions may be approximated as

$$T(0,y,z) = 0 ; \quad T(x,0,z) = 0 ; \quad T(x,y,c) = 0$$

$$T(a,y,z) = 0 ; \quad T(x,b,z) = 0 ; \quad T(x,y,0) = f(x,y) \quad (5.160a,b,c,d,e,f)$$

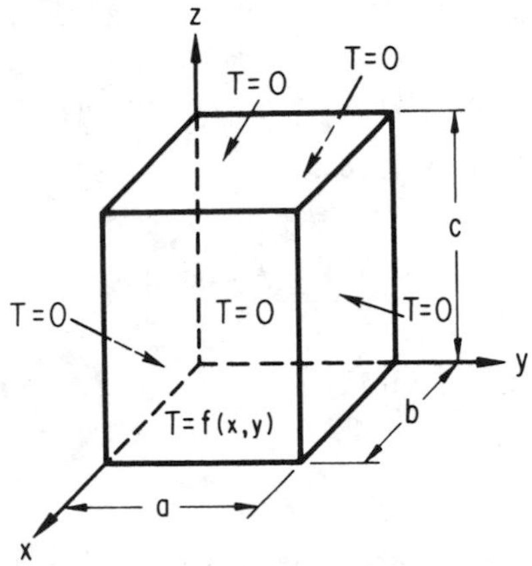

Fig. 5.13 Rectangular parallelepiped.

If we assume a product solution of the relevant equation

$$\frac{\partial^2 T}{\partial x^2} + \frac{\partial^2 T}{\partial y^2} + \frac{\partial^2 T}{\partial z^2} = 0 \tag{5.161}$$

in the form

$$T(x,y,z) = X(x)Y(y)Z(z) \tag{5.162}$$

then this equation can be separated as

$$-\frac{1}{X}\frac{d^2X}{dx^2} = \frac{1}{Y}\frac{d^2Y}{dy^2} + \frac{1}{Z}\frac{d^2Z}{dz^2} \tag{5.163}$$

The left-hand member of Eq. (5.163) is independent of both y and z, and the right-hand member is independent of x. Therefore, the right- and left-hand members of Eq. (5.163) can be equal to each other if and only if they are equal to the same constant. This constant, in view of the homogeneous boundary conditions in the x-direction, is a positive real constant, that is, λ^2. Hence, we have

$$\frac{d^2X}{dx^2} + \lambda^2 X = 0 \tag{5.164}$$

Similarly,

$$-\frac{1}{Y}\frac{d^2Y}{dy^2} = \frac{1}{Z}\frac{d^2Z}{dz^2} - \lambda^2 = \beta^2 \tag{5.165}$$

or

$$\frac{d^2Y}{dy^2} + \beta^2 Y = 0 \tag{5.166}$$

and

$$\frac{d^2Z}{dz^2} - (\lambda^2 + \beta^2)Z = 0 \tag{5.167}$$

General solution of Eq. (5.161) can then be written as

$$T(x,y,z) = (A_1 \cos\lambda x + A_2 \sin\lambda x)(B_1 \cos\beta y + B_2 \sin\beta y)$$

$$\cdot \left[C_1 \; e^{z\sqrt{\lambda^2+\beta^2}} + C_2 \; e^{-z\sqrt{\lambda^2+\beta^2}} \right] \tag{5.168}$$

The boundary condition at z = c yields

$$C_2 = -C_1 \; e^{2c\sqrt{\lambda^2+\beta^2}}$$

and the boundary conditions at x=0 and at y=0 give $A_1 = B_1 = 0$. Thus, substituting these constants into Eq. (5.168) we obtain

$$T(x,y,z) = A \sin\lambda x \sin\beta y \sinh\sqrt{\lambda^2+\beta^2}\,(c-z) \tag{5.169}$$

where $A \equiv 2A_2B_2C_1 \; e^{c\sqrt{\lambda^2+\beta^2}}$. The boundary conditions at x = a and x = b yield

$$\sin\lambda a = 0 \rightarrow \lambda_m = \frac{m\pi}{a}, \; m=1,2,3,\ldots \tag{5.170}$$

$$\sin\beta b = 0 \rightarrow \beta_n = \frac{n\pi}{b}, \; n=1,2,3,\ldots \tag{5.171}$$

Hence, the solution of Eq. (5.161) which will satisfy all the boundary conditions (5.160) can be written as

$$T(x,y,z) = \sum_{m=1}^{\infty} \sum_{n=1}^{\infty} A_{mn} \sin\frac{m\pi x}{a} \sin\frac{n\pi y}{b}$$

$$\cdot \sinh\left[\sqrt{\left(\frac{m\pi}{a}\right)^2 + \left(\frac{n\pi}{b}\right)^2}\,(c-z) \right] \tag{5.172}$$

Application of the boundary condition (5.160f) gives

$$f(x,y) = \sum_{m=1}^{\infty} \sum_{n=1}^{\infty} A_{mn} \sin\frac{m\pi x}{a} \sin\frac{n\pi y}{b} \sinh\left[\sqrt{\left(\frac{m\pi}{a}\right)^2 + \left(\frac{n\pi}{b}\right)^2}\; c \right] \tag{5.173a}$$

or

$$f(x,y) = \sum_{m=1}^{\infty} \sum_{n=1}^{\infty} B_{mn} \sin\frac{m\pi x}{a} \sin\frac{n\pi y}{b} \tag{5.173b}$$

where we have introduced

$$B_{mn} = A_{mn} \sinh \left[\sqrt{\left(\frac{m\pi}{a}\right)^2 + \left(\frac{n\pi}{b}\right)^2}\ c \right] \tag{5.173c}$$

Here the constants B_{mn} are the coefficients of the *double Fourier sine series expansion* of $f(x,y)$ over the rectangular domain $0 < x < a$, $0 < y < b$.

If both sides of Eq. (5.173b) are multiplied by $\sin\frac{p\pi x}{a} \sin\frac{q\pi y}{b}$, where p and q are arbitrary integers, and the result is integrated over the rectangular domain $0 < x < a$, $0 < y < b$, there follows

$$\int_0^a \int_0^b f(x,y) \sin \frac{p\pi x}{a} \sin \frac{q\pi y}{b}\, dx\, dy$$

$$= \sum_{m=1}^{\infty} \sum_{n=1}^{\infty} B_{mn} \int_0^a \int_0^b \sin \frac{p\pi x}{a} \sin \frac{q\pi y}{b} \sin \frac{m\pi x}{a} \sin \frac{n\pi y}{b}\, dx\, dy \tag{5.174}$$

The double integrals in this result can be written as the product

$$\left[\int_0^a \sin \frac{p\pi x}{a} \sin \frac{m\pi x}{a}\, dx \right] \left[\int_0^b \sin \frac{q\pi y}{b} \sin \frac{n\pi y}{b}\, dy \right]$$

This product vanishes unless p=m and q=n, in which case it has the value $(a/2)\cdot(b/2) = ab/4$. Thus the double series in Eq. (5.174) reduces to a single term for which m=p and n=q, leading to the result

$$A_{mn} = \frac{4}{ab} \left\{ \sinh \left[\sqrt{\left(\frac{m\pi}{a}\right)^2 + \left(\frac{n\pi}{b}\right)^2}\ c \right] \right\}^{-1}$$

$$\cdot \int_0^a \int_0^b f(x,y) \sin \frac{m\pi x}{a} \sin \frac{n\pi y}{b}\, dx\, dy \tag{5.175}$$

Finally, substitution of Eq. (5.175) into Eq. (5.172) yields the temperature distribution.

5.6 HEAT TRANSFER RATES

In the preceding sections temperature distributions in several representative problems have been obtained. With the temperature distribution known, heat transfer rate across any surface A can be calculated by using the Fourier's law of heat conduction:

$$\dot{Q}_n = -\int_A k \frac{\partial T}{\partial n}\, dA \tag{5.176}$$

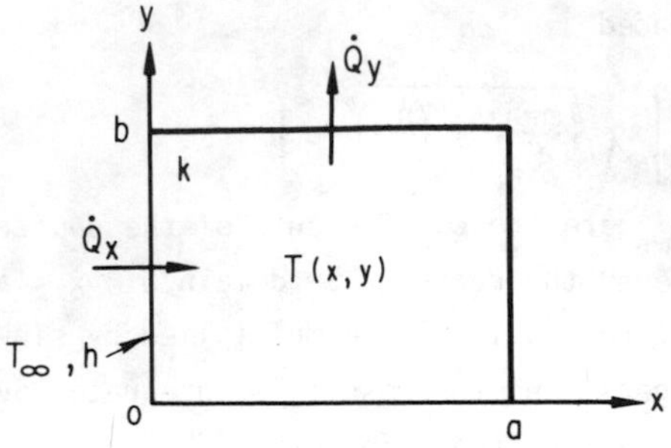

Fig. 5.14 Two-dimensional fin.

Across a surface A exposed to a fluid at temperature T_∞ with a finite heat transfer coefficient h, the heat transfer rate may also be calculated by the use of the Newton's law of cooling,

$$\dot{Q} = \int_A h(T_w - T_\infty)dA \tag{5.177}$$

For example, the rate of heat transfer per unit width across the surface at y = b of the fin in Fig. 5.14 will be

$$\dot{Q}_y = -\int_0^a k \frac{\partial T(x,b)}{\partial y} dx \tag{5.178}$$

and the rate of heat transfer per unit width of the fin across the surface at x=0 is

$$\dot{Q}_x = -\int_0^b k \frac{\partial T(0,y)}{\partial x} dy \tag{5.179}$$

Similarly, in Fig. 5.13 the rate of heat transfer in the z-direction across the surface at z=0 is given by

$$\dot{Q}_z = -\int_0^a \int_0^b k \frac{\partial T(x,y,0)}{\partial z} dx\, dy \tag{5.180}$$

Furthermore, the rate of heat transfer across the surface at z=L of the finite cylinder in Fig. 5.15 is

$$\dot{Q}_z = -\int_0^{r_0} k \frac{\partial T(r,L)}{\partial z} 2\pi r\, dr \tag{5.181}$$

If h and T_∞ are given, for example, Eqs. (5.179) and (5.181) can equally be

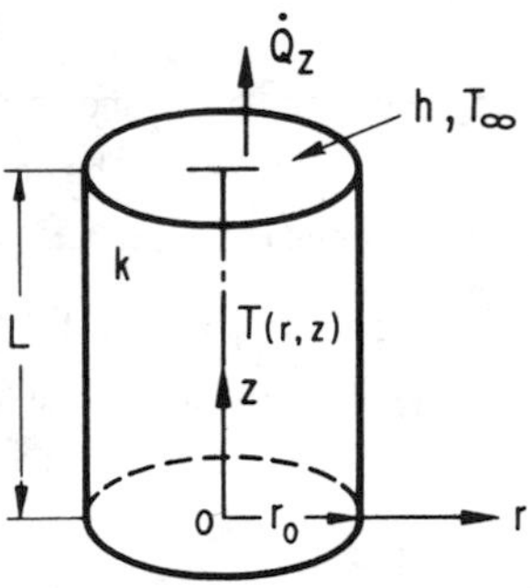

Fig. 5.15 Solid cylinder.

replaced by

$$\dot{Q}_x = \int_0^b h[T_\infty - T(0,y)]\, dy \tag{5.182}$$

$$\dot{Q}_z = \int_0^{r_0} h[T(r,L) - T_\infty]\, 2\pi r\, dr \tag{5.183}$$

Equations (5.182) and (5.183) can be used only if h is finite.

REFERENCES

1. Arpacı, V. S., *Conduction Heat Transfer*, Addison-Wesley Publishing Co., Inc., Reading, Massachusetts, 1966.
2. Hildebrand, F. B., *Advanced Calculus for Applications*, 2nd ed., Prentice-Hall, Inc., Englewood Cliffs, New Jersey, 1976.
3. Rohsenow, W. M., Class Notes on *Advanced Heat Transfer*, M.I.T., U.S.A., 1958.
4. Özışık, M. N., *Boundary Value Problems of Heat Conduction*, International Textbook Co., Scranton, Pennsylvania, 1968.
5. Özışık, M. N., *Heat Conduction*, John Wiley and Sons, Inc., New York, 1980.
6. Greenberg, M. D., *Foundations of Applied Mathematics*, Prentice-Hall, Inc., Englewood Cliffs, New Jersey, 1978.
7. Carslaw, H. S., and Jaeger, J. C., *Conduction of Heat in Solids*, 2nd ed., Oxford University Press, 1965.
8. Churchill, R. V., *Operational Mathematics*, 3rd ed., McGraw-Hill Book Co., New York, 1972.
9. Schneider, P. J., *Conduction Heat Transfer*, Addison-Wesley Publishing Co., Inc., Reading, Massachusetts, 1965.
10. Luikov, A. V., *Analytical Heat Diffusion Theory*, Academic Press, New York, 1968.

PROBLEMS

5.1 Three surfaces of a long square bar are maintained at 0°C, while the fourth surface is at 100°C. Calculate the temperature at the centerline of the bar under steady-state conditions.

5.2 Applying the method of separation of variables obtain the steady-state temperature distribution $T(x,y)$ in the two-dimensional rectangular fin shown in Fig. 5.16.

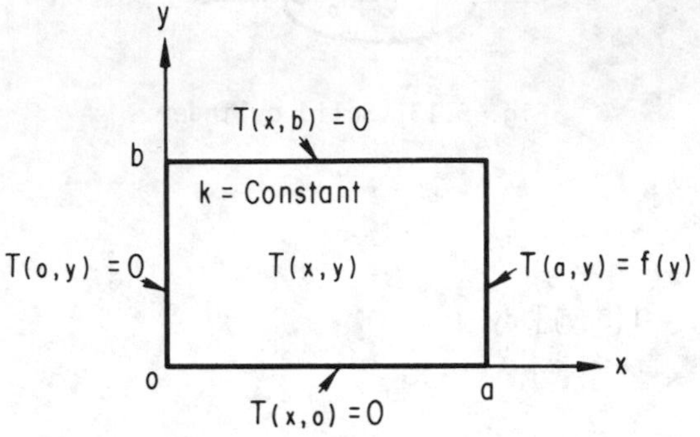

Fig. 5.16 Two-dimensional rectangular fin in Problem 5.2.

5.3 Using the temperature distribution found in Problem 5.2, obtain the steady-state temperature distribution $T(x,y)$ in the two-dimensional rectangular fin shown in Fig. 5.17.

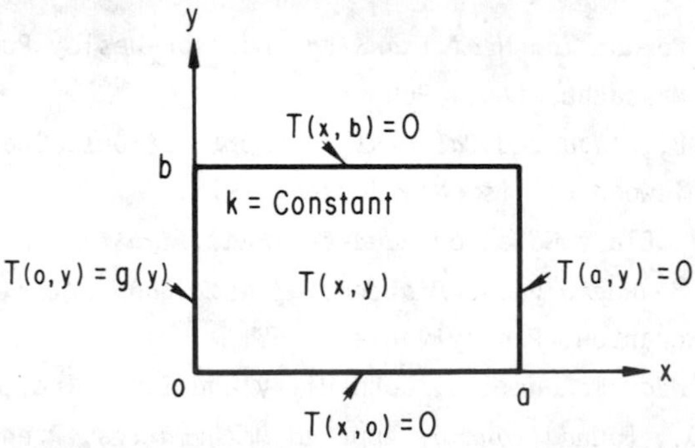

Fig. 5.17 Two-dimensional rectangular fin in Problem 5.3.

5.4 Consider a two-dimensional rectangular fin of constant thermal conductivity, which has a thickness a in the x-direction and is semi-infinite in the y-direction. Obtain an expression for the steady-state temperature distribution $T(x,y)$ in this fin under the following boundary conditions.

$$T(0,y) = 0; \qquad T(x,\infty) = 0$$

$$\frac{\partial T(a,y)}{\partial x} = 0; \qquad T(x,0) = f(x)$$

What is the heat transfer rate at $y = 0$?

5.5 Obtain an expression for the temperature distribution $T(x,y)$ for steady-state heat transfer in the two-dimensional rectangular fin shown in Fig. 5.18.

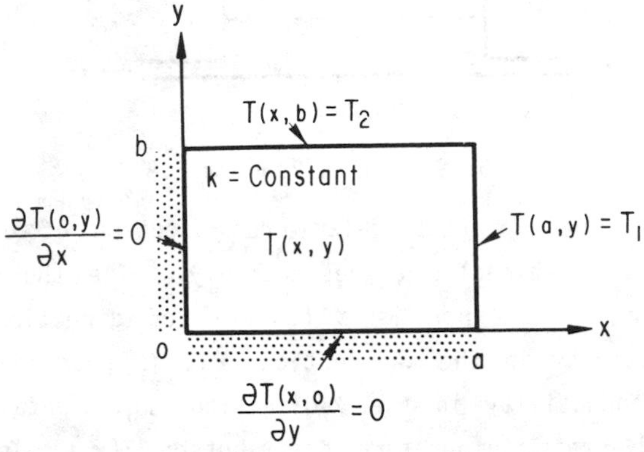

Fig. 5.18 Two-dimensional rectangular fin in Problem 5.5.

5.6 The temperature is maintained at 0°C along the three surfaces of a rectangular bar while the fourth surface at $y = 0$ is held at constant temperature T_0. Calculate $T(x,y)/T_0$ at $x/a = 0.25$, $y/b = 0.25$. Take $b/a = 2$ (see Fig. 5.2).

5.7 Develop an expression for the steady-state temperature distribution $T(x,y)$ in a long rectangular bar for the following boundary conditions:

$$T(x,b) = T_1; \qquad T(x,0) = T_2$$

$$T(a,y) = T_3; \qquad T(0,y) = T_4$$

where T_1, T_2, T_3 and T_4 are all constant temperatures.

5.8 A jet engine combustion chamber wall is cooled by water flowing in an annular jacket around the combustion chamber. Fins running spirally around the chamber in the annular space are cast integral with the combustion chamber and guide the cooling water. As an approximation, the fins may be assumed to be long straight finite rectangular fins as shown in Fig. 5.19. Obtain an expression for the steady-state temperature distribution $T(x,y)$ in the fins and an expression for the heat transfer rate per unit depth into the base of one fin from the combustion gases.

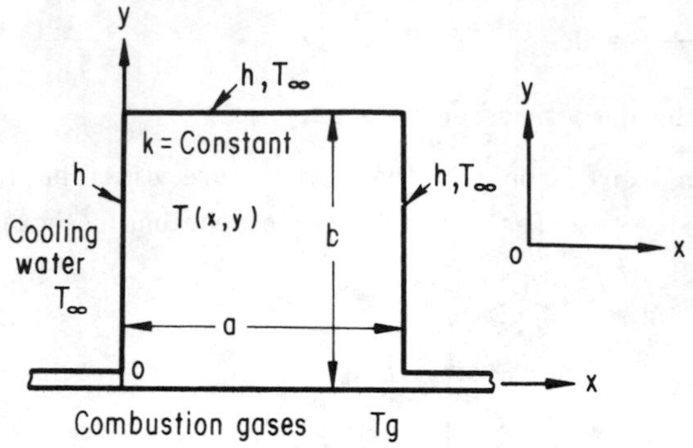

Fig. 5.19 Fin in Problem 5.8.

5.9 Consider a long bar of rectangular cross section of dimensions a and b in the x- and y-directions, respectively. The thermal conductivity of the material of the bar is a function of direction such that the thermal conductivity in the x-direction, k_x, is four times greater than the thermal conductivity in the y-direction, k_y. Obtain an expression for the steady-state temperature distribution T(x,y) in the bar under the following boundary conditions:

$$T(0,y) = T_1; \qquad T(a,y) = T_1$$

$$T(x,0) = T_1; \qquad T(x,b) = T_2$$

where T_1 and T_2 are constant temperatures.

5.10 Obtain an expression for the steady-state temperature distribution T(x,y) in the long rod of triangular cross section shown in Fig. 5.20. Assume that the thermal conductivity of the material of the rod is constant.

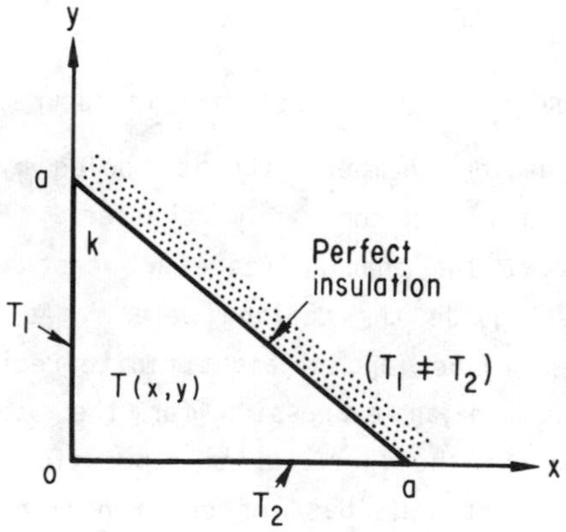

Fig. 5.20 Infinitely long rod of triangular cross section in Problem 5.10.

5.11 Obtain an expression for the steady-state temperature distribution in the long rod of triangular cross section shown in Fig. 5.21. Assume that the thermal conductivity of the material of the rod is constant.

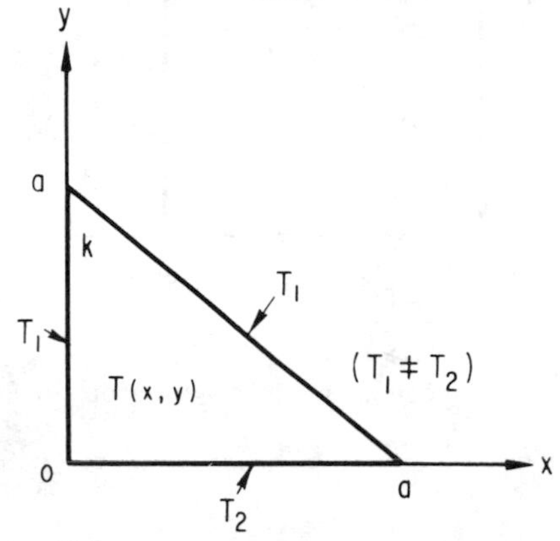

Fig. 5.21 Infinitely long rod of triangular cross section in Problem 5.11.

5.12 Consider the long rod of rectangular cross section shown in Fig. 5.22, in which internal energy is generated at a rate $\dot{q}(x)$ per unit volume. Assuming that the thermal conductivity of the material of the rod is constant, obtain an expression for the steady-state temperature distribution $T(x,y)$ under the boundary conditions given in the figure in the following ways:

(a) By separating the problem into two simple ones as

$$T(x,y) = \psi(x,y) + \phi(y)$$

(b) By seeking a solution as

$$T(x,y) = \sum_{n=0}^{\infty} A_n(y) \cos \frac{(2n+1)\pi}{2a} x, \qquad \frac{dA_n(o)}{dy} = A_n(b) = 0$$

Hint: Also expand

$$\frac{\dot{q}(x)}{k} = \sum_{n=0}^{\infty} B_n \cos \frac{(2n+1)\pi}{2a} x \text{ in } 0 < x < a$$

and substitute these two expansions into the differential equation of the problem to relate the unknown coefficients $A_n(y)$ to the known constants B_n through a set of ordinary differential equations.

(c) By seeking a solution as

$$T(x,y) = \sum_{m=0}^{\infty} \sum_{n=0}^{\infty} A_{mn} \cos \frac{(2m+1)\pi}{2a} x \cos \frac{(2n+1)\pi}{2b} y$$

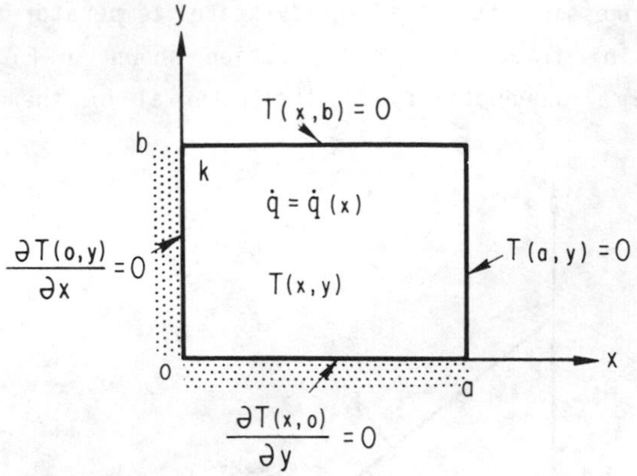

Fig. 5.22 Long rod of rectangular cross section in Problem 5.12.

Hint: Also expand

$$\frac{\dot{q}(x)}{k} = \sum_{m=0}^{\infty} \sum_{n=0}^{\infty} B_{mn} \cos \frac{(2m+1)\pi}{2a} x \cos \frac{(2n+1)\pi}{2b} y$$

and relate the unknown constants A_{mn} to the known constants B_{mn} through the use of the differential equation of the problem.

5.13 Obtain the temperature distribution T(x,y) for a steady-state heat conduction problem formulated as

$$\frac{\partial^2 T}{\partial x^2} + \frac{\partial^2 T}{\partial y^2} + \frac{\dot{q}(x,y)}{k} = 0$$

$$\frac{\partial T(0,y)}{\partial x} = 0; \qquad T(a,y) = 0$$

$$\frac{\partial T(x,0)}{\partial y} = 0; \qquad T(x,b) = 0$$

Hint: See the hint in Part (c) of Problem 5.12.

5.14 Determine the steady-state temperature distribution T(r,z) in the solid cylinder of constant thermal conductivity k shown in Fig. 5.23 under the following boundary conditions:

$$T(r_0,z) = T_0$$

$$T(r,0) = 0$$

$$T(r,L) = 0$$

where r_0 and L are the radius and height of the cylinder, respectively.

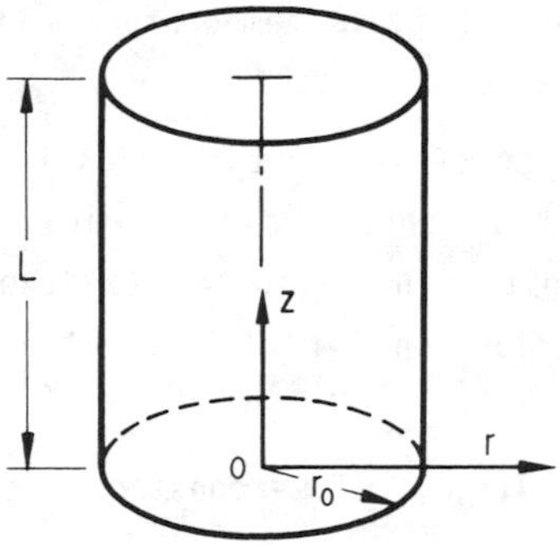

Fig. 5.23 Finite cylinder in Problem 5.14.

5.15 Determine the steady-state temperature distribution T(r,z) in the solid cylinder of Problem 5.14 for the following boundary conditions:

$$T(r_0,z) = 0$$

$$T(r,0) = 0$$

$$T(r,L) = f(r)$$

5.16 (a) Develop an expression for the steady-state temperature distribution T(r,z) in the solid cylinder of Problem 5.14 for the following boundary conditions:

$$T(r_0,z) = f_1(z)$$

$$T(r,0) = f_2(r)$$

$$T(r,L) = f_3(r)$$

(b) If $f_1(z) = T_0$ and $f_2(r) = f_3(r) = T_1$, where T_0 and T_1 are uniform and constant temperatures, find the steady-state temperature distribution in the cylinder in terms of $\theta = T - T_0$ and $\phi = T - T_1$.

5.17 Obtain an expression for the steady-state temperature distribution T(r,z) in a solid cylinder of length L, radius r_0 and of constant thermal conductivity k for the following boundary conditions:

$$T(r,0) = f(r)$$

$$\frac{\partial T(r_0,z)}{\partial r} = -\frac{h}{k}\,[T(r_0,z) - T_\infty]$$

$$\frac{\partial T(r,L)}{\partial z} = -\frac{h}{k}\,[T(r,L) - T_\infty]$$

where T_∞ is the surrounding fluid temperature and h is the heat transfer coefficient and they are both constants.

5.18 Consider a solid cylinder of length L, radius r_0 and of constant thermal conductivity k. Internal energy is generated in this cylinder at a constant rate $\dot{q}$ per unit volume. Obtain an expression for the steady-state temperature distribution T(r,z) under the following boundary conditions:

$$T(r_0,z) = T(r,0) = T(r,L) = T_0 = \text{constant}$$

5.19 Obtain the temperature distribution T(r,z) for a steady-state heat conduction problem formulated as

$$\frac{\partial^2 T}{\partial r^2} + \frac{1}{r}\frac{\partial T}{\partial r} + \frac{\partial^2 T}{\partial z^2} + \frac{\dot{q}(r,z)}{k} = 0$$

$$T(0,z) = \text{finite}; \qquad T(r_0,z) = 0$$

$$T(r,0) = 0; \qquad T(r,L) = 0$$

5.20 Obtain an expression for the steady-state temperature distribution T(r,z) in the semi-infinite solid cylinder shown in Fig. 5.24. Assume that the thermal conductivity of the material of the solid is constant.

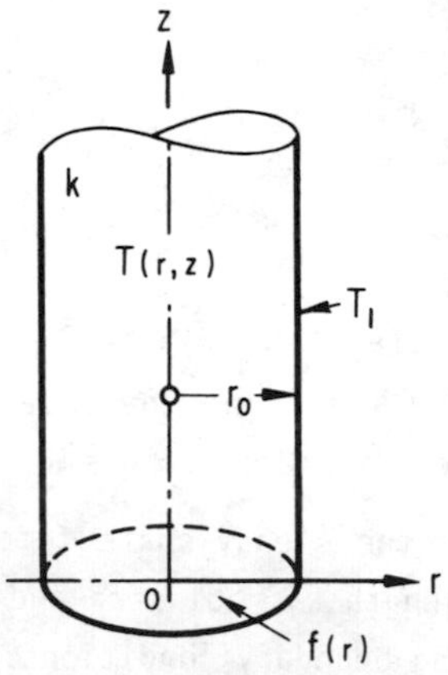

Fig. 5.24 Semi-infinite cylinder of Problem 5.20.

5.21 Give an expression for the temperature distribution in Problem 5.20 for the particular case of $f(r) = T_2$ where $T_1 \neq T_2$.

5.22 Obtain an expression for the steady-state temperature distribution T(r,z) in the semi-infinite cylinder of Problem 5.20 under the following boundary conditions:

$$T(r_0,z) = T_1$$

$$\frac{\partial T(r,0)}{\partial z} = -\frac{q_w}{k}$$

where q_w represents the constant heat flux applied at the circular surface at $z = 0$.

5.23 Determine the steady-state temperature distribution $T(r,z)$ in a solid rod of radius r_0, height H and of constant thermal conductivity k under the following boundary conditions:

$$\frac{\partial T(r_0,z)}{\partial r} = 0$$

$$T(r,0) = T_1$$

$$T(r,H) = T_2$$

5.24 (a) Determine the two-dimensional steady-state temperature distribution $T(r,\phi)$ inside the sector $0 < \phi < \alpha$, $0 < r < r_0$ of a circular plate of constant thermal conductivity if the temperature is maintained at zero along the straight edges and at a prescribed distribution $T(r_0,\phi) = f(\phi)$ along the curved edge.

(b) What will be the temperature distribution if, in particular, $f(\phi) = T_1 =$ constant?

5.25 Determine the two-dimensional steady-state temperature distribution $T(r,\phi)$ inside the sector $0 < \phi < \alpha$, $0 < r < r_0$ of a circular plate of constant thermal conductivity if the temperature is maintained at a prescribed constant value T_0 along the straight edge at $\phi = 0$ and at zero along the straight edge at $\phi = \alpha$ and the curved edge when $0 < \phi < \alpha$.

5.26 Use the Poisson integral formula (5.86) to solve the problem defined by Eqs. (5.62) and (5.63a,b,c,d).

5.27 Use the Poisson integral formula (5.86) to solve the following two-dimensional steady-state heat conduction problem:

$$\frac{\partial^2 T}{\partial r^2} + \frac{1}{r}\frac{\partial T}{\partial r} + \frac{1}{r^2}\frac{\partial^2 T}{\partial \phi^2} = 0, \qquad 0 < r < r_0, \qquad 0 < \phi < \pi$$

$$T(r_0,\phi) = f(\phi)$$

$$\frac{\partial T(r,0)}{\partial \phi} = \frac{\partial T(r,\pi)}{\partial \phi} = 0$$

5.28 Determine the two-dimensional steady-state temperature distribution

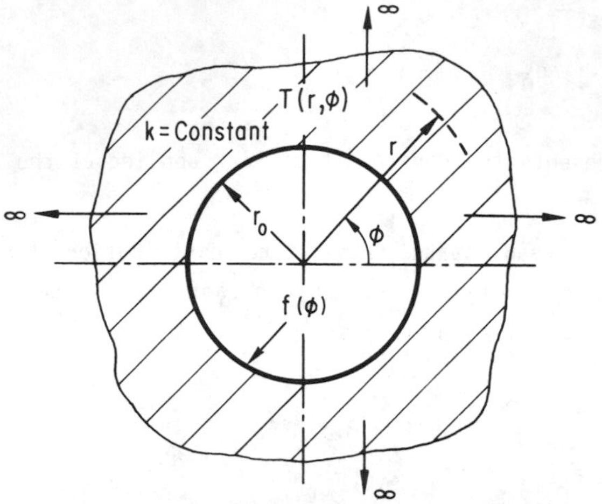

Fig. 5.25 A large plate with a circular hole in it.

$T(r,\phi)$ inside a large plate with a circular hole of radius r_0 as illustrated in Fig. 5.25, where the circular surface at r_0 is held at a prescribed temperature $f(\phi)$.

5.29 Find the first three coefficients in the expansion of the function

$$f(x) = \begin{cases} 0, & -1 < x < 0 \\ x, & 0 < x < 1 \end{cases}$$

in a series of Legendre polynomials over the interval (-1,1).

5.30 Find the first three coefficients in the expansion of the function

$$f(\theta) = \begin{cases} \cos\theta, & 0 < \theta < \pi/2 \\ 0 \quad , & \pi/2 < \theta < \pi \end{cases}$$

in a series of the form

$$f(\theta) = \sum_{n=0}^{\infty} A_n P_n(\cos\theta), \qquad 0 < \theta < \pi$$

5.31 (a) The temperature of the surface of a solid sphere of constant thermal conductivity k and radius r_0 is maintained at $T = T_0(1 - \cos\theta)$, where θ is the polar angle. Find the steady-state temperature distribution inside the sphere.

(b) If, instead, the temperature is maintained at a constant value T_0 over the upper hemisphere $(0 < \theta < \pi/2)$ and at zero over the remainder of the surface, determine the first three terms in the series solution.

5.32 Determine the steady-state temperature distribution $T(r, \phi, z)$ in the solid cylinder of Problem 5.14 for the following boundary conditions:

$$T(r_0,\phi,z) = f(\phi)$$

$$T(r,\phi,o) = 0$$

$$T(r,\phi,L) = 0$$

6

Unsteady Heat Conduction Solutions with Separation of Variables

6.1 INTRODUCTION

In this chapter we discuss the solution of unsteady heat conduction problems by the method of separation of variables in terms of some representative problems.

In our discussion of heat conduction, we have so far considered steady-state problems where the temperature distribution depends only on the space coordinates. A great majority of engineering problems, however, involve thermal conditions which change with time. For example, the boundary conditions may vary periodically, suddenly (step-wise), or irregularly. The initial temperature distribution may not be in thermal equilibrium. Internal energy generation may be started or stopped suddenly, or the rate of internal energy generation may be time dependent. The temperature at each point in such systems changes with time accordingly. Some examples of engineering interest include the temperature distribution in a turbine wheel or a turbine blade during the start-up and shut-down of the turbine when it is subjected to sudden changes in gas temperature; the temperature distribution in a piece of steel during the quench-hardening operation when it is removed from a furnace and plunged suddenly into a cold bath of air, water or oil; the temperature distribution in the cylinder wall of a reciprocating internal combustion engine which is subjected to a periodically varying gas temperature; the temperature distribution in the wall of the combustion chamber of a jet engine or a rocket motor during start-up, etc. From the knowledge of the temperature distribution at any instant the thermal stress distribution can be evaluated and from the knowledge of how the temperature at a point varies with time, the metallurgical conditions may be determined. Such unsteady or transient problems can be divided into two groups; namely, *non-periodic* and *periodic* problems.

The first group includes processes involving heating or cooling of bodies placed in a medium of given thermal state such as heating of steel ingots, cooling of iron bars in steelworks, hardening of steel by quenching in an oil bath, starting-up or shutting-down of a nuclear reactor or a furnace, etc.

The second group includes processes such as the heating process of regenerators whose packings are periodically heated by flue gases and cooled by air, daily periodic variation of heat transfer from the sun to the earth's surface, temperature fluctuations in the walls of internal combustion engines, etc.

In non-periodic transient problems, the temperature at any point within the body under consideration changes as some general non-linear function of time, while in periodic transient cases the temperature undergoes a periodic change which is either regular or irregular but always cyclic. A regular periodic variation is characterized by a harmonic function and an irregular periodic variation by any function which is cyclic but not necessarily harmonic.

In any heating or cooling problem, the heat transfer process between a body and the fluid surrounding it is influenced by both the internal resistance of the body and the surface resistance. In Chapter 3, we defined the ratio of internal resistance to surface resistance as Biot number, and discussed how to interpret the spatial variation of temperature in a solid in terms of the Biot number under steady-state conditions (see Fig. 3.11). We now extend this interpretation to spatial temperature distribution of unsteady-state problems. In a thin or a small body with a large thermal conductivity, the internal resistance may be negligible compared to the surface resistance. If this is the case, that is, if $Bi \to 0$, then the spatial variation of temperature can be neglected as illustrated in Fig. 6.1. Such a system is called a *lumped-heat-capacity system*.

When the surface resistance is negligible compared to the internal resistance, that is, when $Bi \to \infty$, the surface (boundary) temperature will be equal to the surrounding fluid temperature as illustrated in Fig. 6.2. This is usually the case with boiling, condensation and highly turbulent flows.

When the internal and surface resistances are comparable, that is, when

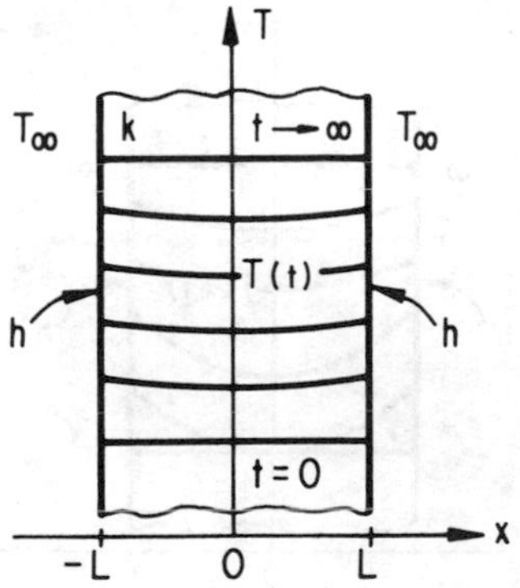

Fig. 6.1 Temperature distribution in a flat plate with $Bi = hL/k \to 0$.

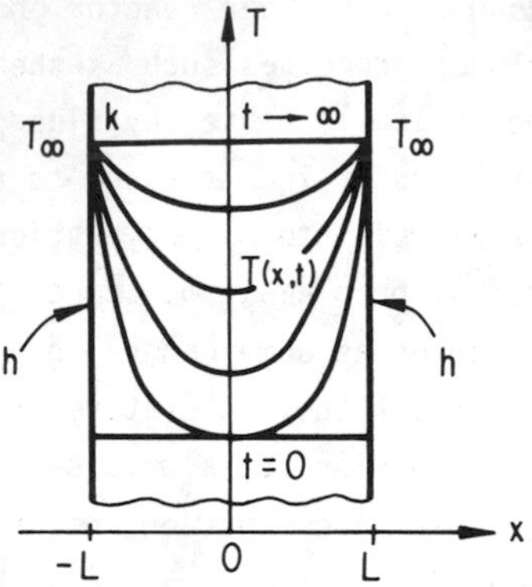

Fig. 6.2 Temperature distribution in a flat plate with Bi = $hL/k \to \infty$.

Bi≈1, the situation illustrated in Fig. 6.3 results. The last two cases correspond to what is known as *distributed systems*.

Since the elimination of spatial variations of temperature considerably simplifies the problems, we shall first study lumped-heat-capacity systems and then consider distributed systems. In distributed systems, we shall discuss first some representative non-periodic heating and cooling problems and then a periodic problem.

6.2 LUMPED-HEAT-CAPACITY SYSTEMS

The simplest example of transient cooling or heating is the case in which the internal resistance of a solid body, into or out of which heat transfer takes place, is negligibly small compared to the resistance to the heat flow at the boundary of the body. This would be the case, for example, if the thermal conductivity k of the body is infinitely large. Actually, the assumption is that

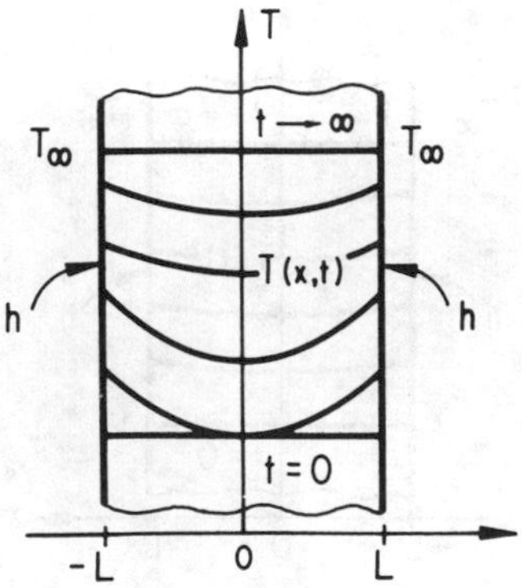

Fig. 6.3 Temperature distribution in a flat plate with Bi = $hL/k \approx 1$.

the heat transferred to the body is instantaneously and uniformly distributed throughout the body, resulting in uniform temperature within the system. Such systems, as already mentioned, are called lumped-heat-capacity systems. They are obviously idealized systems, since k of even the best conductor is not infinitely large. In fact, a temperature gradient must exist in a material if heat is to be conducted into or out of the material. However, investigations of the temperature distributions in bodies at any time during transient heating or cooling processes show that only small temperature differences exist if the Biot number is very small, that is, if $Bi = h(V/A)/k < 0.1$ where V is the volume and A is the total heat transfer area.

If a long thin steel bar, for example, is immersed in a cool pan of water, the lumped-heat-capacity method of analysis may be used if we can justify the assumption of uniform bar temperature at any time during the cooling process. Clearly, the temperature distribution in the bar will depend on the thermal conductivity of the bar material and the heat transfer conditions from the surface of the bar to the surrounding fluid, that is, the heat transfer coefficient h.

In order to make a lumped analysis the whole bar is taken to be the system shown in Fig. 6.4. The first law of thermodynamics as applied to this system yields

$$\frac{dE}{dt} = -Aq \tag{6.1}$$

where q is the rate of heat loss from the bar per unit surface area and A is the total heat transfer area of the bar. If we use the definition of specific heat, the left-hand side of Eq. (6.1) becomes

$$\frac{dE}{dt} = \rho c V \frac{dT}{dt} \tag{6.2}$$

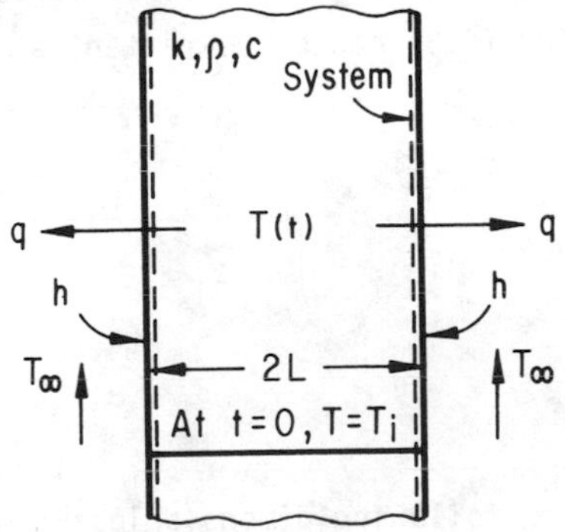

Fig. 6.4 Long thin steel bar; lumped-heat-capacity system.

where V is the volume of the bar. The heat flux term q on the right-hand side of Eq. (6.1) can be written as

$$q = h(T-T_\infty) \tag{6.3}$$

Combining Eqs. (6.1), (6.2) and (6.3) yields

$$\rho c V \frac{dT}{dt} = -Ah(T-T_\infty) \tag{6.4}$$

which is a first-order ordinary differential equation, for which we have the following initial condition:

$$T(0) = T_i \tag{6.5}$$

Equation (6.4) can be rewritten as

$$\frac{dT}{T-T_\infty} = -\frac{hA}{\rho c V}\,dt \tag{6.6}$$

Solution of this equation is

$$\ln(T-T_\infty) = -\frac{hA}{\rho c V}\,t + C_1 \tag{6.7}$$

Using the initial condition (6.5) yields

$$C_1 = \ln(T_i-T_\infty)$$

so that the solution of Eq. (6.4) becomes

$$\frac{T(t)-T_\infty}{T_i-T_\infty} = e^{-\left(\frac{hA}{\rho c V}\right)t} \tag{6.8}$$

It has been assumed that the fluid temperature T_∞, heat transfer coefficient h, and the thermophysical properties are constants. This result is quite useful in calculating the response rates of thermocouples. Defining a characteristic length as $L = V/A$, Eq. (6.8) can be rewritten as

$$\frac{T(t)-T_\infty}{T_i-T_\infty} = e^{-\left(\frac{hL}{k}\right)\left(\frac{\alpha t}{L^2}\right)} \tag{6.9a}$$

or

$$\frac{T(t)-T_\infty}{T_i-T_\infty} = e^{-Bi.Fo} \tag{6.9b}$$

where we have introduced the following dimensionless numbers:

$\frac{hL}{k}$ = Bi, *Biot* Number,

$\frac{\alpha t}{L^2}$ = Fo, *Fourier* Number.

The physical significance of Fourier number will be discussed in Section 6.3.1.

Equation (6.8) is applicable for any geometry as long as the assumption Bi < 0.1 holds. For simple geometries, L is given by

Sphere : $L = \frac{r}{3}$ (r = radius of sphere)

Cylinder: $L = \frac{r}{2}$ (Height > r) (r = radius of cylinder)

Cube : $L = \frac{a}{6}$ (a = side of cube)

Example 6.1: A cylindrical steel rod, 3 cm in diameter and 15 cm long, is first heated to 650°C in a furnace and then immersed in an oil bath maintained at 40°C for a heat treatment process. The heat transfer coefficient can be taken as 110 $W/m^2 \cdot K$. The density, specific heat and thermal conductivity of the steel are ρ = 7.83 g/cm^3, c = 465 J/kg·K, and k = 45 W/m·K, respectively. Calculate the time required for the rod to reach 345°C assuming a uniform temperature variation with time (lumped-heat-capacity system).

Solution: From the given data we have

$$Bi = \frac{hL}{k} = \frac{h\ r/2}{k} = \frac{110 \times 1.5\text{x}10^{-2}}{45} = 0.0367$$

Hence, the lumped-heat-capacity system analysis is a valid assumption. Furthermore,

$$\alpha = \frac{k}{\rho c} = \frac{45 \times 3600}{7830 \times 465} = 0.0445 \text{ m/hr}$$

and

$$\frac{T-T_\infty}{T_i-T_\infty} = \frac{345 - 40}{650 - 40} = 0.5$$

Substituting these values into Eq. (6.9b) we calculate the Fourier number and then the corresponding time:

$$Fo = \frac{1}{Bi} \ln \frac{T_i-T_\infty}{T-T_\infty} = \frac{1}{0.0367} \ln 2 = 18.887$$

$$t = \frac{Fo\text{x}L^2}{\alpha} = \frac{Fo\text{x}(r/2)^2}{\alpha}$$

$$= \frac{18.887 \text{ x}(1.5 \times 10^{-2})^2}{0.0445} = 0.095 \text{ hr} = 5.73 \text{ min.}$$

6.3. ONE-DIMENSIONAL DISTRIBUTED SYSTEMS

In engineering situations which can be idealized either as infinite plates with no temperature gradients along the y- and z-coordinates (*i.e.*, $\partial T/\partial y = \partial T/\partial z = 0$) as illustrated in Fig. 6.5 or as perfectly insulated rods as shown in Fig. 6.6, the unsteady-state temperature distribution will be one dimensional and, when the thermophysical properties (k, ρ, c) are constant, will satisfy the following diffusion equation:

$$\frac{\partial^2 T}{\partial x^2} = \frac{1}{\alpha}\frac{\partial T}{\partial t} \tag{6.10}$$

Assume that the following initial and boundary conditions exist:

$$T(x,0) = f(x) \tag{6.11a}$$

$$T(0,t) = 0 \tag{6.11b}$$

$$T(L,t) = 0 \tag{6.11c}$$

Since the boundary conditions are homogeneous, the assumption of the existence of a product solution of the form $T(x,t) = \psi(x)\Gamma(t)$ yields the following general solution

$$T(x,t) = e^{-\alpha\lambda^2 t}(A \sin \lambda x + B \cos \lambda x) \tag{6.12}$$

for the differential equation (6.10), where the sign of the separation constant λ^2 is consistent with the fact that the temperature distribution will asymptotically approach zero as the time increases indefinitely. Imposing the boundary conditions gives

$$T(x,t) = \sum_{n=1}^{\infty} A_n\, e^{-\alpha\lambda_n^2 t} \sin \lambda_n x \tag{6.13}$$

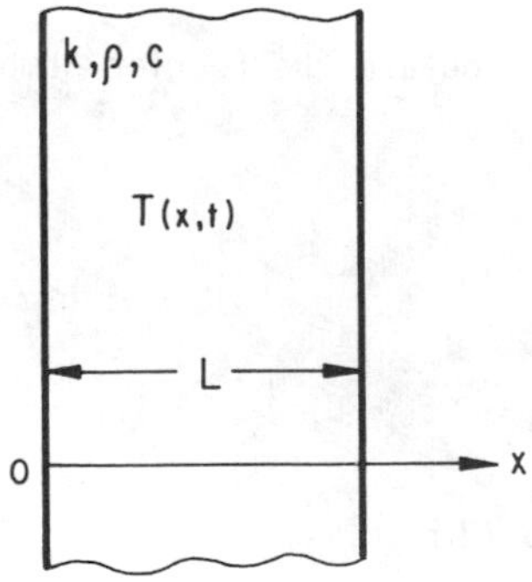

Fig. 6.5 One-dimensional plate.

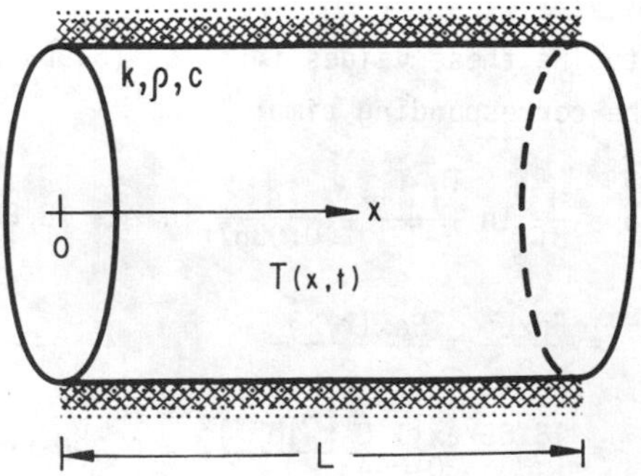

Fig. 6.6 One-dimensional insulated rod.

where

$$\lambda_n = \frac{n\pi}{L}, \qquad n = 1,2,3...$$

Introducing the initial condition (6.11a) into Eq. (6.13) we get

$$f(x) = \sum_{n=1}^{\infty} A_n \sin \frac{n\pi}{L} x \qquad (6.14)$$

which is the Fourier sine expansion of f(x) over the interval (0,L). By making use of Table 4.1, the coefficients A_n are obtained as

$$A_n = \frac{2}{L}\int_0^L f(x)\sin \frac{n\pi}{L} x\, dx \qquad (6.15)$$

Thus, the solution for T(x,t) is given by

$$T(x,t) = \frac{2}{L}\sum_{n=1}^{\infty} e^{-\left(\frac{n\pi}{L}\right)^2 \alpha t} \sin \frac{n\pi}{L} x \int_0^L f(x') \sin \frac{n\pi}{L} x' dx' \qquad (6.16)$$

Let us now consider the following initial and boundary conditions:

$$T(x,0) = f(x) \qquad (6.17a)$$

$$T(0,t) = T_1 \qquad (6.17b)$$

$$T(L,t) = T_2 \qquad (6.17c)$$

Since the problem is linear, the superposition technique can be used to obtain the solution of Eq. (6.10) as:

$$T(x,y) = T_s(x) + T_t(x,t) \qquad (6.18)$$

Here $T_s(x)$ is the steady-state solution of the problem and satisfies

$$\frac{d^2T_s}{dx^2} = 0 \qquad (6.19a)$$

$$T_s(0) = T_1, \qquad T_s(L) = T_2 \qquad (6.19b,c)$$

and $T_t(x,t)$ is the transient solution and satisfies

$$\frac{\partial^2 T_t}{\partial x^2} = \frac{1}{\alpha}\frac{\partial T_t}{\partial t} \qquad (6.20a)$$

$$T_t(x,0) = f(x) - T_s(x) \qquad (6.20b)$$

$$T_t(0,t) = T_t(L,t) = 0 \tag{6.20c,d}$$

The solution of the steady-state problem (6.19) is given by

$$T_s(x) = (T_2 - T_1)\frac{x}{L} + T_1 \tag{6.21}$$

and the solution of the transient problem (6.20) is

$$T_t(x,t) = \sum_{n=1}^{\infty} A_n e^{-\left(\frac{n\pi}{L}\right)^2 \alpha t} \sin\frac{n\pi}{L}x \tag{6.22}$$

Hence, the solution for $T(x,t)$ becomes

$$T(x,t) = T_1 + (T_2-T_1)\frac{x}{L} + \sum_{n=1}^{\infty} A_n e^{-\left(\frac{n\pi}{L}\right)^2 \alpha t} \sin\frac{n\pi}{L}x \tag{6.23}$$

Applying the initial condition (6.17a) we get

$$f(x) = T_1 + (T_2-T_1)\frac{x}{L} + \sum_{n=1}^{\infty} A_n \sin\frac{n\pi}{L}x \tag{6.24a}$$

which is the Fourier sine expansion of

$$F(x) = f(x) - T_1 - (T_2-T_1)\frac{x}{L} \tag{6.24b}$$

over the interval $(0,L)$. Thus, the expansion coefficients A_n are given by

$$A_n = \frac{2}{L}\int_0^L F(x)\sin\frac{n\pi}{L}x\,dx$$

$$= \frac{2}{n\pi}\left[(-1)^n T_2 - T_1\right] + \frac{2}{L}\int_0^L f(x)\sin\frac{n\pi}{L}x\,dx \tag{6.25}$$

We can now conclude that, to be able to solve a linear time-dependent problem by the method of separation of variables, the differential equation and the boundary conditions have to be homogeneous. Non-homogeneities, however, can be avoided by the principle of superposition. If the non-homogeneity in a boundary condition is due to a non-zero heat flux, then the nature of the problem may require a superposition of the form

$$T(x,t) = T_1(x,t) + T_2(x) + T_3(t) \tag{6.26}$$

instead of the form (6.18) (see Problem 6.19).

6.3.1 Cooling (or Heating) of an Infinite Flat Plate

The determination of the temperature variation through a relatively thin sheet or slab of material which is cooled (or heated) on both sides can be approximated as a one-dimensional problem, and such problems are frequently encountered in engineering applications. Solutions can be obtained for several different combinations of the initial and boundary conditions. The most useful and illustrative of these is possibly the case of a flat plate initially at a uniform temperature T_i which is subjected to the same cooling (or heating) conditions on both sides as illustrated in Fig. 6.7. This plate is considered to be sufficiently large in the y- and z-directions compared to its thickness 2L in the x-direction. The heat transfer coefficient h between the surfaces of the plate and the surrounding fluid on both sides is assumed to be constant. It is further assumed that the surrounding fluid temperature remains constant at T_∞ during the whole cooling (or heating) process.

Under these conditions heat flow through the plate will be one-dimensional. If the thermo-physical properties (k, ρ, c) are also assumed to be constant, the temperature distribution function defined as $\theta(x,t) = T(x,t) - T_\infty$ will then satisfy

$$\frac{\partial^2\theta}{\partial x^2} = \frac{1}{\alpha}\frac{\partial\theta}{\partial t} \tag{6.27}$$

with the initial condition

$$\theta(x,0) = T_i - T_\infty \equiv \theta_i \tag{6.28a}$$

and the boundary conditions

$$\frac{\partial\theta(0,t)}{\partial x} = 0 \tag{6.28b}$$

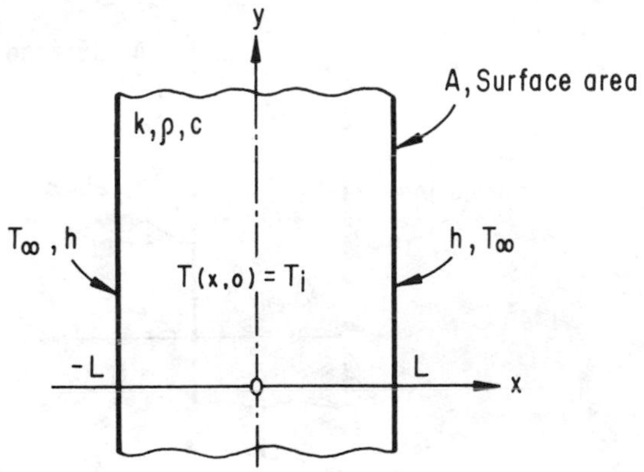

Fig. 6.7 Transient cooling (or heating) of a flat plate of thickness 2L.

$$k \frac{\partial\theta(L,t)}{\partial x} + h\,\theta(L,t) = 0 \tag{6.28c}$$

It may be noticed that the problem under consideration is mathematically the same as the problem of a slab of the same material but thickness L, which is insulated on one side and cooled (or heated) on the other, as illustrated in Fig. 6.8.

Since the above problem is linear with two homogeneous boundary conditions, assuming a product solution of the form $\theta(x,t) = X(x)\Gamma(t)$ it can be shown that

$$\theta(x,t) = \sum_{n=1}^{\infty} A_n e^{-\lambda_n^2 \alpha t} \cos \lambda_n x \tag{6.29}$$

where the characteristic values λ_n are the positive roots of

$$\cot \lambda L = \frac{\lambda L}{Bi} \tag{6.30}$$

with $Bi = hL/k$. Since Eq. (6.30) is a transcendental equation the characteristic values can be determined numerically or graphically, as discussed in Section 5.2.1 (see Fig. 5.7). The constants A_n in the solution (6.29) are determined from the application of the initial condition (6.28a); that is,

$$\theta_i = \sum_{n=1}^{\infty} A_n \cos \lambda_n x \tag{6.31}$$

which is a Fourier cosine expansion of θ_i with the values of λ_n determined from the characteristic-value equation (6.30). Hence, from Table 4.1 we have

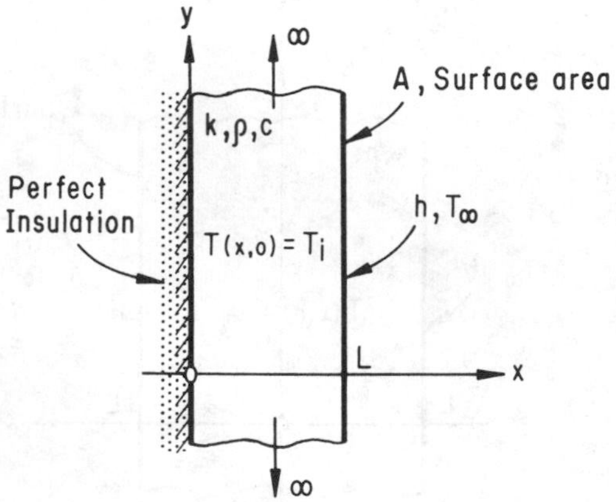

Fig. 6.8 Transient cooling (or heating) of a flat plate insulated on one side.

$$A_n = \frac{2\lambda_n}{\lambda_n L + \sin \lambda_n L \cos \lambda_n L} \int_0^L \theta_i \cos \lambda_n x dx \tag{6.32a}$$

which reduces to

$$A_n = \frac{2\, \theta_i \sin \lambda_n L}{\lambda_n L + \sin \lambda_n L \cos \lambda_n L} \tag{6.32b}$$

The solution is then given by

$$\frac{\theta(x,t)}{\theta_i} = \frac{T(x,t)-T_\infty}{T_i-T_\infty} = 2 \sum_{n=1}^{\infty} \frac{\sin \lambda_n L \cos \lambda_n x}{\lambda_n L + \sin \lambda_n L \cos \lambda_n L} e^{-\lambda_n^2 \alpha t} \tag{6.33}$$

For the case in which $h \to \infty$, that is, $Bi \to \infty$, the boundary condition (6.28c) becomes $T = T_\infty$ or $\theta = 0$ at $x = L$. This condition leads to the characteristic-value equation $\cos \lambda L = 0$, so that the characteristic values are given by

$$\lambda_n = \frac{2n-1}{L} \frac{\pi}{2}, \quad n = 1,2,3... \tag{6.34}$$

Then, the solution becomes

$$\frac{\theta(x,t)}{\theta_i} = \frac{4}{\pi} \sum_{n=1}^{\infty} \frac{(-1)^{n+1}}{(2n-1)} e^{-\left(\frac{2n-1}{2L}\right)^2 \pi^2 \alpha t} \cos \frac{(2n-1)\pi}{2L} x \tag{6.35}$$

It is often necessary to know the instantaneous rate of heat transfer to or from the plate, and also the total amount of heat which has been transferred over a period of time after the transient has started. The instantaneous rate of heat loss from both the surfaces is given by

$$\dot{Q} = -2kA \frac{\partial T(L,t)}{\partial x} \tag{6.36}$$

By the use of Eq. (6.33) we obtain

$$\dot{Q} = 4kA\theta_i \sum_{n=1}^{\infty} \frac{\lambda_n \sin^2 \lambda_n L}{\lambda_n L + \sin \lambda_n L \cos \lambda_n L} e^{-\lambda_n^2 \alpha t} \tag{6.37}$$

The total amount of heat transfer from the plate up to time t, after the transient period has started, is given by

$$Q = \int_0^t \dot{Q}\, dt$$

$$= 4kA\theta_i \int_0^t \sum_{n=1}^{\infty} \frac{\lambda_n \sin^2\lambda_n L}{\lambda_n L + \sin \lambda_n L \cos \lambda_n L} e^{-\lambda_n^2 \alpha t'} dt' \tag{6.38a}$$

When the integral over the time variable is evaluated we get

$$Q = 4\rho cA\theta_i \sum_{n=1}^{\infty} \frac{\sin^2\lambda_n L}{\lambda_n(\lambda_n L + \sin \lambda_n L \cos \lambda_n L)} (1-e^{-\lambda_n^2 \alpha t}) \tag{6.38b}$$

This result can also be written as

$$\frac{Q}{Q_i} = 2 \sum_{n=1}^{\infty} \frac{1}{\lambda_n L} \frac{\sin^2\lambda_n L}{\lambda_n L + \sin \lambda_n L \cos \lambda_n L} (1-e^{-\lambda_n^2 \alpha t}) \tag{6.38c}$$

where $Q_i = 2AL\rho c(T_i - T_\infty)$ which is the initial energy content of the plate with reference to the surrounding fluid temperature T_∞.

In the case of the plate of thickness L, which is cooled (or heated) on one side and insulated on the other, Eqs. (6.37) and (6.38b) must be divided by 2, an also in Eq. (6.38c) $Q_i = AL\rho c(T_i - T_\infty)$.

The series in the above results converge rather rapidly and satisfactory accuracy can be achieved by considering only a few terms (except for very small values of time) of the series.

The above results will apply only for the particular plate under consideration with the given thermal conductivity, heat transfer coefficient and the surrounding fluid temperature. Different sets of L, k and h values yield different sets of characteristic values which will require re-evaluation of the constant in Eq. (6.29). Obviously a more general approach would be desirable. This is possible since Eqs. (6.33) and (6.38c) may be regarded as functions of three dimensionless groups or variables. If solutions over wide ranges of these dimensionless groups are presented in the form of tables or charts, the answers to a specific problem can then be determined readily from the appropriate table or chart. For example, if we let $\lambda_n L = \gamma_n$ then the characteristic-value equation (6.30) can be written as

$$\cot \gamma_n = \frac{\gamma_n}{Bi} \tag{6.39}$$

Therefore,

$$\gamma_n = f(bi) = f\left(\frac{hL}{k}\right) \tag{6.40}$$

On the other hand, Eq. (6.33) becomes

$$\frac{\theta(x,t)}{\theta_i} = \frac{T(x,t)-T_\infty}{T_i-T_\infty} = 2\sum_{n=1}^{\infty} \frac{\sin\gamma_n \cos\left(\gamma_n \frac{x}{L}\right)}{\gamma_n + \sin\gamma_n \cos\gamma_n} e^{-\gamma_n^2 Fo} \tag{6.41}$$

where the grouping $Fo = \alpha t/L^2$ is the so-called Fourier number. Its magnitude is a measure of the degree of cooling or heating effects generated through the plate. If, for example, α/L^2 is small, a large value of t is required before significant temperature changes through the body occur. In view of Eq. (6.41), we now conclude that the dimensionless temperature ratio $\theta(x,t)/\theta_i$ is a function of the Biot number Bi, the Fourier number Fo, and the dimensionless position x/L; that is,

$$\frac{T(x,t)-T_\infty}{T_i-T_\infty} = \psi\left(Bi, Fo, \frac{x}{L}\right) = \psi\left(\frac{hL}{k}, \frac{\alpha t}{L^2}, \frac{x}{L}\right) \tag{6.42}$$

Similarly, Eq. (6.38c) can be written as

$$\frac{Q}{Q_i} = 2\sum_{n=1}^{\infty} \frac{1}{\gamma_n} \frac{\sin^2\gamma_n}{\gamma_n + \sin\gamma_n \cos\gamma_n} \left(1-e^{-\gamma_n^2 Fo}\right) \tag{6.43}$$

Therefore,

$$\frac{Q}{Q_i} = \phi(Bi, Fo) = \phi\left(\frac{hL}{k}, \frac{\alpha t}{L^2}\right) \tag{6.44}$$

Several sets of charts presenting solutions to these equations have already been published. We shall describe here those presented by Heisler [1] and Gröber [3] since they are the most complete and better known.

Figure 6.9 is a chart for the determination of the variation of the midplane temperature T_c of an infinite plate of thickness 2L or the temperature of the insulated surface of an infinite plate of thickness L. In this figure the dimensionless temperature ratio $(T_c-T_\infty)/(T_i-T_\infty)$ is plotted versus Fourier number $F_o = \alpha t/L^2$ for values of 1/Bi up to 100. Temperatures at other locations are obtained by multiplying the temperature ratios from Fig. 6.9 by the position-correction factors taken from Fig. 6.10.

It should be noted that the curve $1/Bi = k/hL = 0$ in Fig. 6.9 corresponds to the negligible surface resistance ($h \to \infty$). This is equivalent to the case in which the temperature of the surfaces of the plate is suddenly changed and held at the surrounding fluid temperature T_∞.

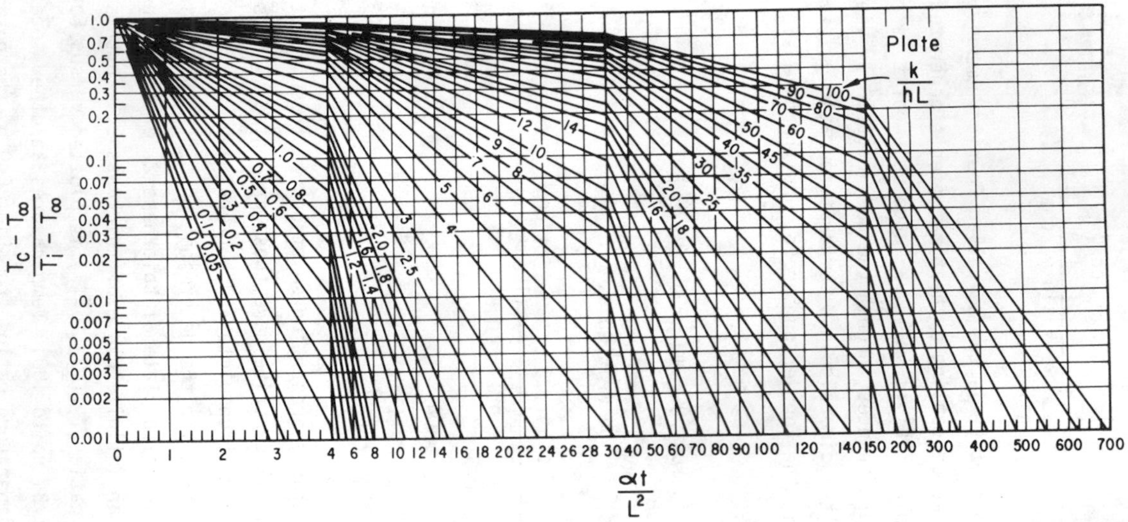

Fig. 6.9 Temperature-time history at the midplane of an infinite plate of thickness 2L [1].

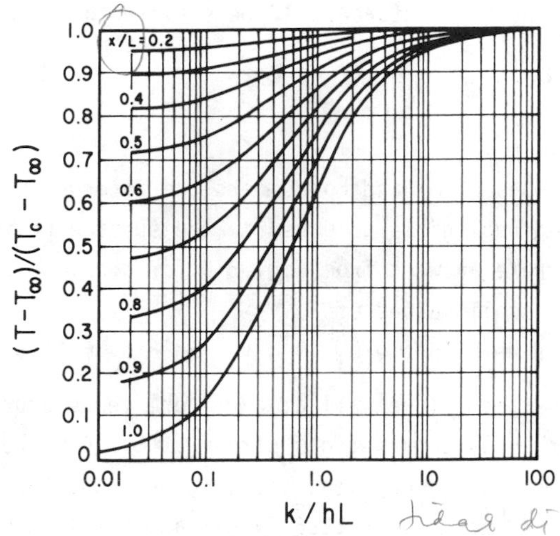

Fig. 6.10 Position-correction chart for an infinite plate of thickness 2L [1].

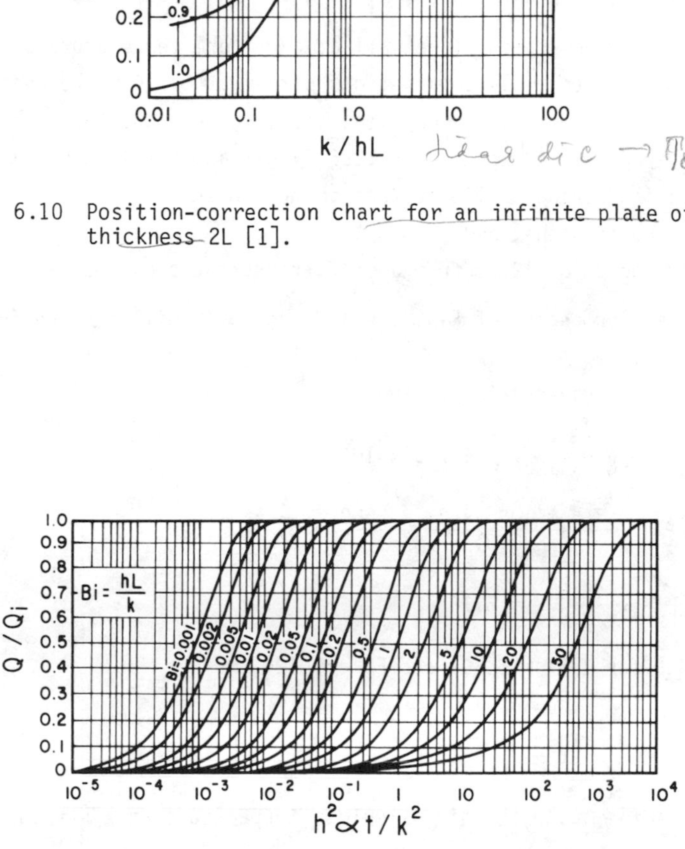

Fig. 6.11 Dimensionless heat loss Q/Q_i of an infinite plate of thickness 2L [3].

For problems in which k/hL is greater than 100 (and Fo > 0.2) Heisler [1] showed that the midplane temperature could be satisfactorily represented by

$$\frac{T_c - T_\infty}{T_i - T_\infty} = e^{-\left(\frac{hL}{k}\right)\left(\frac{\alpha t}{L^2}\right)} \tag{6.45}$$

That is, for these values of Bi and Fo numbers the internal resistance to heat flow may be considered negligible compared to the surface resistance. It is difficult to obtain accurate answers from Fig. 6.9 for short heating or cooling periods, that is, for small values of Fo.

The dimensionless heat loss Q/Q_i is given in Fig. 6.11.

Example 6.2: A 60-cm-thick plane wall at a uniform temperature of 21°C is suddenly exposed on both sides to a hot gas stream at 577°C. The heat transfer coefficient is 10 W/m²·K. The density, specific heat and thermal conductivity of the wall are ρ = 2600 kg/m³, c = 1256 J/kg·K and k = 15.5 W/m·K. After 27 hours of heating period, calculate the temperatures

a) at the midplane, and

b) at a depth of 12 cm from one of the surfaces of the wall.

Solution: Heisler charts of Fig. 6.9 and Fig. 6.10 will be used for the solution of this problem.

a) From the given data we have

$$\alpha = \frac{k}{\rho c} = \frac{15.5}{2600 \times 1256} = 4.746 \times 10^{-6} \text{ m}^2/\text{s}$$

$$\text{Fo} = \frac{\alpha t}{L^2} = \frac{4.746 \times 10^{-6} \times 27 \times 3600}{(0.3)^2} = 5.126$$

$$\frac{1}{\text{Bi}} = \frac{k}{hL} = \frac{15.5}{10 \times 0.3} = 5.167$$

From Fig. 6.9 for 1/Bi = 5.167 and Fo = 5.126, we get

$$\frac{T_c - T_\infty}{T_i - T_\infty} \cong 0.4 \Rightarrow T_c \cong 355°\text{C}$$

b) From Fig. 6.10 the dimensionless temperature at x/L = 18/30 = 0.6 is

$$\frac{T - T_\infty}{T_c - T_\infty} \cong 0.97 \Rightarrow T \cong 361°\text{C}$$

6.3.2 Cooling (or Heating) of an Infinite Solid Cylinder

Equally important as the plate configuration we have just discussed is a solid circular cylinder. In order to insure one-dimensional heat conduction so that the transient problem will involve only two independent variables, it will be assumed that the cylinder is infinitely long and that axial symmetry exists. Thus, the heat flow can be considered to be in the radial direction only.

Assume that the cylinder under consideration is heated initially to some known axially symmetric temperature distribution f(r) and then suddenly at t = 0 placed in contact with a fluid at temperature T_∞ as shown in Fig. 6.12. The heat transfer coefficient h at the surface of the cylinder is constant. In addition, it is also assumed that the temperature of the surrounding fluid remains constant at T_∞ during the whole cooling (or heating) process for $t > 0$.

Under the above conditions and if the thermo-physical properties (k, ρ, c) are constant, the formulation of the problem in terms of $\theta(r,t) = T(r,t) - T_\infty$ is given by the following heat conduction equation

$$\frac{\partial^2\theta}{\partial r^2} + \frac{1}{r}\frac{\partial\theta}{\partial r} = \frac{1}{\alpha}\frac{\partial\theta}{\partial t} \tag{6.46}$$

together with the initial condition

$$\theta(r,0) = f(r) - T_\infty \equiv F(r) \tag{6.47a}$$

and the boundary conditions

$$\theta(0,t) \neq \infty \; ; \; k\frac{\partial\theta(r_0,t)}{\partial r} + h\theta(r_0,t) = 0 \tag{6.47b,c}$$

Since the boundary conditions are homogeneous, the assumption of the existence of a product solution of the form $\theta(r,t) = R(r)\Gamma(t)$ yields the following solution

$$\theta(r,t) = \sum_{n=1}^{\infty} A_n e^{-\lambda_n^2\alpha t} J_0(\lambda_n r) \tag{6.48}$$

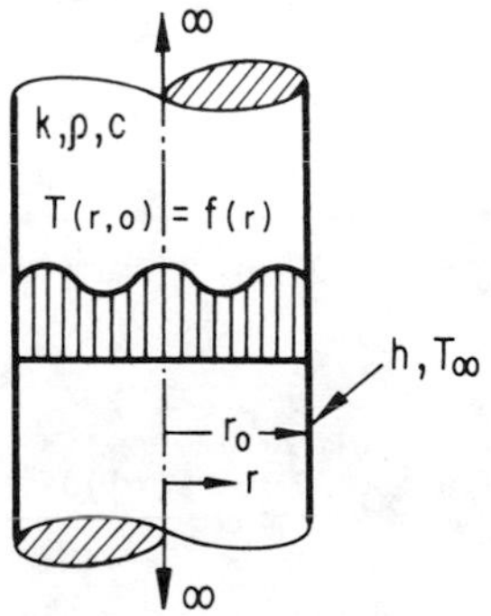

Fig. 6.12 Infinitely long solid cylinder.

where the characteristic values λ_n are the positive roots of the following transcendental equation

$$-\lambda k J_1(\lambda r_0) + h J_0(\lambda r_0) = 0 \tag{6.49}$$

which is obtained by applying the boundary condition (6.47c). The characteristic value equation (6.49) can be rearranged in the following form

$$\frac{J_1(\lambda r_0)}{J_0(\lambda r_0)} = \frac{Bi}{\lambda r_0} \tag{6.50}$$

where $Bi = hr_0/k$. Equation (6.50) is a transcendental equation and can be solved numerically to obtain the characteristic values λ_n. The first six characteristic values are tabulated in Reference [13] for several values of the Biot number. Applying the initial condition (6.47a) we obtain

$$F(r) = \sum_{n=1}^{\infty} A_n J_0(\lambda_n r) \tag{6.51}$$

which is a Fourier-Bessel expansion of F(r) with the values of λ_n determined from Eq. (6.50). Hence, from Table 4.2, we get

$$A_n = \frac{1}{N_n} \int_0^{r_0} F(r) J_0(\lambda_n r) r dr \tag{6.52}$$

where

$$N_n = \frac{r_0^2}{2} \left[1 + \frac{1}{\lambda_n^2}\left(\frac{h}{k}\right)^2\right] J_0^2(\lambda_n r_0) \tag{6.53a}$$

which, by the use of Eq. (6.49), can be rearranged as

$$N_n = \frac{r_0^2}{2} \left[J_0^2(\lambda_n r_0) + J_1^2(\lambda_n r_0)\right] \tag{6.53b}$$

Thus, the solution of the problem becomes

$$\theta(r,t) = T(r,t) - T_\infty = \frac{2}{r_0^2} \sum_{n=1}^{\infty} \frac{J_0(\lambda_n r)}{[J_0^2(\lambda_n r_0) + J_1^2(\lambda_n r_0)]}$$

$$\cdot\, e^{-\lambda_n^2 \alpha t} \int_0^{r_0} F(r') J_0(\lambda_n r') r' dr' \tag{6.54}$$

As in the previous flat plate case, a circular bar initially heated to a uniform temperature is of practical value. So, with $f(r) = T_i$ = constant, the above expression (6.54) becomes

$$\frac{\theta(r,t)}{\theta_i} = \frac{T(r,t)-T_\infty}{T_i-T_\infty}$$

$$= \frac{2}{r_0}\sum_{n=1}^{\infty}\frac{1}{\lambda_n}\frac{J_1(\lambda_n r_0)\,J_0(\lambda_n r)}{[J_0^2(\lambda_n r_0) + J_1^2(\lambda_n r_0)]}\,e^{-\lambda_n^2\alpha t} \tag{6.55}$$

This result can also be written as

$$\frac{\theta(r,t)}{\theta_i} = \frac{T(r,t)-T_\infty}{T_i-T_\infty}$$

$$= 2\sum_{n=1}^{\infty}\frac{1}{\gamma_n}\frac{J_1(\gamma_n)J_0(\gamma_n\frac{r}{r_0})}{[J_0^2(\gamma_n) + J_1^2(\gamma_n)]}\,e^{-\gamma_n^2 Fo} \tag{6.56}$$

where $\gamma_n = \lambda_n r_0$ and $Fo = \alpha t/r_0^2$.

It is seen from Eqs. (6.50) and (6.56) that the dimensionless temperature distribution $\theta(r,t)/\theta_i$ is a function of the Fourier number Fo, the Biot number Bi and the dimensionless position ratio r/r_0, that is,

$$\frac{T(r,t)-T_\infty}{T_i-T_\infty} = \psi(Bi, Fo, \frac{r}{r_0}) = \psi\left(\frac{hr_0}{k}\,,\,\frac{\alpha t}{r_0^2}\,,\,\frac{r}{r_0}\right) \tag{6.57}$$

Charts similar to those for the infinite plates have also been prepared for long solid cylinders, and are given in Figs. 6.13 and 6.14. Therefore, the transient temperature variations in an infinitely long cylinder which is initially at a uniform temperature T_i and suddenly exposed (at $t = 0$) to convective cooling or heating can be determined by the procedure same as that used for the infinite plate.

The total heat flow from a length L of the cylinder over a time interval (0,t) can be evaluated as

$$Q = -2\pi r_0\,Lk\int_0^t \frac{\partial}{\partial r}T(r_0,t')dt' \tag{6.58}$$

Substituting Eq. (6.55) into Eq. (6.58), we obtain

$$\frac{Q}{Q_i} = 4\sum_{n=1}^{\infty}\frac{J_1^2(\gamma_n)}{[J_0^2(\gamma_n) + J_1^2(\gamma_n)]}\,(1-e^{-\gamma_n^2 Fo}) \tag{6.59a}$$

where $Q_i = \pi r_0^2 L\rho c(T_i-T_\infty)$ is the initial energy content of the cylinder relative

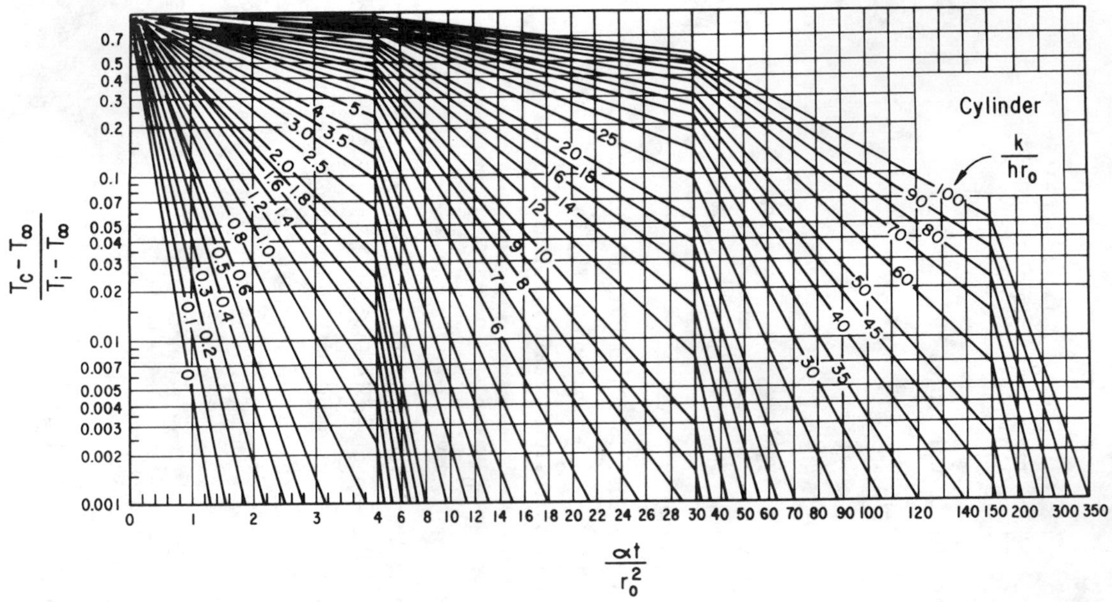

Fig. 6.13 Temperature-time history at the centerline of an infinitely long cylinder of radius r_0[1].

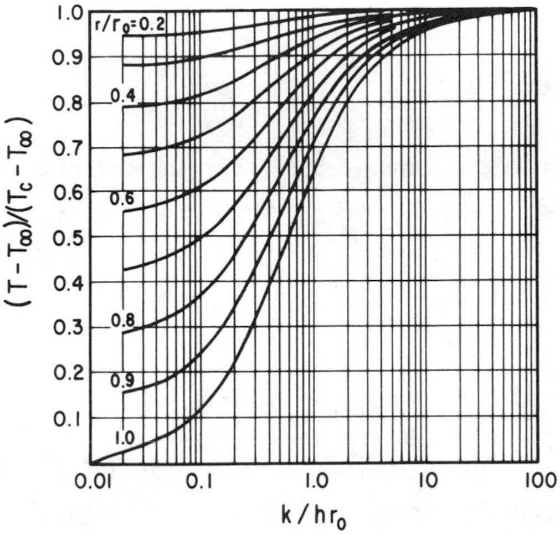

Fig. 6.14 Position-correction chart for an infinitely long cylinder of radius r_0 [1].

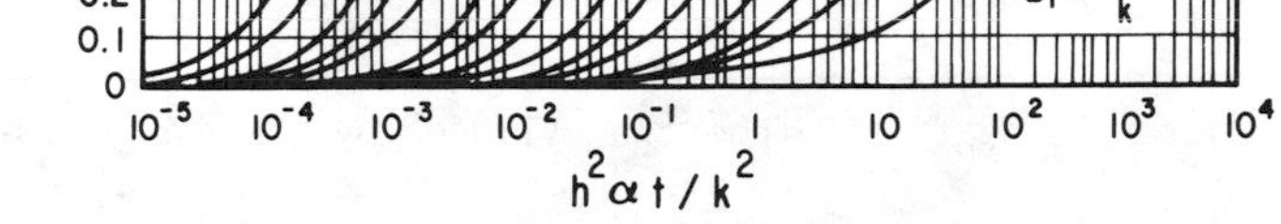

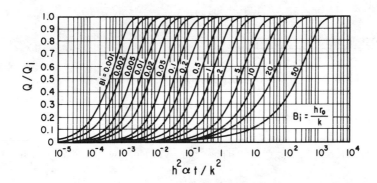

Fig. 6.15 Dimensionless heat loss Q/Q_i of a solid cylinder of radius r_0 [3].

to the surrounding fluid temperature T_∞. Taking into account that $J_0(\gamma_n)/J_1(\gamma_n) = \gamma_n/Bi$, Eq. (6.59a) can also be written as

$$\frac{Q}{Q_i} = 4 \sum_{n=1}^{\infty} \frac{Bi^2}{\gamma_n^2(\gamma_n^2 + Bi^2)} (1-e^{-\gamma_n^2 Fo}) \tag{6.59b}$$

It is seen from Eqs. (6.50) and (6.59b) that the dimensionless heat flow can be represented as

$$\frac{Q}{Q_i} = \phi(Bi, Fo) = \phi\left(\frac{hr_0}{k}, \frac{\alpha t}{r_0^2}\right) \tag{6.60}$$

In Fig. 6.15, the dimensionless heat loss Q/Q_i is given as a function of $Bi^2 Fo = (h/k)^2 \alpha t$ for several values of Bi.

Example 6.3: A steel cylinder, 20 cm in diameter and 5 m in length, is initially at 500°C. It is suddenly immersed in an oil bath maintained at 20°C. The properties of steel are: $\rho = 7700$ kg/m³, $c = 500$ J/kg·K and $k = 41$ W/m·K. Assuming a heat transfer coefficient of 1200 W/m²·K, calculate, after the cylinder has been exposed to the cooling process for 12 min,

a) the temperature at the centerline, and
b) the heat lost per unit length of the cylinder.

Solution: From the given data, we have

$$\alpha = \frac{k}{\rho c} = \frac{41}{7700 \times 500} = 1.065 \times 10^{-5} m^2/s$$

$$Bi = \frac{1200 \times 0.1}{41} = 2.93, \quad 1/Bi = 0.341$$

$$Fo = \frac{1.065 \times 10^{-5} \times 12 \times 60}{(0.1)^2} = 0.767$$

a) For 1/Bi = 0.341 and Fo = 0.767, Fig. 6.13 yields

$$\frac{T_c - T_\infty}{T_i - T_\infty} \cong 0.12 \Rightarrow T_c \cong 78°C$$

b) We calculate the heat lost from the cylinder by the use of Fig. 6.15. Since

$$\frac{h^2 \alpha t}{k^2} = \frac{(1200)^2 \times 1.065 \times 10^{-5} \times 12 \times 60}{(41)^2} = 6.57$$

from Fig. 6.15 for Bi = 2.93 we have

$$\frac{Q}{Q_i} \cong 0.85$$

For unit length

$$Q_i = \rho c \pi r_0^2 (T_i - T_\infty) = 7700 \times 500 \times \pi \times (0.1)^2 \times (500-20)$$

$$= 5.805 \times 10^7 \text{ J}$$

Thus, the heat lost per unit length is

$$Q = 0.85 Q_i = 0.85 \times 5.805 \times 10^7 \text{J} = 4.934 \times 10^7 \text{J}$$

6.3.3 Cooling (or Heating) of a Solid Sphere

Consider the cooling (or heating) of a solid sphere of radius r_0 in a medium of constant temperature T_∞, with a constant heat transfer coefficient h. At time t = 0, the temperature distribution at all points of the sphere is assumed to be given.

Heat conduction problems in spherical coordinate systems usually involve spherical symmetry so that the derivatives of the temperature with respect to ϕ and θ variables vanish. The general heat conduction equation in the absence of internal heat generation and for constant thermo-physical properties then reduces to

$$\frac{\partial^2 T}{\partial r^2} + \frac{2}{r}\frac{\partial T}{\partial r} = \frac{1}{\alpha}\frac{\partial T}{\partial t} \tag{6.61}$$

This diffusion equation can be reduced to the cartesian form by changing the dependent variable from T to U such that

$$U(r,t) = T(r,t)r \tag{6.62}$$

Substituting T(r,t) from Eq. (6.62) into Eq. (6.61) it can be shown that U satisfies

$$\frac{\partial^2 U}{\partial r^2} = \frac{1}{\alpha}\frac{\partial U}{\partial t} \tag{6.63}$$

Let us now assume that the solid sphere under consideration has an initial temperature distribution given by f(r). The formulation of the problem in terms of $\theta(r,t) = T(r,t) - T_\infty$ is given by

$$\frac{\partial^2 \theta}{\partial r^2} + \frac{2}{r}\frac{\partial \theta}{\partial r} = \frac{1}{\alpha}\frac{\partial \theta}{\partial t}$$

$$\theta(r,0) = f(r) - T_\infty$$

$$\theta(0,t) \neq \infty \qquad (6.64a,b,c,d)$$

$$k\frac{\partial \theta(r_0,t)}{\partial r} + h\theta(r_0,t) = 0$$

In terms of $U(r,t) = \theta(r,t)r$, this problem becomes

$$\frac{\partial^2 U}{\partial r^2} = \frac{1}{\alpha}\frac{\partial U}{\partial t}$$

$$U(r,0) = r[f(r)-T_\infty]$$

$$U(0,t) = 0 \tag{6.65a,b,c,d}$$

$$\frac{\partial U(r_0,t)}{\partial r} = \left[\frac{1}{r_0} - \frac{h}{k}\right] U(r_0,t)$$

Since the boundary conditions (6.65c,d) are homogeneous, the assumption of the existence of a product solution of the form $U(r,t) = R(r)\cdot\Gamma(t)$ yields the following solution:

$$U(r,t) = \sum_{n=1}^{\infty} A_n e^{-\alpha\lambda_n^2 t} \sin\lambda_n r \tag{6.66}$$

where the characteristic values λ_n are the positive roots of

$$\tan \lambda r_0 = -\frac{\lambda r_0}{(Bi-1)} \quad , \quad Bi = \frac{hr_0}{k} \tag{6.67}$$

which is obtained by applying the boundary condition (6.65d). The characteristic-value equation (6.67) is a transcendental equation and can be solved numerically to obtain the characteristic values λ_n. The use of the initial condition (6.65b) yields

$$r[f(r)-T_\infty] = \sum_{n=1}^{\infty} A_n \sin \lambda_n r \tag{6.68}$$

where the constants A_n, from Table 4.1, are given by

$$A_n = \frac{2\lambda_n \int_0^{r_0} r[f(r)-T_\infty]\sin \lambda_n r \, dr}{[\lambda_n r_0 - \sin \lambda_n r_0 \cos \lambda_n r_0]} \tag{6.69}$$

Hence, the solution for $\theta(r,t)$ becomes

$$\theta(r,t) = \frac{1}{r}\sum_{n=1}^{\infty} A_n e^{-\alpha\lambda_n^2 t} \sin\lambda_n r \tag{6.70}$$

with A_n given by Eq. (6.69).

When $f(r) = T_i$ = constant, the constants A_n become

$$A_n = 2(T_i - T_\infty)\frac{(\sin\lambda_n r_o - \lambda_n r_o \cos\lambda_n r_o)}{\lambda_n(\lambda_n r_o - \sin\lambda_n r_o \cos\lambda_n r_o)} \tag{6.71}$$

Then the transient temperature distribution in the sphere becomes

$$\frac{\theta(r,t)}{\theta_i} = \frac{T(r,t)-T_\infty}{T_i-T_\infty}$$

$$= 2\sum_{n=1}^{\infty}\frac{\sin\lambda_n r_o - \lambda_n r_o\cos\lambda_n r_o}{\lambda_n r_o - \sin\lambda_n r_o\cos\lambda_n r_o}\;\frac{\sin\lambda_n r}{\lambda_n r}\,e^{-\alpha\lambda_n^2 t} \tag{6.72}$$

As in the previous cases, Eq. (6.72) can be represented as

$$\frac{T(r,t)-T_\infty}{T_i-T_\infty} = \psi(\mathrm{Bi},\,\mathrm{Fo},\,\frac{r}{r_o}) = \psi\left(\frac{hr_o}{k},\,\frac{\alpha t}{r_o^2},\,\frac{r}{r_o}\right) \tag{6.73}$$

Charts similar to those for the infinite plate and cylinder have also been prepared for the solid sphere from Eq. (6.73). These charts are given in Figs. 6.16 and 6.17.

The total heat transfer Q from the sphere over the time interval (0,t) can be shown to be

$$\frac{Q}{Q_i} = 6\sum_{n=1}^{\infty}\frac{1}{\gamma_n^3}\frac{\sin\gamma_n - \gamma_n\cos\gamma_n}{\gamma_n - \sin\gamma_n\cos\gamma_n}\left(1-e^{-\gamma_n^2\mathrm{Fo}}\right) \tag{6.74}$$

where $\gamma_n = \lambda_n r_o$, $Q_i = {}^4\!/_3\,\pi r_i^3\rho c(T_i - T_\infty)$ and $\mathrm{Fo} = \alpha t/r_o^2$. An examination of Eq. (6.74), together with Eq. (6.67), reveals that

$$\frac{Q}{Q_i} = \phi(\mathrm{Bi},\,\mathrm{Fo}) = \phi\left(\frac{hr_o}{k},\,\frac{\alpha t}{r_o^2}\right) \tag{6.75}$$

In Fig. 6.18, the dimensionless heat loss Q/Q_i is given as a function of $(\mathrm{Bi})^2\mathrm{Fo}$ for several values of Bi.

Example 6.4: A steel sphere, 3 in. in diameter, is suddenly immersed in a water bath maintained at a uniform temperature of 35°C for hardening purposes after it has been initially heated to 900°C in a furnace. How long will it take for the surface of the sphere to cool down to 200°C? The heat transfer coefficient h is 60 $W/m^2 \cdot K$ and the properties of the steel are ρ = 8100 kg/m^3, c = 465 J/kg·K, and k = 44 W/m·K.

Solution: From the given data, we have

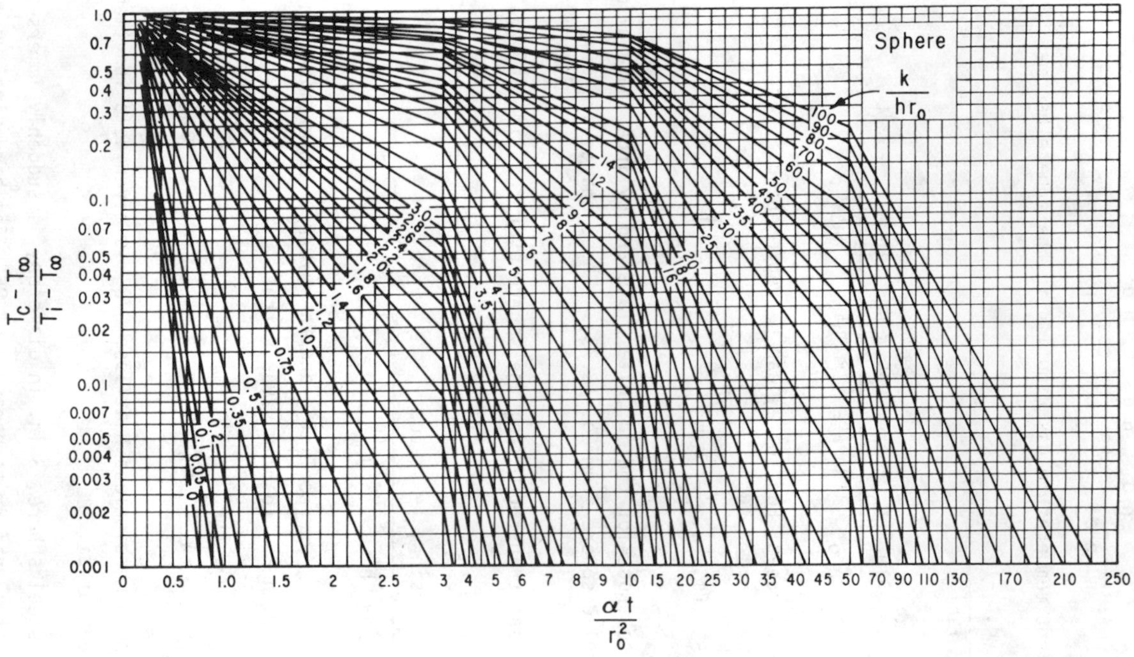

Fig. 6.16 Center temperature in a solid sphere of radius r_0 [1].

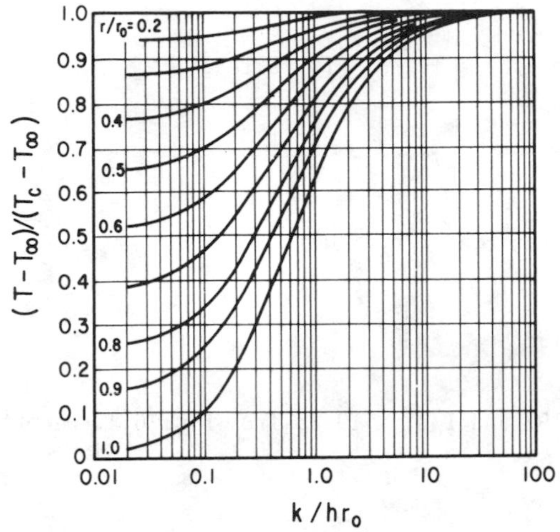

Fig. 6.17 Temperature as a function of center temperature for a sphere of radius r_0 [1].

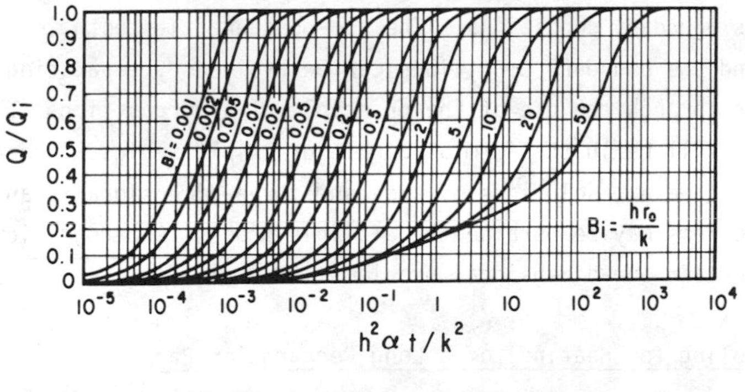

Fig. 6.18 Dimensionless heat loss Q/Q_i of a sphere of radius r_0 [3].

$$\alpha = \frac{k}{\rho c} = \frac{44}{8100 \times 465} = 1.168 \times 10^{-5} \ m^2/s$$

$$1/Bi = \frac{k}{hr_0} = \frac{44 \times 100}{60 \times 1.5 \times 2.54} = 19.25$$

Thus, at $r/r_0 = 1$ with $1/Bi = 19.25$, Fig. 6.18 gives

$$\frac{T_s - T_\infty}{T_c - T_\infty} \cong 0.965 \Rightarrow T_c - T_\infty \cong 171°C$$

Thus, the temperature ratio at $r/r_0 = 0$ is

$$\frac{T_c - T_\infty}{T_i - T_\infty} = \frac{171}{865} \cong 0.198$$

Then, with $\theta_c/\theta_i = 0.198$ and $1/Bi = 19.25$ from Fig. 6.17 we obtain

$$\frac{\alpha t}{r_0^2} \cong 11$$

So that the time required for the surface of the sphere to cool to 200°C is

$$t = \frac{11 \times (1.5 \times 2.54 \times 10^{-2})^2}{1.168 \times 10^{-5} \times 3600} = 0.38hr = 22.8 \ min$$

6.4 MULTI-DIMENSIONAL SYSTEMS

Two- and three-dimensional unsteady-state problems can be solved by the method of separation of variables in the same way as the one-dimensional problems if it is linear and consists of a homogeneous differential equation together with homogeneous boundary conditions. The non-homogeneities both in the differential equation and the boundary conditions can be avoided by separating the problem into simple ones through the use of the principle of superposition. The results, in such cases, will be in the form of double or triple series. Under certain special conditions, however, solutions of two- and three-dimensional transient heat conduction problems may be obtained by a simple product superposition of the solutions of certain one-dimensional problems.

6.4.1 Cooling (or Heating) of a Long Rectangular Bar

Consider the cooling (or heating) of the long rectangular bar shown in Fig. 6.19. The bar has a thickness 2a in the x-direction and height 2b in the y-direction. It is initially at a uniform temperature T_i, and suddenly at $t = 0$ placed in a fluid of constant temperature T_∞, which remains at this value during

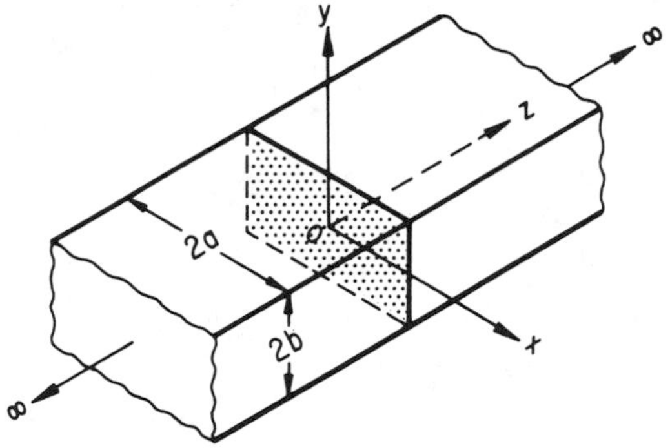

Fig. 6.19 Long rectangular bar.

the whole cooling (or heating) period. Let the heat transfer coefficient on all the surfaces be the same constant h.

If the thermo-physical properties (k, ρ, c) are assumed to be constant, then the problem can be formulated in terms of the temperature function $\theta(x,y,t) = T(x,y,t) - T_\infty$ as follows: The differential equation to be solved is

$$\frac{\partial^2\theta}{\partial x^2} + \frac{\partial^2\theta}{\partial y^2} = \frac{1}{\alpha}\frac{\partial\theta}{\partial t} \tag{6.76}$$

with the initial condition

$$\theta(x,y,0) = T_i - T_\infty \equiv \theta_i \tag{6.77a}$$

and the boundary conditions

$$\frac{\partial\theta(0,y,t)}{\partial x} = 0 \; ; \qquad \frac{\partial\theta(x,0,t)}{\partial y} = 0 \tag{6.77b,c}$$

$$\frac{\partial\theta(a,y,t)}{\partial x} = -\frac{h}{k}\,\theta(a,y,t) \; ; \qquad \frac{\partial\theta(x,b,t)}{\partial y} = -\frac{h}{k}\,\theta(x,b,t) \tag{6.77d,e}$$

With these initial and boundary conditions it can be shown that the solution for the temperature distribution is given by

$$\frac{\theta(x,y,t)}{\theta_i} = \frac{T(x,y,t)-T_\infty}{T_i-T_\infty} = 4\sum_{n=1}^{\infty}\sum_{m=1}^{\infty} e^{-\alpha(\lambda_n^2+\beta_m^2)t} \cdot \frac{(\sin\lambda_n a\cos\lambda_n x)(\sin\beta_m b\cos\beta_m y)}{(\lambda_n a + \sin\lambda_n a\cos\lambda_n a)(\beta_m b + \sin\beta_m b\cos\beta_m b)} \tag{6.78}$$

where λ_n and β_m are the positive roots of

$$\cot \lambda a = \frac{\lambda k}{h} \tag{6.79a}$$

and

$$\cot \beta b = \frac{\beta k}{h} \tag{6.79b}$$

The solution (6.78) can be rewritten as

$$\frac{\theta(x,y,t)}{\theta_i} = 2 \sum_{n=1}^{\infty} \frac{\sin \lambda_n a \cos \lambda_n x}{(\lambda_n a + \sin \lambda_n a \cos \lambda_n a)} e^{-\lambda_n^2 \alpha t}$$

$$\cdot\ 2 \sum_{m=1}^{\infty} \frac{\sin \beta_m b \cos \beta_m y}{(\beta_m b + \sin \beta_m b \cos \beta_m b)} e^{-\beta_m^2 \alpha t} \tag{6.80}$$

If Eq. (6.80) is compared with Eq. (6.33), it is seen that

$$\left(\frac{\theta(x,y,t)}{\theta_i}\right)_{\substack{2a \times 2b \\ \text{bar}}} = \left(\frac{\theta(x,t)}{\theta_i}\right)_{\substack{2a \\ \text{plate}}} \times \left(\frac{\theta(y,t)}{\theta_i}\right)_{\substack{2b \\ \text{plate}}} \tag{6.81a}$$

or

$$\left(\frac{T(x,y,t)-T_\infty}{T_i-T_\infty}\right)_{\substack{2a \times 2b \\ \text{bar}}} = \left(\frac{T(x,t)-T_\infty}{T_i-T_\infty}\right)_{\substack{2a \\ \text{plate}}} \times \left(\frac{T(y,t)-T_\infty}{T_i-T_\infty}\right)_{\substack{2b \\ \text{plate}}} \tag{6.81b}$$

Thus, the solution of the transient heat conduction problem for the dimensionless temperature distribution in the infinitely long bar shown in Fig. 6.19 is the product of the dimensionless temperature distributions in two infinitely long plates whose intersection forms the bar in question as illustrated in Fig. 6.20.

The Heisler chart of Section 6.3.1 (Fig. 6.9) can therefore be used to find the temperature at any point in the two-dimensional bar. It should be remembered that the length L used in the formulas and charts for the flat plate is the half thickness of the plate so that L will have to be replaced by a and b before the results of the previous section are applied to the bar problem set forth in this section.

The Heisler chart in Fig. 6.9 would be applicable even if the heat transfer coefficient h were different on the x and y surfaces but the same on each pair of parallel surfaces; that is, if h in Eq. (6.77d), say, were h_1 and h in Eq. (6.77e) were h_2.

Although we have demonstrated the validity of the product solution after we obtained the solution (6.78) for the bar, this can also be shown from the

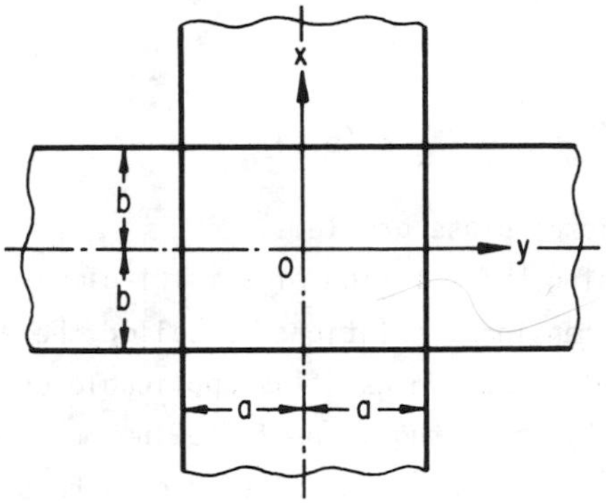

Fig. 6.20 The rectangular bar formed by the intersection of two infinite plates.

formulation of the problem. Thus, for example, if we reformulate the problem in terms of $\phi = \theta/\theta_i$, we get

$$\frac{\partial^2\phi}{\partial x^2} + \frac{\partial^2\phi}{\partial y^2} = \frac{1}{\alpha}\frac{\partial\phi}{\partial t} \tag{6.82}$$

$$\phi(x,y,0) = 1 \tag{6.83a}$$

$$\frac{\partial\phi(0,y,t)}{\partial x} = 0 \; ; \qquad \frac{\partial\phi(a,y,t)}{\partial x} = -\frac{h}{k}\,\phi(a,y,t) \tag{6.83b,c}$$

$$\frac{\partial\phi(x,0,t)}{\partial y} = 0 \; ; \qquad \frac{\partial\phi(x,b,t)}{\partial y} = -\frac{h}{k}\,\phi(x,b,t) \tag{6.83d,e}$$

Now, letting

$$\phi(x,y,t) = \phi_1(x,t)\cdot\phi_2(y,t) \tag{6.84}$$

it can readily be shown that $\phi_1(x,t)$ is the solution of

$$\frac{\partial^2\phi_1}{\partial x^2} = \frac{1}{\alpha}\frac{\partial\phi_1}{\partial t} \tag{6.85a}$$

$$\phi_1(x,0) = 1 \tag{6.85b}$$

$$\frac{\partial\phi_1(0,t)}{\partial x} = 0 \; ; \qquad \frac{\partial\phi_1(a,t)}{\partial x} = -\frac{h}{k}\,\phi_1(a,t) \tag{6.85c,d}$$

and $\phi_2(y,t)$ is the solution of

$$\frac{\partial^2\phi_2}{\partial y^2} = \frac{1}{\alpha}\frac{\partial\phi_2}{\partial t} \tag{6.86a}$$

$$\phi_2(y,0) = 1 \tag{6.86b}$$

$$\frac{\partial\phi_2(0,t)}{\partial y} = 0 \; ; \qquad \frac{\partial\phi_2(b,t)}{\partial y} = -\frac{h}{k}\,\phi_2(b,t) \tag{6.86c,d}$$

where both are one-dimensional plate problems.

This method of obtaining the solution of a multi-dimensional unsteady-state problem in terms of one-dimensional solutions is called the Newman method. It should, however, be used with caution as it is applicable only to certain special cases. In general, to apply the method, the following must hold:

(i) The problem has to be linear, consisting of a homogeneous differential equation together with homogeneous boundary conditions.

(ii) The initial temperature distribution has to be uniform or be in a form that can be factored into a product of functions, each involving only one of the space variables.

(iii) If one (or more) of the one-dimensional problems are for the infinite plate case and, therefore, the Heisler chart in Fig. 6.9 is to be used, then the problem has to have thermal symmetry in the related direction; that is, heat transfer coefficient should be same on each pair of parallel surfaces.

Other examples of the Newman method are given in the following sections.

6.4.2 Cooling (or Heating) of a Parallelepiped and a Finite Cylinder

The product superposition principle just illustrated above for the two-dimensional transient heat conduction in a rectangular bar can be extended to other configurations as well. Consider, for example, the rectangular parallelepiped of dimensions 2a x 2b x 2c shown in Fig. 6.21, which is initially at a uniform temperature T_i and suddenly at t = 0 is cooled (or heated) on all sides

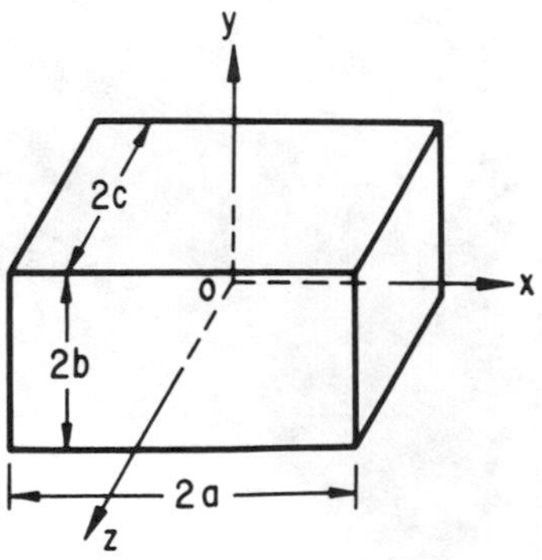

Fig. 6.21 Rectangular parallelepiped.

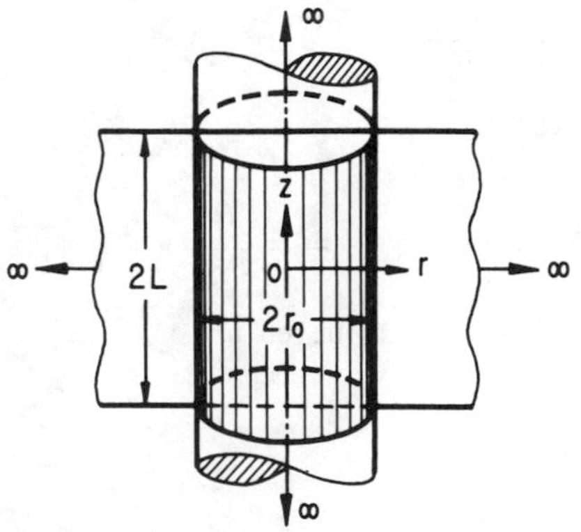

Fig. 6.22 Finite cylinder.

by a fluid at a constant temperature T_∞ with a constant heat transfer coefficient h. As in the preceeding case, it can easily be shown that the dimensionless temperature at any point (x,y,z) is given by

$$\left(\frac{T(x,y,z,t)-T_\infty}{T_i-T_\infty}\right)_{\substack{\text{Rectangular}\\\text{parallelepiped}}} = \left(\frac{T(x,t)-T_\infty}{T_i-T_\infty}\right)_{\substack{2a\\\text{plate}}} \times \left(\frac{T(y,t)-T_\infty}{T_i-T_\infty}\right)_{\substack{2b\\\text{plate}}} \times \left(\frac{T(z,t)-T_\infty}{T_i-T_\infty}\right)_{\substack{2c\\\text{plate}}} \tag{6.87}$$

Similarly, the solution for a short cylinder of radius r_0 and height 2L can be shown to be

$$\left(\frac{T(r,z,t)-T_\infty}{T_i-T_\infty}\right)_{\substack{\text{Finite}\\\text{cylinder}}} = \left(\frac{T(z,t)-T_\infty}{T_i-T_\infty}\right)_{\substack{2L\\\text{plate}}} \times \left(\frac{T(r,t)-T_\infty}{T_i-T_\infty}\right)_{\substack{\text{Infinite}\\\text{cylinder}}} \tag{6.88}$$

as illustrated in Fig. 6.22.

Example 6.5: A brass cylinder, 5 cm in diameter and 5 cm long, is initially at a uniform temperature of 316°C. It is suddenly immersed in a tank of water maintained at 21°C. After 15 seconds of cooling calculate the temperature at the center, and at a radial position of 2 cm and a distance of 2 cm from one end of the cylinder. The heat transfer coefficient h can be taken as 2000 W/m²·K. The properties of brass are given as: ρ = 8520 kg/m³, c = 385 J/kg·K, k = 111 W/m·K.

Solution: The two-dimensional temperature distribution in the cylinder is given by Eq. (6.88); that is, the Heisler charts for an infinite plate and an infinite cylinder are to be used together. From the given data we obtain

$$\alpha = \frac{k}{\rho c} = \frac{111}{8520 \times 385} = 3.38 \times 10^{-5}\ \text{m}^2/\text{s}$$

For both a flat plate of thickness 5 cm and a long cylinder of radius 2.5 cm we have

$$\text{Fo} = \frac{\alpha t}{L^2} = \frac{3.38 \times 10^{-5} \times 15}{(0.025)^2} = 0.811$$

$$1/\text{Bi} = \frac{k}{hL} = \frac{111}{2000 \times 0.025} = 2.22$$

Using Fig. 6.9, with 1/Bi = 2.22 and Fo = 0.811, we get

$$\left(\frac{T_c - T_\infty}{T_i - T_\infty}\right)_{\substack{2L\\ \text{plate}}} \cong 0.8$$

For 1/Bi = 2.22 and Fo = 0.811, Fig. 6.14 gives

$$\left(\frac{T_c - T_\infty}{T_i - T_\infty}\right)_{\substack{2r_o\\ \text{cylinder}}} \cong 0.6$$

Hence, the dimensionless temperature at the center of the finite cylinder is

$$\left(\frac{T_c - T_\infty}{T_i - T_\infty}\right)_{\substack{\text{finite}\\ \text{cylinder}}} \cong 0.8 \times 0.6 \simeq 0.48$$

which yields

$$T_c \simeq 163°C$$

The temperature at $r/r_o = 0.8$ and $z/L = 0.2$: From Fig. 6.10, for $z/L = 0.2$ and $1/Bi = 2.22$, we have

$$\left(\frac{T_z - T_\infty}{T_c - T_\infty}\right)_{\substack{2L\\ \text{plate}}} \cong 0.985$$

For the infinite cylinder from Fig. 6.14, for $r/r_o = 0.8$ and $1/Bi = 2.22$ we get

$$\left(\frac{T_r - T_\infty}{T_c - T_\infty}\right)_{\substack{\text{infinite}\\ \text{cylinder}}} \cong 0.87$$

The dimensionless temperature at $r/r_o = 0.8$ and $z/L = 0.2$ is then

$$\left(\frac{T - T_\infty}{T_c - T_\infty}\right)_{\substack{\text{finite}\\ \text{cylinder}}} \cong 0.985 \times 0.87 \simeq 0.857$$

which yields

$$T \simeq 142°C$$

6.4.3 Semi-Infinite Body

In practice, a number of cases can be idealized as semi-infinite solids, in which, by definition, at a given time there will be regions where the initial temperature remains unchanged after a temperature change occurs on one of its surfaces. A thick plate, for example, can be treated as a semi-infinite solid if the transient temperature response of the plate is to be investigated for short periods of time after a change of temperature of one of its surfaces (such as during a welding process). Another typical example is the earth. After a change of the surface temperature (following the weather conditions), there will always be some regions below the surface where the temperature is not affected by the change at the surface.

Consider such a semi-infinite solid as shown in Fig. 6.23, which is initially at a uniform temperature T_i. The surface at x = 0 is suddenly exposed at t = 0 to a fluid at temperature T_∞ with a constant heat transfer coefficient h. Assuming constant thermo-physical properties (k,ρ,c), the problem can be formulated in terms of $\theta = T-T_\infty$ as

$$\frac{\partial^2\theta}{\partial x^2} = \frac{1}{\alpha}\frac{\partial\theta}{\partial t} \tag{6.89}$$

$$\theta(x,0) = T_i-T_\infty \equiv \theta_i \tag{6.90a}$$

$$\frac{\partial\theta(0,t)}{\partial x} = \frac{h}{k}\,\theta(0,t) \quad ; \qquad \theta(\infty,t) = \theta_i \tag{6.90b,c}$$

This problem cannot be solved easily with the method we have used in the preceeding sections. It can, however, be solved readily by the use of Fourier or Laplace transforms that we shall discuss in the following chapters, and the solution is given by (see Problem 8.7)

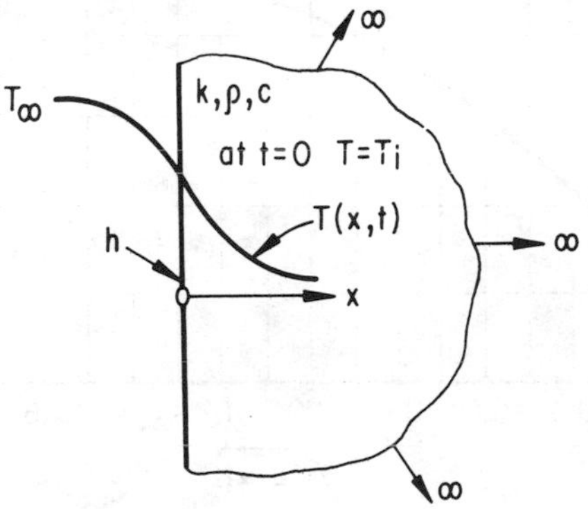

Fig. 6.23 Semi-infinite solid.

$$\frac{\theta(x,t)}{\theta_i} = \frac{T(x,t)-T_\infty}{T_i-T_\infty} = \text{erf}\,\frac{x}{2\sqrt{\alpha t}}$$

$$+ \exp\left[\frac{hx}{k} + \left(\frac{h}{k}\right)^2 \alpha t\right] \text{erfc}\left[\frac{x}{2\sqrt{\alpha t}} + \frac{h}{k}\sqrt{\alpha t}\right] \tag{6.91}$$

where the *error function* erf and the *complementary error function* erfc are defined as

$$\text{erf}\chi = 1 - \text{erfc}\chi = \frac{2}{\sqrt{\pi}}\int_0^\chi e^{-\beta^2}\, d\beta \tag{6.92}$$

The values of erf χ are available in literature. A short table of the values of erf χ is given in Appendix C.

If $h \to \infty$, then the surface temperature $T_w \to T_\infty$ and Eq. (6.91) reduces to

$$\frac{T(x,t)-T_w}{T_i-T_w} = \text{erf}\,\frac{x}{2\sqrt{\alpha t}} \tag{6.93}$$

In Section 8.8, this solution is obtained by Laplace transforms. Figure 6.24 gives the solution (6.93) in graphical form. Similarly, the solution (6.91) is given in graphical form in Figs. 6.25a and 6.25b.

Example 6.6: On a cold winter day, after a sudden change in weather conditions, the surface temperature of the roads in a city drops to -5°C. These cold weather conditions continue for 20 days. After this period calculate the temperature of the ground at a depth of 1 m. The temperature of the ground before the weather change was 5°C. The thermal diffusivity of the earth can be assumed to be 0.0025 m²/hr.

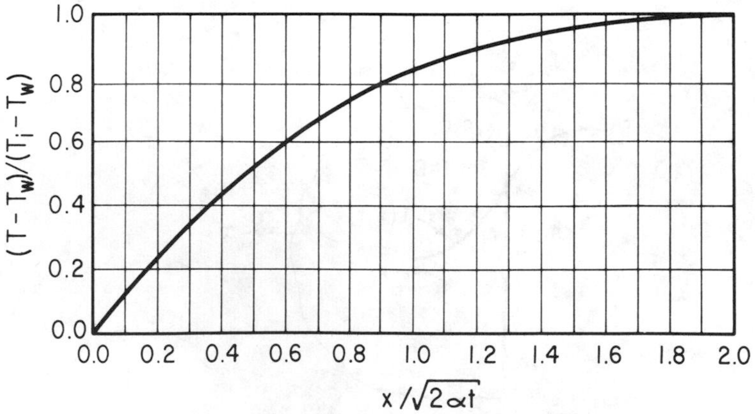

Fig. 6.24 Temperature-time history in a semi-infinite solid ($h \to \infty$), [2].

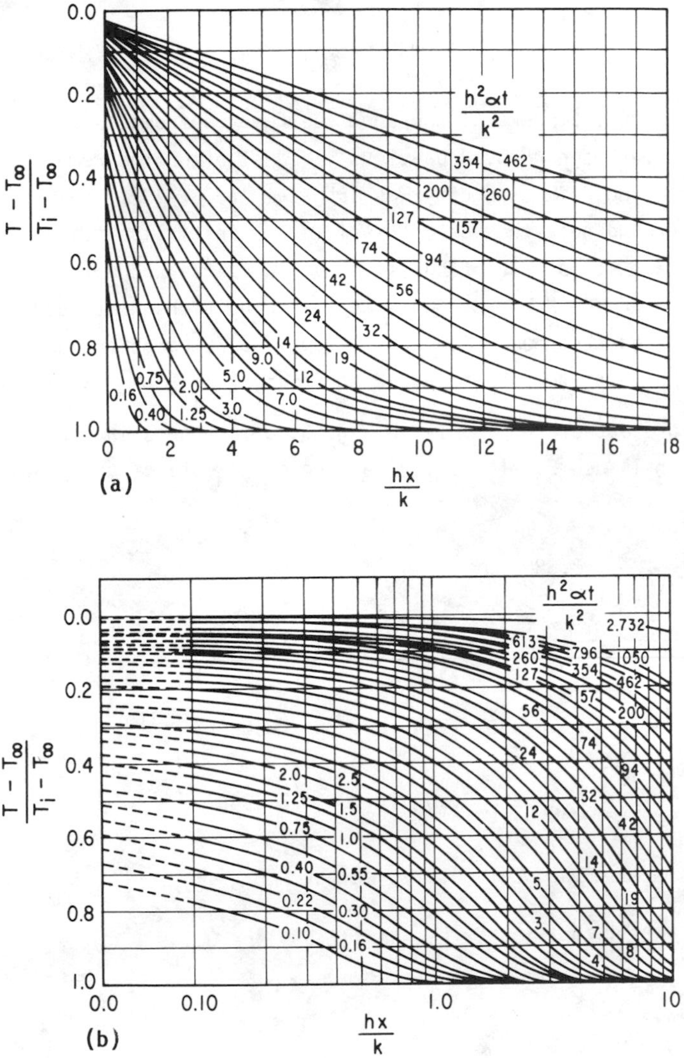

Fig. 6.25 Temperature-time history in a semi-infinite solid (h = finite), [2].

Solution: Equation (6.93) applies for this problem:

$$\frac{T+5}{5-(-5)} = \text{erf}\left[\frac{1}{2(0.0025 \times 20 \times 24)^{\frac{1}{2}}}\right]$$

$$\frac{T+5}{10} = \text{erf}\left[0.456\right]$$

The value of the error function can be taken from Appendix C or from Fig. 6.24.

Thus

$$\frac{T+5}{10} = 0.48 \Rightarrow T = -0.2°C$$

Example 6.7: In the city of Ankara, the snow was removed from parts of the city where it had never been removed previously. Just after the removal, the weather conditions changed and the city was subjected to a -15°C environmental temperature. The ground near the earth's surface was initially at a uniform temperature of 4°C and it is made of coarse gravel with the properties: α = 0.0005 m^2/hr and k = 0.52 W/m·K. Cold weather conditions continued for 2 days. Calculate the earth's temperature at a depth of 25 cm at the end of the cold period. Assume that the heat transfer coefficient is 6 W/m^2·K.

Solution: We may use either Eq. (6.91) or Fig. 6.25 for the solution of this problem. Here we shall use Eq. (6.91). From the given data of the problem we have

$$Bi = \frac{hx}{k} = \frac{6 \times 0.25}{0.52} = 2.885$$

$$\frac{x}{2\sqrt{\alpha t}} = \frac{0.25}{2 \times \sqrt{0.0005 \times 48}} = 0.807$$

$$\left(\frac{h}{k}\right)^2 \alpha t + \frac{hx}{k} = \left(\frac{6}{0.52}\right)^2 \times 0.0005 \times 48 + 2.885 = 6.080$$

$$\frac{x}{2\sqrt{\alpha t}} + \frac{h}{k}\sqrt{\alpha t} = 0.807 + \frac{6}{0.52}\left(0.0005 \times 48\right)^{\frac{1}{2}} = 2.594$$

Substituting these values into Eq. (6.91), we get

$$\frac{T-4}{-15-4} = 1 - \text{erf}(0.807) - e^{6.080}\,[1 - \text{erf}(2.594)] = 0.149$$

which yields

$$T = 1.17°C$$

6.4.4 Cooling (or Heating) of Semi-Infinite Cylinders, Bars and Plates

Consider the cooling (or heating) of a semi-infinite cylinder, a semi-infinite rectangular bar and a semi-infinite plate in a surrounding of constant temperature T_∞. Let the heat transfer coefficient at the base and the surfaces be the same or satisfy the Newman condition. At the beginning of the cooling (or heating) process at t = 0 all points are at a uniform temperature T_i. The temperature distributions for the transient heat conduction problems shown in Figs. 6.26a,b,c can be obtained in dimensionless form as a product of the dimensionless

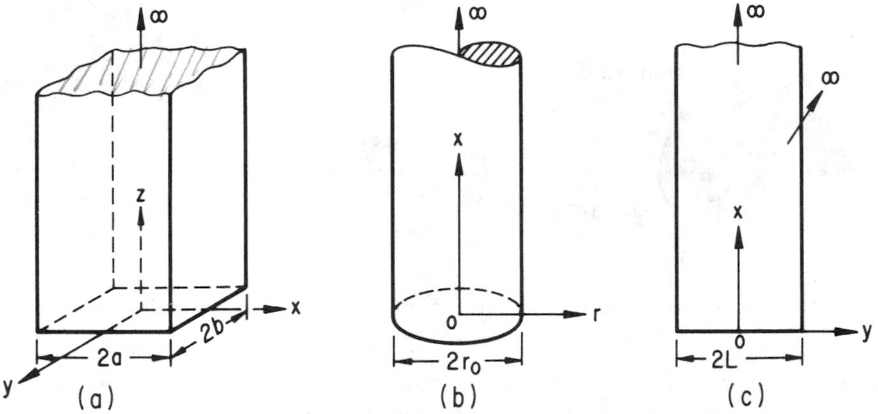

Fig. 6.26 Semi-infinite rectangular bar, cylinder and plate.

temperature distributions for the previous appropriate one-dimensional and semi-infinite solids, whose intersection forms the semi-infinite solid in question.

For example, the dimensionless temperature distribution in the semi-infinite rectangular bar of Fig. 6.26a can be shown to be

$$\left(\frac{T(x,y,z,t)-T_\infty}{T_i-T_\infty}\right)_{\substack{\text{semi-infinite}\\ \text{bar}}}$$

$$=\left(\frac{T(x,t)-T_\infty}{T_i-T_\infty}\right)_{\substack{2a\\ \text{plate}}} \times \left(\frac{T(y,t)-T_\infty}{T_i-T_\infty}\right)_{\substack{2b\\ \text{plate}}} \times \left(\frac{T(z,t)-T_\infty}{T_i-T_\infty}\right)_{\substack{\text{semi-infinite}\\ \text{body}}} \tag{6.94a}$$

The dimensionless temperature in the semi-infinite cylinder of Fig. 6.26b is

$$\left(\frac{T(r,x,t)-T_\infty}{T_i-T_\infty}\right)_{\substack{\text{semi-infinite}\\ \text{cylinder}}}$$

$$=\left(\frac{T(x,t)-T_\infty}{T_i-T_\infty}\right)_{\substack{\text{semi-infinite}\\ \text{body}}} \times \left(\frac{T(r,t)-T_\infty}{T_i-T_\infty}\right)_{\substack{r_0\\ \text{cylinder}}} \tag{6.94b}$$

and the dimensionless temperature in the semi-infinite plate of Fig. 6.26c is

$$\left(\frac{T(x,y,t)-T_\infty}{T_i-T_\infty}\right)_{\substack{\text{semi-infinite}\\ \text{plate}}}$$

$$=\left(\frac{T(x,t)-T_\infty}{T_i-T_\infty}\right)_{\substack{\text{semi-infinite}\\ \text{body}}} \times \left(\frac{T(y,t)-T_\infty}{T_i-T_\infty}\right)_{\substack{2L\\ \text{plate}}} \tag{6.94c}$$

Example 6.8: A semi-infinite steel plate of thickness 5 cm is initially at a temperature of 500°C. It is suddenly immersed in an oil bath maintained at 20°C. The properties of steel are ρ = 7833 kg/m^3, c = 465 J/kg•K and k = 45 W/m•K. Assuming a heat transfer coefficient of 1200 W/m^2•K, calculate the temperature on the axis of the plate 20 cm from the end 6 minutes after the cooling process has started.

Solution: For the solution of this problem we will take the product of an infinite plate and a semi-infinite solid in accordance with Eq. (6.94c). From the given data we have

$$\alpha = \frac{k}{\rho c} = \frac{45}{7833 \times 465} = 1.235 \times 10^{-5} \text{ m}^2/\text{s}$$

For the semi-infinite solid, Fig. 6.25 will be used. The parameters to be used with x = 20 cm are the following:

$$\text{Bi} = \frac{hx}{k} = \frac{1200 \times 0.20}{45} = 5.33$$

$$\left(\frac{h}{k}\right)^2 \alpha t = \left(\frac{1200}{45}\right)^2 \times 1.235 \times 10^{-5} \times 6 \times 60 = 3.16$$

From Fig. 6.25 we obtain

$$\left(\frac{\theta}{\theta_i}\right)_{\substack{\text{semi-infinite}\\ \text{solid}}} \cong 0.984$$

For the infinite plate with L = 2.5 cm we have

$$\frac{k}{hL} = \frac{45}{1200 \times 0.025} = 1.5$$

$$\frac{\alpha t}{L^2} = \frac{1.235 \times 10^{-5} \times 6 \times 60}{(0.025)^2} = 7.11$$

From Fig. 6.9 we get

$$\left(\frac{\theta}{\theta_i}\right)_{\substack{\text{infinite}\\ \text{plate}}} \cong 0.025$$

Combining the above results according to Eq. (6.94c) yields the solution for the semi-infinite plate:

$$\left(\frac{\theta}{\theta_i}\right)_{\substack{\text{semi-infinite}\\ \text{plate}}} = \frac{T - T_\infty}{T_i - T_\infty} \cong 0.984 \times 0.025 \simeq 0.0246$$

which gives

$$T \simeq 32°C$$

6.5 PERIODIC SURFACE TEMPERATURE CHANGE

The problem to be considered in this section is the determination of the temperature distribution in the semi-infinite solid shown in Fig. 6.27 which has a surface temperature varying periodically.

The thickness of the semi-infinite solid in the x-direction is so large that the temperature-time variation at any location x within the solid will depend only on the condition imposed on the surface at $x = 0$.

It will be assumed that the periodic temperature variations have occurred for a sufficiently long period of time so that, at points below the surface, succeeding cycles are identical. This is equivalent to neglecting the heating-up period that would result if a uniformly cold solid was suddenly subjected to a periodic surface temperature variation. The type of conduction just described is referred to as quasi-steady conduction. By this it is meant that the unsteady temperature changes repeat themselves steadily.

In practice, the daily temperature variation of buildings and the earth

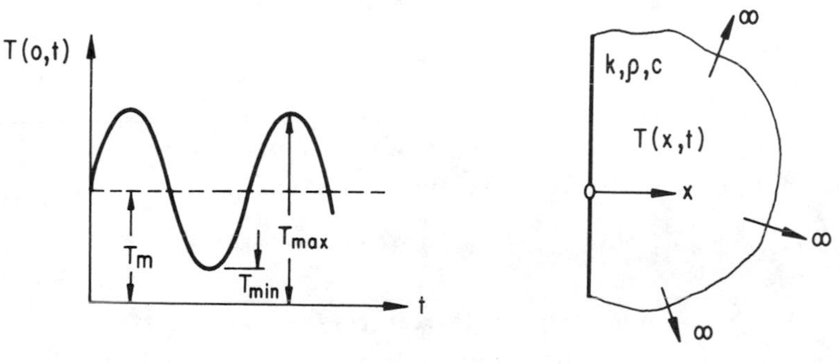

Fig. 6.27 Periodic surface temperature variation of a semi-infinite solid.

exposed to solar radiation or the temperature variation of the cylinder walls of internal combustion engines form this type of problems.

The boundary condition at the surface of the solid at x = 0 will be taken to be a sinusoidal surface temperature variation as shown in Fig. 6.27; that is,

$$T(0,t) = T_m + (\Delta T)_0 \sin(2\pi nt) \tag{6.95}$$

where

$$T_m = \frac{T_{max}+T_{min}}{2} \tag{6.96a}$$

and

$$(\Delta T)_0 = \frac{T_{max}-T_{min}}{2} \tag{6.96b}$$

are the mean and the amplitude of the surface temperatures, respectively. In addition, n is the frequency of oscillations. So, 1/n is the period of the surface temperature variation.

Defining a temperature variable as $\theta = T - T_m$, the periodic boundary condition (6.95) can be rewritten as

$$\theta(0,t) = (\Delta t)_0 \sin(2\pi nt) \tag{6.97}$$

If the thermo-physical properties (k,ρ,c) are assumed to be constant, then the temperature variable θ will satisfy the following one-dimensional unsteady heat conduction equation:

$$\frac{\partial^2\theta}{\partial x^2} = \frac{1}{\alpha}\frac{\partial\theta}{\partial t} \tag{6.98}$$

Now, consider an auxiliary problem with the differential equation

$$\frac{\partial^2\tilde{\theta}}{\partial x^2} = \frac{1}{\alpha}\frac{\partial\tilde{\theta}}{\partial t} \tag{6.99}$$

and the boundary condition

$$\tilde{\theta}(0,t) = (\Delta T)_0 \cos(2\pi nt) \tag{6.100}$$

Let us define a (complex) temperature function as

$$\theta_c(x,t) = \tilde{\theta}(x,t) + i\,\theta(x,t) \tag{6.101}$$

where $i = \sqrt{-1}$. The complex function $\theta_c(x,t)$ can be shown to satisfy the following differential equation and the boundary condition:

$$\frac{\partial^2\theta_c}{\partial x^2} = \frac{1}{\alpha}\frac{\partial\theta_c}{\partial t} \tag{6.102}$$

$$\theta_c(0,t) = (\Delta T)_0\, e^{2i\pi nt} \tag{6.103}$$

Assume a product solution in the form

$$\theta_c(x,t) = X(x)e^{2i\pi nt} \tag{6.104}$$

Substitution of Eq. (6.104) into Eqs. (6.102) and (6.103) yields

$$\frac{d^2X}{dx^2} - 2i\frac{\pi n}{\alpha}X(x) = 0 \tag{6.105}$$

$$X(0) = (\Delta T)_0 \tag{6.106a}$$

Since $\theta \neq \infty$ and $\tilde{\theta} \neq \infty$ as $x \to \infty$, we also have

$$X(\infty) \neq \infty \tag{6.106b}$$

Noting that

$$\sqrt{2i\frac{\pi n}{\alpha}} = \pm(1+i)\sqrt{\frac{\pi n}{\alpha}} \tag{6.107}$$

the solution of Eq. (6.105) can be written as

$$X(x) = A_1\, e^{-(1+i)\sqrt{\frac{\pi n}{\alpha}}\,x} + A_2\, e^{(1+i)\sqrt{\frac{\pi n}{\alpha}}\,x} \tag{6.108}$$

Use of the condition (6.106b) gives $A_2 = 0$. Thus

$$X(x) = A_1\, e^{-(1+i)\sqrt{\frac{\pi n}{\alpha}}\,x} \tag{6.109}$$

Hence, the product solution (6.104) gives

$$\theta_c(x,t) = A_1\, e^{-\sqrt{\frac{\pi n}{\alpha}}\,x}\exp\left[i\left(2\pi nt - \sqrt{\frac{\pi n}{\alpha}}\,x\right)\right] \tag{6.110}$$

Since $e^{i\beta} = \cos\beta + i\sin\beta$, Eq. (6.110) can also be written as

$$\theta_c(x,t) = A_1 e^{-\sqrt{\frac{\pi n}{\alpha}}\,x}\left[\cos\left(2n\pi t - \sqrt{\frac{\pi n}{\alpha}}\,x\right) + i\sin\left(2n\pi t - \sqrt{\frac{n\pi}{\alpha}}\,x\right)\right] \tag{6.111}$$

Hence, from the definition (6.101) we see that

$$\theta(x,t) = A_1 e^{-\sqrt{\frac{\pi n}{\alpha}}x} \sin(2n\pi t - \sqrt{\frac{n\pi}{\alpha}}x) \tag{6.112}$$

Application of the boundary condition (6.97) yields $A_1 = (\Delta T)_0$. Thus,

$$\theta(x,t) = (\Delta T)_0 e^{-\sqrt{\frac{\pi n}{\alpha}}x} \sin(2n\pi t - \sqrt{\frac{n\pi}{\alpha}}x) \tag{6.113}$$

Comparison of this result with the imposed variation at the surface, x = 0, shows that at a given distance from the surface the temperature variation is a periodic variation of the same period but of an amplitude that decreases exponentially with distance as illustrated in Fig. 6.28. That is, the amplitude at a distance x is

$$(\Delta T)_x = (\Delta T)_0 e^{-\sqrt{\frac{n\pi}{\alpha}}x} \tag{6.114}$$

Also, the period at a distance x is the same as that at the surface, but lags by a phase difference equal to

$$\Delta t = \frac{x}{2}\left(\frac{1}{\alpha\pi n}\right)^{\frac{1}{2}} \tag{6.115}$$

Figure 6.29 illustrates the temperature distribution within the semi-infinite solid at two different times.

The temperature at any given location x will be maximum when

$$\sin\left[2\pi nt - \sqrt{\frac{\pi n}{\alpha}}x\right] = 1$$

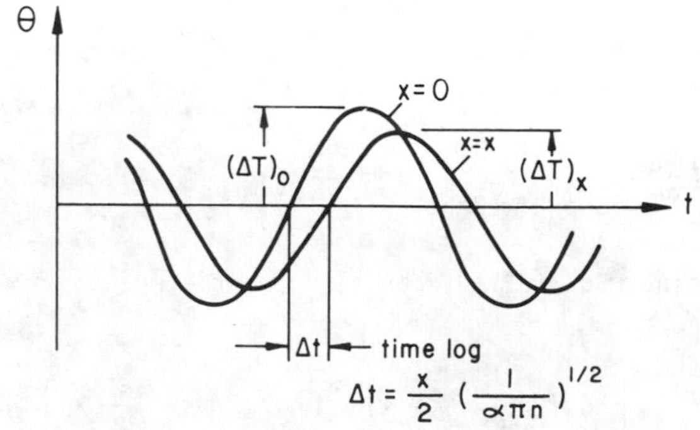

Fig. 6.28 Temperature variation at the surface and at a depth x in a semi-infinite solid subjected to a periodic surface temperature.

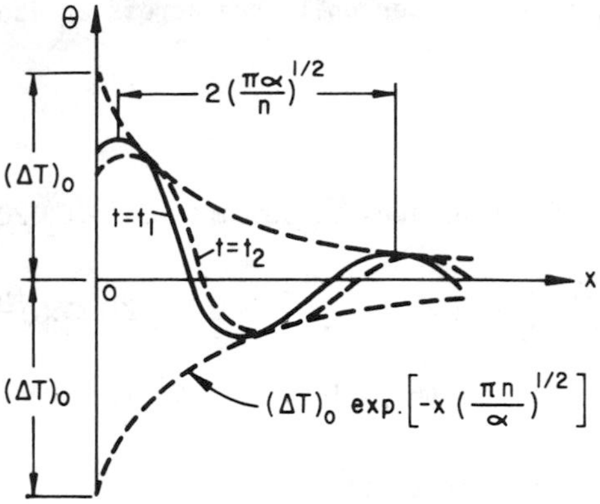

Fig. 6.29 Penetration of the surface temperature oscillations through a semi-infinite solid.

or

$$2\pi nt - \sqrt{\frac{\pi n}{\alpha}}\, x = \frac{(2m-1)\pi}{2}, \qquad m = 1,2,3,\ldots$$

Hence, the time at which the temperature at a given location x is maximum is

$$t = \frac{2m-1}{4n} + \frac{x}{2}\sqrt{\frac{1}{\alpha\pi n}} \tag{6.116}$$

From the solution (6.113) we see that for any particular value of time, the wave-length is given by

$$x_0 = 2\sqrt{\frac{\pi\alpha}{n}} \tag{6.117}$$

From the wave theory we know that the velocity of propagation is equal to the wave-length divided by the period. Hence, the propagation velocity of the temperature waves through the semi-infinite solid is given by

$$V(\text{velocity of wave}) = \frac{2(\pi\alpha/n)^{\frac{1}{2}}}{1/n} = 2(\pi\alpha n)^{\frac{1}{2}} \tag{6.118}$$

Thus, the wave at t_2 is displaced from the wave at t_1 by a distance

$$(t_2-t_1)2(\pi\alpha n)^{\frac{1}{2}} \tag{6.119}$$

In Eq. (6.113), the exponential factor has no effect on the phase difference, wavelength or propagation velocity, but results in a rapid decrease of the amplitude of the waves with increasing distance x.

The instantaneous heat flow rate per unit area across an isothermal surface at a distance x is

$$q = -k \frac{\partial T}{\partial x} = -k \frac{\partial \theta}{\partial x} \qquad (6.120)$$

Substituting the temperature distribution (6.113) into Eq. (6.120) we obtain

$$q = k(\Delta T)_0 \, e^{-x(n\pi/\alpha)^{\frac{1}{2}}} \sqrt{\frac{n\pi}{\alpha}} \left[\cos\left(2\pi nt - x\sqrt{\frac{n\pi}{\alpha}}\right) + \sin\left(2\pi nt - x\sqrt{\frac{n\pi}{\alpha}}\right)\right] \qquad (6.121)$$

The heat flow rate per unit area at the surface (x = 0) is, therefore, given by

$$q_0 = k(\Delta T)_0 \sqrt{\frac{n\pi}{\alpha}} \quad (\cos 2\pi nt + \sin 2\pi nt) \qquad (6.122a)$$

or

$$q_0 = k(\Delta T)_0 \sqrt{\frac{2n\pi}{\alpha}} \quad [\sin (2\pi nt + \frac{\pi}{4})] \qquad (6.122b)$$

It is seen from Eq. (6.122b) that the surface heat flow rate is also a periodic function of time. The phase lag between the heat flow and temperature function is 1/8n.

Example 6.9: On winter days the surface temperature of the ground in Ankara varies sinusoidally between -8°C and 10°C with a period of 24 hours. Calculate the depth above which a water pipe must not be placed to prevent it from freezing Take $\alpha = 0.0025$ m²/hr.

Solution: From the given data we have

$$(\Delta T)_0 = \frac{T_{max} - T_{min}}{2} = \frac{10-(-8)}{2} = \frac{18}{2} = 9°C$$

$$T_m = \frac{T_{max} + T_{min}}{2} = \frac{10-8}{2} = 1°C$$

$$n = \frac{1}{24}\ hr^{-1} \Rightarrow \sqrt{\frac{\pi n}{\alpha}} = 7.236\ m^{-1}$$

Since T_m = 1°C, for a water pipe at a depth x not to freeze the amplitude must be

$$(\Delta T)_x = 1°C$$

Thus, Eq. (6.114) gives

$$x = \frac{\ln(\Delta T)_0/(\Delta T)_x}{\sqrt{(\pi n/\alpha)}} = \frac{\ln(9/1)}{7.236} = 0.304 \text{ m}$$

$$x = 30.4 \text{ cm}$$

REFERENCES

1. Heisler, M. P., "Temperature Charts for Induction and Constant Temperature Heating," *Trans. ASME*, Vol. 69, pp. 227-236, 1947.
2. Arpacı, V. S., *Conduction Heat Transfer*, Addison-Wesley Pub. Co., Reading, Massachusetts, 1966.
3. Gröber, H., Erk, S., and Grigull, U., *Fundamentals of Heat Transfer*, McGraw-Hill Book Co., New York, 1961.
4. Holman, J. P., *Heat Transfer*, 5th ed., McGraw-Hill Book Co., New York, 1981.
5. Rohsenow, W. M., Class Notes on Advanced Heat Transfer, M.I.T., 1958.
6. Jacob, M., *Heat Transfer*, Vol. 1, John Wiley and Sons, Inc., New York, 1949.
7. Chapman, A. J., *Heat Transfer*, 3rd ed., McMillan Pub. Co., New York, 1974.
8. Özışık, M. N., *Boundary Value Problems of Heat Conduction*, International Textbook Co., Scranton, Pennsylvania, 1968.
9. Özışık, M. N., *Heat Conduction*, John Wiley and Sons, Inc., New York, 1980.
10. Luikov, A. V., *Analytical Heat Diffusion Theory*, Academic Press, New York, 1968.
11. Carslaw, H. S., and Jaeger, J. C., *Conduction of Heat in Solids*, 2nd ed., Oxford University Press, 1965.
12. Giedt, W. H., *Principles of Engineering Heat Transfer*, D. Van Nostrand Co., Inc., New York, 1957.
13. Janhke, E., Emde, F., and Losch, F., *Tables of Higher Functions*, McGraw-Hill Book Co., New York, 1960.

PROBLEMS

6.1 Consider a solid body of volume V and surface area A, surrounded by a coolant maintained at T_∞. The solid body is initially at temperature T_∞. For times $t \geq 0$, heat is generated in the solid at a rate per unit volume according to

$$\dot{q} = \dot{q}_0 \; e^{-\beta t}$$

where $\dot{q}_0$ and β are given constants. Let the heat transfer coefficient h between the solid body and the coolant be constant. Assuming constant thermophysical properties, and neglecting the spatial variation of the temperature, obtain an expression for the temperature of the solid body as a function of time for $t > 0$.

6.2 A slab extending from $x = 0$ to $x = L$ is perfectly insulated at $x = 0$ and held at a constant temperature T_w at $x = L$. Assuming constant thermo-physical properties, obtain an expression for the temperature distribution $T(x,t)$ in the slab for a uniform initial temperature distribution T_i.

6.3 A slab extending from $x = 0$ to $x = L$ and infinite in extent in the y- and z-directions is initially at 0°C. The surface temperatures at $x = 0$ and $x = L$ are suddenly raised to T_1 and T_2 at $t = 0$ respectively, and held constant at these values for times $t > 0$. The thermo-physical properties of the material of the slab are constant.

(a) Write the differential equation, and the initial and boundary conditions for the temperature distribution in the slab.

(b) Solve the problem formulated in part (a) and obtain the temperature distribution.

6.4 The wall of the combustion chamber of a jet engine may be assumed to be a large flat plate of thickness L. Initially the wall is at a uniform temperature equal to the temperature T_∞ of the surrounding atmosphere. As a first approximation it may be assumed that, when combustion begins, the gas inside the chamber flows at a high velocity resulting in a very high surface heat transfer coefficient, so that the temperature of the inner face of the wall immediately takes on the flame temperature T_f. Assuming that the heat transfer coefficient h between the wall and the surrounding atmosphere is a constant, obtain an expression for the temperature distribution in the chamber wall. The thermo-physical properties of the wall material may be taken as constants.

6.5 An infinitely long rod of radius r_o is initially at a uniform temperature T_i. At time $t = 0$, its surface temperture is changed to T_w and is subsequently maintained at this value for $t > 0$. Assuming constant thermo-physical properties; obtain

(a) an expression for the temperature distribution in the rod as a function of the radial coordinate r and time;

(b) an expression for the heat flux through the surface at $r = r_o$ as a function of time; and

(c) an expression for the average temperature of the rod as a function of time.

6.6 A flat plate of thickness L in the x-direction and of infinite extent in the y- and z-directions is initially at a uniform temperature T_i. At time t = 0, the temperature of the surface at x = 0 is changed to T_w and subsequently kept at this value for t > 0, while the other surface at x = L is maintained at T_i. Assuming constant thermo-physical properties, obtain

(a) an expression for the temperature distribution in the plate as a function of position and time;

(b) an expression for the heat flux through each surface as a function of time; and

(c) an expression for the average temperature of the plate as a function of time.

6.7 A large steel plate, 1.5 cm thick and initially at a uniform temperature of 24°C, is placed in a furnace maintained at 930°C. The mean heat transfer coefficient for combined convection and radiation may be taken as 90 $W/m^2 \cdot K$. Estimate the time required for the midplane of the plate to reach 540°C and the corresponding surface temperature.

6.8 The cylindrical combustion chamber of a rocket engine is made of 1-cm-thick stainless steel. The heat transfer coefficient on the gas side is 4000 $W/m^2 \cdot K$, and the temperature of the combustion gases is 2320°C. The temperature of the engine prior to firing is 28°C. Estimate the minimum possible combustion period if the temperature anywhere in the combustion wall is not to exceed 1100°C. The wall thickness may be assumed small in comparison with the diameter of the combustion chamber.

6.9 If the combustion chamber of the rocket engine discussed in Problem 6.8 is surrounded by a jacket through which liquid oxygen is circulated at -185°C, calculate the gain in the permissible firing time. Assume that the combustion chamber wall is precooled to -185°C and that the heat transfer coefficient on the oxygen side is 9800 $W/m^2 \cdot K$.

6.10 A large slab of hard rubber, 1 in. thick and initially at 20°C, is placed between two heated steel plates maintained at 140°C. The heating is to be discontinued when the temperature at the midplane of the slab reaches 130°C. The rubber has a thermal conductivity of 0.15 $W/m \cdot K$ and a thermal diffusivity of 8.8×10^{-8} m^2/s. The thermal contact resistance between the steel plates and the rubber is negligible.

(a) Calculate the length of the heating period.

(b) Calculate the temperature of the rubber at a distance 1/4 in. from one of the steel plates at the end of the heating period.

(c) Calculate the time required for the temperature at a distance

1/4 in. from one of the steel plates to reach 130°C.

6.11 A long cylindrical steel bar, 20 cm in diameter, is heated to 980°C and then quenched in an oil bath maintained at 40°C. The heat transfer coefficient can be assumed to be 500 $W/m^2 \cdot K$. How long will it take for the temperature of the centerline of the cylinder to reach 260°C?

6.12 Stainless steel circular bars, each 12 cm in diameter, are to be quenched in a large oil bath maintained at 38°C. The initial temperature of the bars is 870°C. The maximum temperature within the bars at the end of the quenching process will have to be 200°C. How long must the bars be kept in the oil bath if

(a) the bars are infinitely long; or

(b) the length of the bars is twice the diameter?

The properties of the stainless steel are $k = 41$ W/m·K, $\rho = 7865$ kg/m^3, $c = 460$ J/kg·K.

6.13 In an aluminum mill, aluminum bars, 2.5 cm x 5 cm in cross section, are extruded at 500°C as illustrated in Fig. 6.30. In order to handle a large quantity of extruded aluminum, it is necessary to prepare them for shipping soon after extrusion. This necessitates rapid cooling. A cooling system has been proposed wherein an extruded bar passes through a water spray bath after it leaves the extrusion die. At the end of the water bath the temperature of the cooled bar must be low enough to permit handling. The water spray is available at a temperature of 25°C. The heat transfer coefficient between the cooling water and the surface of the bar is 5000 $W/m^2 \cdot K$. The properties of aluminum are $k = 230$ W/m·K, $\rho = 2707$ kg/m^3 and $c = 896$ J/kg·K.

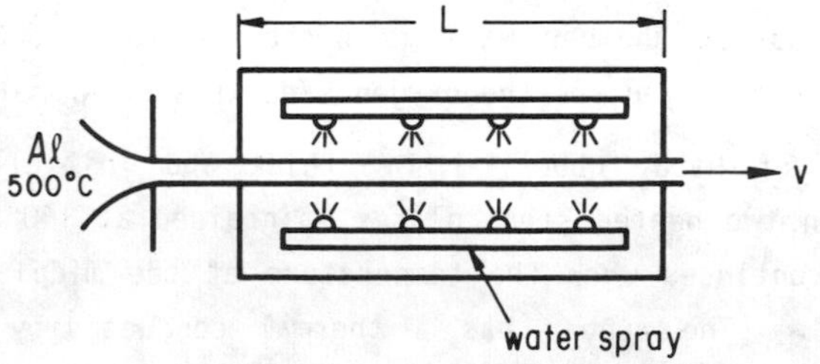

Fig. 6.30 Figure for Problem 6.13.

(a) If the bar is extruded at a velocity of 0.5 m/s, determine the length L of the cooling tank required to reduce the centerline temperature of the bar to 150°C.

(b) As the bar emerges from the cooling tank the surfaces will be cooler than the center. What is the maximum surface temperature that would occur on the extruded bar after the cooling process.

6.14 A solid steel sphere 8 cm in diameter, is heated in a furnace to a uniform temperature of 800°C. It is to be quenched by immersing in a well-stirred lead bath, maintained at the melting point, until the center reaches 510°C. At this point the sphere will be removed from the lead bath and quenched in oil. Calculate the required period of immersion in the lead bath. State and discuss the assumptions to be made. The properties of steel are k = 36.4 W/m·K, = 7753 kb/m^3, and c = 486 J/kg·K, and the melting point of lead at atmospheric pressure is 327°C.

6.15 A solid steel sphere, 10 cm in diamter, is immersed in a tank of oil maintained at 10°C. The initial temperature of the sphere is 260°C. The heat transfer coefficient h can be taken as 250 W/m^2 K. How long will it take for the center of the sphere to cool to 150°C.

6.16 A long rectangular steel rod, 10 cm x 20 cm in cross section, is initially at a uniform temperature of 16°C. At time t = 0, the temperature of its surfaces is raised to 100°C by immersing in boiling water and subsequently maintained constant at this value. Calculate the temperature at the axis of the rod after 40 seconds has elapsed.

6.17 A solid rod of radius r_o and height H is perfectly insulated against radial heat flow. The end of the rod at z = 0 is initially maintained at T_1 while the other end at z = H is at T_2. At time t = 0, the end at z = 0 is insulated against heat flow and kept insulated for times $t > 0$ while the other end is still maintained at T_2. Determine the subsequent change in the temperature distribution in the rod for times $t > 0$. Assume that the therm-physical properties of the rod are constant.

6.18 A long solid cylinder of radius r_o is initially at a uniform temperature T_i. For times $t \geq 0$, internal energy is generated in the rod at a constant rate q per unit volume while the peripheral surface at $r = r_o$ is maintained at T_i. Obtain an expression for the temperature distribution in the cylinder for times $t > 0$. Assume that the thermo-physical properties (k, ρ, c) of the material of the cylinder are constant.

6.19 An infinitely long solid cylinder of radius r_o is initially at a uniform temperature T_i. For times $t \geq 0$, a constant heat flux q_w is applied to the peripheral surface. Obtain an expression for the temperature distribution in the cylinder for $t > 0$. Assume that the thermo-physical properties (k,ρ,c) of the material of the rod are constant.

6.20 A hollow sphere $r_1 \le r \le r_2$ is initially at a uniform temperature T_i. For times t > 0, the inner boundary surface at $r = r_1$ is kept at zero temperature and the outer boundary surface at $r = r_2$ is subjected to convection into a medium at zero temperature with a constant heat transfer coefficient h. Determine the temperature distribution in the solid for times t > 0.

6.21 Consider a two-dimensional fin of length L in the x-direction and thickness 2b in the y-direction. The fin is initially at a uniform temperature T_0. At time t = 0, while the temperature of the base at x = 0 is kept at T_0, the temperatures of the outer surfaces at x = L and y = ±b are suddenly changed to T_∞ and subsequently maintained at this constant value for times t > 0. Obtain an expression for the unsteady temperature distribution T(x,y,t) in the fin for t > 0. Assume constant thermo-physical properties.

6.22 A long semi-cylinder, $0 \le r \le r_0$, $0 \le \phi \le \pi$, has an initial temperature distribution given by $F(r,\phi)$. For times t > 0, the boundary surfaces at $r = r_0$, $\phi = 0$ and $\phi = \pi$ are all maintained at a constant temperature T_w. Determine the temperature distribution $T(r,\phi,t)$ in the cylinder for times t > 0. Assume constant thermo-physical properties.

6.23 The surface temperature of a very thick wall changes suddenly from 0°C to 1500°C and then remains constant at the new value. Initially the entire wall was at a temperature of 0°C. The wall is made of concrete with the properties k = 0.814 W/m·K, c = 879 J/kg·K, and $\rho = 1906$ kg/m³.

(a) Calculate the temperature at a depth of 15 cm from the surface after 6 hours has elapsed.

(b) How long will it take for the temperature at a depth of 30 cm to reach the value calculated in part (a)?

6.24 Obtain the quasi-steady temperature distribution t(x,t) in a semi-infinite body when the surface temperature varies according to

$$T(0,t) = T_m + (\Delta T)_0 \cos (2\pi nt)$$

where n is the frequency, and T_m and $(\Delta T)_0$ are the mean and the amplitudes of the surface temperature variation, respectively.

6.25 Using the temperature distribution obtained in Problem 6.24, obtain an expression for the time lag between the temperature waves at the surface and at a depth x.

6.26 On cold days, at a certain locality, temperature of the earth's surface varies sinusoidally between 5°C and 15°C with a period of 24 hours. Calculate the amplitude of the temperature variation at a depth of 10 cm from the earth's surface. Take $\alpha = 0.0036\ m^2/hr$ for the earth's material.

7

Solutions with Integral Transforms

7.1 INTRODUCTION

In the preceding two chapters, we have discussed the method of solution of heat conduction problems by the application of the classical method of separation of variables. When separation is possible, this method turns out to be an effective and simple method to apply. In fact, separation may not even be possible for most of the problems. In this chapter, we study the method of solution of linear heat conduction problems in the rectangular, cylindrical and sperical coordinates by the application of the integral transforms that eliminate the partial derivatives with respect to space variables, such as Fourier and Hankel transforms. The method of Laplace transforms, which is also an integral transform method and is used to remove the partial derivative with respect to time variable, will be studied in the next chapter. With the integral transforms that remove space variable, inversion poses no problem simply because the inversion relation is given together with the definition of the transform at the onset of the problem. They are also equally attractive for both steady and unsteady-state problems.

We have already seen in Chapter 4 that finite integral transforms and the corresponding inversions are obtained by rewriting the Fourier expansion of an arbitrary function in two parts. Based on this approach Eringen [1] and Churchill [2] developed the theory of finite integral transforms by expanding an arbitrary function into an infinite series of the characteristic functions of different Sturm-Liouville systems, and thus derived different finite integral transforms. These transforms, following Eringen, are called *finite Sturm-Liouville transforms*. Finite Fourier transforms, finite Hankel transforms, finite Legendre transforms, etc., are all examples of finite Sturm-Liouville transforms. In the following sections we introduce the method of solution of heat conduction problems using integral transforms in terms of examples. More fundamental treatment of the theory of integral transforms can be found in the book by Sneddon [3] and the application of integral transforms to the solution of linear heat conduction problems are given in the book by Ozisik [4].

7.2 AN INTRODUCTORY EXAMPLE

As an introductory example, let us consider the plane wall shown in Fig. 7.1 which has a thickness L. Let the initial temperature distribution be $T_i(x)$. Assume that heat is generated in this wall at a rate of $\dot{q}(x,t)$ per unit volume for times $t \geq 0$, and dissipated by convection from the boundaries at $x = 0$ and $x = L$ into a surrounding medium whose temperature T_∞ varies with time. We also assume that the thermal conductivity k, thermal diffusivity α are constants and the heat transfer coefficients h_1 and h_2 are very large.

The unsteady temperature distribution $T(x,t)$ in this wall satisfies the following initial-and-boundary-value problem for $t > 0$:

$$\frac{\partial^2 T}{\partial x^2} + \frac{\dot{q}(x,t)}{k} = \frac{1}{\alpha}\frac{\partial T}{\partial t} \tag{7.1a}$$

$$T(x,0) = T_i(x) \tag{7.1b}$$

$$T(0,t) = T(L,t) = T_\infty(t) \tag{7.1c,d}$$

This problem cannot be solved by the method of separation of variables (why?). In order to illustrate the method of solution, we shall solve it by the method of integral transforms. The basic idea behind the application of this method is to eliminate the partial derivative with respect to the space variable x from the problem. We note that x changes from zero to L. In the finite interval (0,L), an integral transform $\overline{T}(\lambda_n,t)$ of the temperature distribution $T(x,t)$ with respect to the space variable x can be defined as

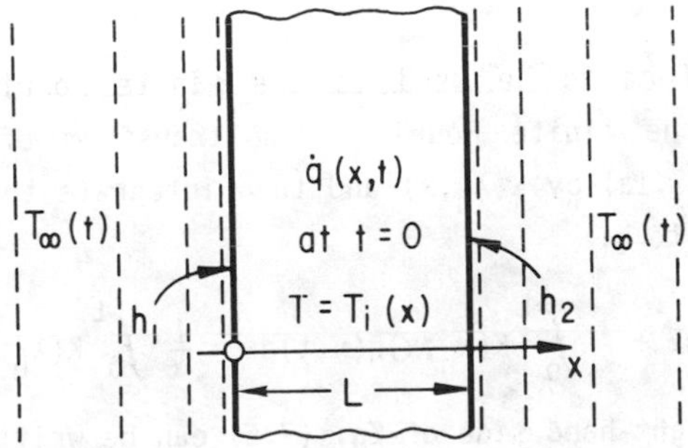

Fig. 7.1 Plane wall.

$$\overline{T}(\lambda_n,t) = \int_0^L K(\lambda_n,x)T(x,t)dx \tag{7.2a}$$

with the inversion

$$T(x,t) = \sum_{n=1}^{\infty} K(\lambda_n,x)\overline{T}(\lambda_n,t) \tag{7.2b}$$

At this stage in order to select a proper kernel $K(\lambda_n,x)$ we proceed as follows: Consider the following auxiliary problem:

$$\frac{\partial^2\psi}{\partial x^2} = \frac{1}{\alpha}\frac{\partial\psi}{\partial t} \tag{7.3a}$$

$$\psi(x,0) = T_i(x) \tag{7.3b}$$

$$\psi(0,t) = \psi(L,t) = 0 \tag{7.3c,d}$$

which is obtained by removing the non-homogeneous terms from the differential equation and the boundary conditions of the original problem (7.1). This problem can be separated into two by letting $\psi(x,t) = Y(x).\Gamma(t)$. Then, $Y(x)$ will satisfy the following characteristic-value problem:

$$\frac{d^2Y}{dx^2} + \lambda^2 Y = 0 \tag{7.4a}$$

$$Y(0) = Y(L) = 0 \tag{7.4b,c}$$

We take the kernel used in Eqs. (7.2a) and (7.2b) to be the normalized characteristic functions of this characteristic-value problem. These normalized characteristic functions are (see Section 4.6.1):

$$K(\lambda_n,x) = \sqrt{\frac{2}{L}}\sin\lambda_n x \tag{7.5a}$$

with

$$\lambda_n = \frac{n\pi}{L}, \qquad n = 1,2,3,... \tag{7.5b}$$

Therefore, the transform to be used is the finite Fourier sine transform.

We now obtain the finite Fourier sine transform of Eq. (7.1a); that is, we multiply Eq. (7.1a) by $K(\lambda_n,x)$ and then integrate the resulting expression over x from zero to L:

$$\int_0^L K(\lambda_n,x)\frac{\partial^2 T}{\partial x^2}dx + \frac{1}{k}\int_0^L K(\lambda_n,x)\dot{q}(x,t)dx = \frac{1}{\alpha}\int_0^L K(\lambda_n,x)\frac{\partial T}{\partial t}dx \tag{7.6}$$

The integral on the right-hand side of Eq. (7.6) can be written as

$$\int_0^L K(\lambda_n,x)\ \frac{\partial T}{\partial t}dx = \frac{d}{dt}\int_0^L K(\lambda_n,x)T(x,t)dt = \frac{d\overline{T}}{dt} \tag{7.7}$$

The first integral on the left-hand side of Eq.(7.6) can be evaluated as follows: Integrate it by parts

$$\int_0^L K(\lambda_n,x)\ \frac{\partial^2 T}{\partial x^2}\ dx = K(\lambda_n,L)\ \frac{\partial T}{\partial x}\Big|_{x=L} - K(\lambda_n,0)\ \frac{\partial T}{\partial x}\Big|_{x=0} - \int_0^L \frac{dK}{dx}\frac{\partial T}{\partial x}\ dx$$

$$= -\int_0^L \frac{dK}{dx}\frac{\partial T}{\partial x}\ dx \tag{7.8a}$$

One more integration by parts yields

$$\int_0^L K(\lambda_n,x)\ \frac{\partial^2 T}{\partial x^2}\ dx = -T(L,t)\ \frac{dK}{dx}\Big|_{x=L} + T(0,t)\ \frac{dK}{dx}\Big|_{x=0} + \int_0^L \frac{d^2K}{dx^2}\ T(x,t)dx \tag{7.8b}$$

Since

$$\frac{dK}{dx} = \sqrt{\frac{2}{L}}\ \lambda_n \cos \lambda_n x$$

and

$$\frac{d^2K}{dx^2} = -\lambda_n^2\ K(\lambda_n,x)$$

Eq. (7.8b) can be written as

$$\int_0^L K(\lambda_n,x)\ \frac{\partial^2 T}{\partial x^2}\ dx = \sqrt{\frac{2}{L}}\ \lambda_n \left[1-(-1)^n\right] T_\infty(t) - \lambda_n^2\ \overline{T}(\lambda_n,t) \tag{7.8c}$$

When Eqs. (7.7) and (7.8c) are substituted into Eq. (7.6) we get

$$\frac{1}{\alpha}\frac{d\overline{T}}{dt} + \lambda_n^2\ \overline{T}(\lambda_n,t) = \sqrt{\frac{2}{L}}\ \lambda_n \left[1-(-1)^n\right] T_\infty(t) + \frac{1}{k}\ \dot{q}(\lambda_n,t) \tag{7.9}$$

where $\dot{q}(\lambda_n,t)$ is the finite Fourier sine transform of $\dot{q}(x,t)$; that is,

$$\overline{\dot{q}}(\lambda_n,t) = \int_0^L K(\lambda_n,x)\ \dot{q}(x,t)\ dx \tag{7.10}$$

Thus, we have reduced the original partial differential equation (7.1a) to an ordinary differential equation (7.9) by eliminating the space variable x. Solution of Eq. (7.9) can be found as follows: Multiply both sides of this equation by $\exp(\alpha\lambda_n^2 t)$ and then integrate the resulting expression from zero to t,

$$\int_0^t e^{\alpha\lambda_n^2 t'}\left[\frac{1}{\alpha}\frac{d\overline{T}}{dt'} + \lambda_n^2\,\overline{T}(\lambda_n,t')\right]dt' = \int_0^t e^{\alpha\lambda_n^2 t'}\;F(\lambda_n,t')\;dt' \tag{7.11}$$

where

$$F(\lambda_n,t') = \sqrt{\frac{2}{L}}\;\lambda_n\left[1-(-1)^n\right]T_\infty(t') + \frac{1}{k}\,\dot{q}(\lambda_n,t') \tag{7.12}$$

Equation (7.11) can be rewritten as

$$\frac{1}{\alpha}\int_0^t \frac{d}{dt'}\left[e^{\alpha\lambda_n^2 t'}\;\overline{T}(\lambda_n,t')\right]dt' = \int_0^t e^{\alpha\lambda_n^2 t'}\;F(\lambda_n,t')\;dt' \tag{7.13}$$

which yields

$$\overline{T}(\lambda_n,t) = e^{-\alpha\lambda_n^2 t}\left[\overline{T}(\lambda_n,0) + \alpha\int_0^t e^{\alpha\lambda_n^2 t'}\;F(\lambda_n,t')\;dt'\right] \tag{7.14}$$

where $\overline{T}(\lambda_n,0)$ is the finite Fourier sine transform of the initial conditon:

$$\overline{T}(\lambda_n,0) = \int_0^L K(\lambda_n,x)\;T_i(x)dx \tag{7.15}$$

Now, $\overline{T}(\lambda_n,t)$ given by Eq. (7.14) can be inverted by using the inversion formula (7.2b) to obtain the temperature distribution $T(x,t)$ as follows:

$$T(x,t) = \sum_{n=1}^{\infty} K(\lambda_n,x)e^{-\alpha\lambda_n^2 t}\left[\overline{T}(\lambda_n,0) + \alpha\int_0^t e^{\alpha\lambda_n t'}\;F(\lambda_n,t')\;dt'\right] \tag{7.16}$$

which can be rewritten as

$$\begin{aligned} T(x,t) = \frac{2}{L}\sum_{n=1}^{\infty} e^{-\alpha(\frac{n\pi}{L})^2 t}\;\sin\frac{n\pi}{L}x\;\Bigg\{&\int_0^L T_i(x')\;\sin\frac{n\pi}{L}x'dx' \\ &+ \alpha\int_0^t e^{\alpha(\frac{n\pi}{L})^2 t'}\Big[\frac{n\pi}{L}\,[1-(-1)^n]\;T_\infty(t') \\ &+ \frac{1}{k}\int_0^L \dot{q}(x',t')\;\sin\frac{n\pi}{L}x'\;dx'\Big]\;dt'\Bigg\} \end{aligned} \tag{7.17}$$

Here it should be noted that this solution is valid only in the interval $0 < x < L$, but not at $x = 0$ and $x = L$. This is due to the fact that the solution (7.17) is a series solution in terms of the functions $\sin\frac{n\pi}{L}x$, $n = 1,2,3,\ldots$, which vanishes both at $x = 0$ and at $x = L$.

The foregoing example illustrates the steps in solving heat conduction problems by integral transforms. The procedure is straightforward. The determination of the kernel of the transform poses no problem. In the examples of the following sections we shall see that the auxiliary characteristic-value problem can immediately be written down merely by looking at the formulation of the original problem.

7.3 FINITE FOURIER TRANSFORMS

The finite Fourier sine and cosine transforms that we developed in Chapter 4 can be used in the analysis of heat conduction problems in rectangular (or spherical) coordinates with simple boundary conditions. We have seen such an application of the finite Fourier sine transform in the previous section. These two transforms were developed by considering the characteristic values and characteristic functions of two special cases of the following Sturm-Liouville problem:

$$\frac{d^2y}{dx^2} + \lambda^2 y = 0 \tag{7.18a}$$

$$\alpha_1 y(0) + \beta_1 \frac{dy(0)}{dx} = 0 \tag{7.18b}$$

$$\alpha_2 y(L) + \beta_2 \frac{dy(L)}{dx} = 0 \tag{7.18c}$$

The finite Fourier sine and cosine transforms were developed by taking $\beta_1 = \beta_2 = 0$ and $\alpha_1 = \alpha_2 = 0$, respectively. Similar transforms can be developed by considering the characteristic values and characteristic functions of different combinations of the boundary conditions of this Sturm-Liouville problem. In fact, there are a total of nine different combinations of the boundary conditions.

Since the system (7.18) is a special case of the Sturm-Liouville system (4.13) with

$$p(x) = 1, \qquad q(x) = 0, \qquad \text{and} \qquad w(x) = 1$$

the characteristic functions $\phi_n(x)$ form a complete orthogonal set with respect to the weight function unity in the interval (0,L). As we did in Section 4.6, the expansion (4.27) of an arbitrary function f(x) in the interval (0,L) into a series of these characteristic functions can be rearranged as the integral transform of the function f(x) in the form as

$$\overline{f}(\lambda_n) = \int_0^L f(x)K(\lambda_n,x)\,dx \tag{7.19a}$$

with the inversion

$$f(x) = \sum_{n=0}^{\infty} \bar{f}(\lambda_n) K(\lambda_n, x) \tag{7.19b}$$

where the kernel $K(\lambda_n, x)$ is defined as

$$K(\lambda_n, x) = \frac{\phi_n(x)}{\sqrt{N_n}} \tag{7.19c}$$

with

$$N_n = \int_0^L \left[\phi_n(x) \right]^2 dx \tag{7.19d}$$

In Chapter 4 (see Table 4.1) the characteristic functions of the system (7.18) were found to be

$$\phi_n(x) = \lambda_n \cos \lambda_n x - H_1 \sin \lambda_n x \tag{7.20}$$

where the characteristic values are the positive zeros of the transcendental equation

$$\tan \lambda L = \frac{\lambda(H_2 - H_1)}{\lambda^2 + H_1 H_2} \tag{7.21}$$

with

$$H_1 = \frac{\alpha_1}{\beta_1} \qquad \text{and} \qquad H_2 = \frac{\alpha_2}{\beta_2}$$

Furthermore, the norm N_n is given by

$$N_n = \frac{1}{2}\left[(\lambda_n^2 + H_1^2)\left(L + \frac{H_2}{\lambda_n^2 + H_2^2} \right) - H_1 \right] \tag{7.22}$$

Now the kernel (7.19c) can be written as

$$K(\lambda_n, x) = \sqrt{2}\, \frac{\lambda_n \cos \lambda_n x - H_1 \sin \lambda_n x}{\left[(\lambda_n^2 + H_1^2)\left(L + \dfrac{H_2}{\lambda_n^2 + H_2^2} \right) - H_1 \right]^{1/2}} \tag{7.23}$$

In this section, we have obtained the kernel of the finite transform for the case where $\alpha_1 \neq 0$, $\alpha_2 \neq 0$, $\beta_1 \neq 0$, and $\beta_2 \neq 0$. The kernels for the remaining eight cases can be obtained following the same procedure. In fact, they can also be obtained easily from Eq. (7.23) as limiting cases. The kernels for all of these nine cases were evaluated and tabulated by Ozisik

[4], and they are summarized here in Table 7.1.

These transforms are called finite Fourier transforms simply because they are derived from the Fourier expansion of an arbitrary function f(x) in the finite interval (0,L).

7.4 FOURIER TRANSFORMS IN THE SEMI-INFINITE INTERVAL

In the previous section, we have introduced Fourier transforms in the finite interval (0,L). These can be used for the solution of heat conduction problems in finite regions. For problems in the semi-infinite region we need transforms in the interval $(0,\infty)$. Such transforms can be obtained from finite Fourier transforms by taking limits as $L \to \infty$.

Consider, for example the finite Fourier sine transform

$$\bar{f}_n = \sqrt{\frac{2}{L}} \int_0^L f(x) \sin \frac{n\pi}{L} x \, dx \tag{7.24a}$$

with the inversion

$$f(x) = \sqrt{\frac{2}{L}} \sum_{n=1}^{\infty} \bar{f}_n \sin \frac{n\pi}{L} x \tag{7.24b}$$

Substitution of Eq. (7.24a) into Eq. (7.24b) results in

$$f(x) = \frac{2}{L} \sum_{n=1}^{\infty} \int_0^L f(x') \sin \frac{n\pi}{L} x' \, dx' \cdot \sin \frac{n\pi}{L} x \tag{7.24c}$$

which is the Fourier sine expansion of f(x) in the interval (0,L). Calling

$$\Delta\beta = \frac{\pi}{L}, \qquad \text{and} \qquad \beta = n\Delta\beta$$

Eq. (7.24c) can be rewritten as

$$f(x) = \frac{2}{\pi} \sum_{n=1}^{\infty} \int_0^L f(x') \sin (n\Delta\beta) x' \, dx' \cdot \sin (n\Delta\beta) x \cdot \Delta\beta \tag{7.25}$$

As $L \to \infty$ (or $\Delta\beta \to 0$), Eq. (7.25) can be written as

$$f(x) = \lim_{\Delta\beta \to 0} \sum_{n=1}^{\infty} g(n\Delta\beta) \cdot \sin (n\Delta\beta) x \cdot \Delta\beta \tag{7.26a}$$

where we have written

$$g(\beta) = \frac{2}{\pi} \int_0^{\infty} f(x') \sin \beta x' \, dx' \tag{7.26b}$$

Table 7.1 Characteristic values and kernels for use in finite Fourier Transforms.

(Integral Transform) $\bar{f}(\lambda_n) = \int_0^L K(\lambda_n,x)f(x)dx$

(Inversion Formula) $f(x) = \sum_{n=1}^{\infty} K(\lambda_n,x)\bar{f}(\lambda_n)$

(Characteristic Value Problem)

$$\frac{d^2y}{dx^2} + \lambda^2 y = 0$$

$$\alpha_1 y(0) + \beta_1 \frac{dy(0)}{dx} = 0$$

$$\alpha_2 y(L) + \beta_2 \frac{dy(L)}{dx} = 0$$

B.C. at x = 0	B.C. at x = L	Kernel* $K(\lambda_n,x)$	Characteristic values λ_n are the positive roots of*
Third kind ($\alpha_1 \neq 0$, $\beta_1 \neq 0$)	Third kind ($\alpha_2 \neq 0$, $\beta_2 \neq 0$)	$\sqrt{2}\,\dfrac{\lambda_n \cos\lambda_n x - H_1 \sin\lambda_n x}{\left[(\lambda_n^2 + H_1^2)\left(L + \dfrac{H_2}{\lambda_n^2 + H_1^2}\right) - H_1\right]^{1/2}}$	$\tan\lambda L = \dfrac{\lambda(H_2 - H_1)}{\lambda^2 + H_1 H_2}$
Third kind ($\alpha_1 \neq 0$, $\beta_1 \neq 0$)	Second kind ($\alpha_2 = 0$, $\beta_2 \neq 0$)	$\sqrt{2\lambda_n}\,\dfrac{\cos\lambda_n(L-x)}{[\lambda_n L + \sin\lambda_n L \cos\lambda_n L]^{1/2}}$	$\lambda\tan\lambda L = -H_1$
Third kind ($\alpha_1 \neq 0$, $\beta_1 \neq 0$)	First kind ($\alpha_2 \neq 0$, $\beta_2 = 0$)	$\sqrt{2\lambda_n}\,\dfrac{\sin\lambda_n(L-x)}{[\lambda_n L - \sin\lambda_n L \cos\lambda_n L]^{1/2}}$	$\lambda\cot\lambda L = H_1$
Second kind ($\alpha_1 = 0$, $\beta_1 \neq 0$)	Third kind ($\alpha_2 \neq 0$, $\beta_2 \neq 0$)	$\sqrt{2\lambda_n}\,\dfrac{\cos\lambda_n x}{[\lambda_n L + \sin\lambda_n L \cos\lambda_n L]^{1/2}}$	$\lambda\tan\lambda L = H_2$

Table 7.1 (Continued).

B.C. at x = 0	B.C. at x = L	Kernel* $K(\lambda_n,x)$	Characteristic values λ_n are the positive roots of*
Second kind ($\alpha_1 = 0, \beta_1 \neq 0$)	Second kind ($\alpha_2 = 0, \beta_2 \neq 0$)	$\sqrt{\frac{2}{L}} \cos\lambda_n x$**	$\sin\lambda L = 0$ ($\lambda_n = \frac{n\pi}{L}$, n=0,1,2...)
Second kind ($\alpha_1 = 0, \beta_1 \neq 0$)	First kind ($\alpha_2 \neq 0, \beta_2 = 0$)	$\sqrt{\frac{2}{L}} \cos\lambda_n x$	$\cos\lambda L = 0$ ($\lambda_n = \frac{2n-1}{L}\frac{\pi}{2}$, n=1,2,3...)
First kind ($\alpha_1 \neq 0, \beta_2 = 0$)	Third kind ($\alpha_2 \neq 0, \beta_2 \neq 0$	$\sqrt{2\lambda_n} \frac{\sin\lambda_n x}{[\lambda_n L - \sin\lambda_n L \cos\lambda_n L]^{1/2}}$	$\lambda\cot\lambda L = -H_2$
First kind ($\alpha_1 \neq 0, \beta_1 = 0$)	Second kind ($\alpha_2 = 0, \beta_2 \neq 0$)	$\sqrt{\frac{2}{L}} \sin\lambda_n x$	$\cos\lambda L = 0$ ($\lambda_n = \frac{2n-1}{L}\frac{\pi}{2}$, n=1,2,3...)
First kind ($\alpha_1 \neq 0, \beta_1 = 0$)	First kind ($\alpha_2 \neq 0, \beta_2 = 0$)	$\sqrt{\frac{2}{L}} \sin\lambda_n x$	$\sin\lambda L = 0$ ($\lambda_n = \frac{n\pi}{L}$, n=1,2,3...)

*$H_1 = \alpha_1/\beta_1$, $H_2 = \alpha_2/\beta_2$.

**When n=0 replace $\sqrt{\frac{2}{L}}$ by $\sqrt{\frac{1}{L}}$.

But the limit (7.26a) is by definition equal to

$$f(x) = \int_0^\infty g(\beta) \sin \beta x \, d\beta \tag{7.26c}$$

Substitution of Eq. (7.26b) into Eq. (7.26c) yields

$$f(x) = \frac{2}{\pi} \int_0^\infty \sin \beta x \left(\int_0^\infty f(x') \sin \beta x' \, dx' \right) d\beta \tag{7.27}$$

Expression (7.27) is called the *Fourier sine integral representation* of f(x). It is valid for x > 0, if f(x) and df/dx are piecewise continuous over each bounded interval of the semi-infinite interval x > 0 and f(x) is absolutely integrable from zero to ∞. Like the Fourier sine expansion, Eq. (7.27) represents f(x) at points of continuity in the interval (0,∞) and gives $\frac{1}{2}[f(x^+) + f(x^-)]$ at points where f(x) has finite jumps. When f(x) is an odd function Eq. (7.27) is valid for all values of x (Why?).

The Fourier sine integral representation of f(x) can be rearranged as the integral transform of f(x) in the interval (0,∞) as

$$\bar{f}(\beta) = \sqrt{\frac{2}{\pi}} \int_0^\infty f(x) \sin \beta x \, dx \tag{7.28a}$$

with the inversion

$$f(x) = \sqrt{\frac{2}{\pi}} \int_0^\infty \bar{f}(\beta) \sin \beta x \, d\beta \tag{7.28b}$$

Equation (7.28a) is known as the *Fourier sine transform* of f(x) in the interval (0,∞) and Eq. (7.28b) is the corresponding inversion formula. The kernel of the tranform is

$$K(\beta,x) = \sqrt{\frac{2}{\pi}} \sin \beta x \tag{7.28c}$$

and it satisfies the following characteristic-value problem

$$\frac{d^2y}{dx^2} + \beta^2 y = 0 \tag{7.29a}$$

$$y(0) = 0 \qquad \text{and} \qquad |y(\infty)| \leq 1 \tag{7.29b,c}$$

for all values of β from zero to ∞. In this case the characteristic values are continuous rather than discrete.

The transform $\bar{f}(\beta)$ exists as a continuous function of the parameter β if f(x) is continuous over each bounded interval of the semi-infinite interval x > 0 and absolutely integrable from zero to ∞.

Table 7.2 Kernels for use in Fourier Transforms in semi-infinite interval.

$$\bar{f}(\beta) = \int_0^\infty K(\beta,x)\, f(x)\, dx \qquad \frac{d^2y}{dx^2} + \beta^2 y = 0$$

$$f(x) = \int_0^\infty K(\beta,x)\, \bar{f}(\beta)\, d\beta \qquad \alpha_1 y(0) + \beta_1 \frac{dy(0)}{dx} = 0$$

$$|y(x)| \leq 1$$

B.C. at x = 0	Kernel $K(\beta,x)$*
Third kind ($\alpha_1 \neq 0,\ \beta_1 \neq 0$)	$\sqrt{\frac{2}{\pi}}\ \frac{\beta \cos \beta x - H \sin \beta x}{\sqrt{\beta^2 + H^2}}$
Second kind ($\alpha_1 = 0$)	$\sqrt{\frac{2}{\pi}} \cos \beta x$
First kind ($\beta_1 = 0$)	$\sqrt{\frac{2}{\pi}} \sin \beta x$

$^*H_1 = \alpha_1/\beta_1$.

Similar integral transforms in the interval $(0,\infty)$ can be developed by considering other series expansions. Table 7.2 gives a summary of such transforms for three different kinds of boundary conditions at $x = 0$ that the kernels satisfy.

7.5 UNSTEADY-STATE PROBLEMS IN RECTANGULAR COORDINATES

In this section we examine the solution of unsteady heat conduction problems in the rectangular coordinate system by the application of Fourier transforms. We have already seen the method of solution by solving an unsteady one-dimensional problem in Section 7.2. For two and three-dimensional unsteady problems the method of solution is again the same: Partial derivatives with respect to space variables are eliminated from the problem by the repeated application of Fourier transforms. Thus an ordinary differential equation with an initial condition is obtained, the solution of which yields the multiple transform of the temperature distribution. The temperature distribution is then obtained by multiple inversion of the transform.

Let us now consider the semi-infinite rectangular strip shown in Fig. 7.2, which is initially at temperature $T_i(x,y)$. The surface at $x = 0$ is insulated. For times $t > 0$ the surfaces at $x = a$ and at $y = 0$ are kept at temperatures $T_1(y,t)$ and $T_2(x,t)$, respectively, which are functions of both

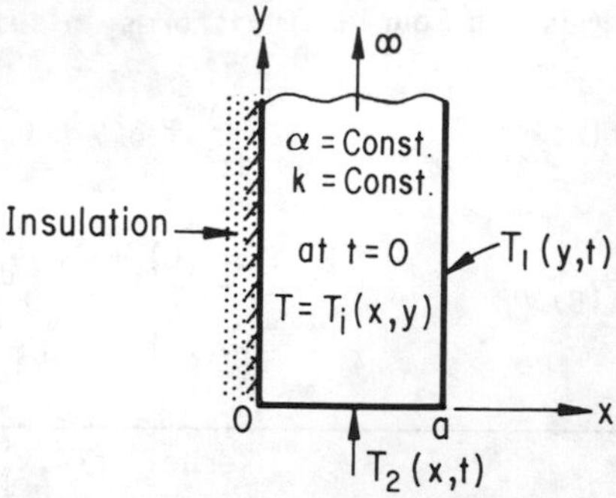

Fig. 7.2 Semi-infinite rectangular strip.

time and position along the surfaces. We wish to find the temperature distribution in the strip for $t > 0$.

Formulation of the problem for $t > 0$ in terms of $T = T(x,y,t)$ is given by

$$\frac{\partial^2 T}{\partial x^2} + \frac{\partial^2 T}{\partial y^2} = \frac{1}{\alpha}\frac{\partial T}{\partial t} \tag{7.30a}$$

$$T(x,y,0) = T_i(x,y) \tag{7.30b}$$

$$\frac{\partial T(0,y,t)}{\partial x} = 0; \qquad T(a,y,t) = T_1(y,t) \tag{7.30c,d}$$

$$T(x,0,t) = T_2(x,t) \tag{7.30e}$$

We note that the range of the space variable x is finite; that is, it changes from zero to a. In this finite interval (0,a), the integral transform $\overline{T}(\lambda_n,y,t)$ of the temperature distribution $T(x,y,t)$ with respect to the space variable x can be defined as

$$\overline{T}(\lambda_n,y,t) = \int_0^a K(\lambda_n,x)\, T(x,y,t)\, dx \tag{7.31a}$$

with the inversion

$$T(x,y,t) = \sum_{n=1}^{\infty} K(\lambda_n,x)\, \overline{T}(\lambda_n,y,t) \tag{7.31b}$$

The kernel $K(\lambda_n,x)$ to be used in this transform can be determined by following the same procedure as in Section 7.2. Just by looking at the problem (7.30), especially the partial derivative with respect to the variable x and the types of the boundary conditions in the x-direction, we see that the kernel should be the normalized characteristic function of the following characteristic-value problem:

$$\frac{d^2y}{dx^2} + \lambda^2 y(x) = 0 \tag{7.32a}$$

$$\frac{dy(0)}{dx} = 0 \qquad \text{and} \qquad y(a) = 0 \tag{7.32b}$$

The kernel and the characteristic values can be taken from Table 7.1. They are,

$$K(\lambda_n, x) = \sqrt{\frac{2}{a}} \cos \lambda_n x \tag{7.33a}$$

with

$$\lambda_n = \frac{(2n-1)\pi}{2a}, \qquad n = 1,2,3,\ldots \tag{7.33b}$$

We now obtain the transform of Eq. (7.30a) with respect to variable x as

$$\int_0^a K(\lambda_n, x)\left[\frac{\partial^2 T}{\partial x^2} + \frac{\partial^2 T}{\partial y^2}\right] dx = \frac{1}{\alpha}\int_0^a K(\lambda_n, x)\frac{\partial T}{\partial t}\, dx \tag{7.34a}$$

which can be rewritten in the form

$$\int_0^a K(\lambda_n, x)\frac{\partial^2 T}{\partial x^2}\, dx + \frac{\partial^2 \overline{T}}{\partial y^2} = \frac{1}{\alpha}\frac{\partial \overline{T}}{\partial t} \tag{7.34b}$$

The integral on the left-hand side can be evaluated by integrating it by parts twice, resulting in

$$\int_0^a K(\lambda_n, x)\frac{\partial^2 T}{\partial x^2}\, dx = \sqrt{\frac{2}{a}}\ \lambda_n(-1)^n T_1(y,t) - \lambda_n^2 \overline{T}(\lambda_n, y, t,) \tag{7.35}$$

Substitution of Eq. (7.35) into Eq. (7.34b) gives

$$\frac{\partial^2 \overline{T}}{\partial y^2} - \lambda_n^2 \overline{T}(\lambda_n, y, t) + (-1)^n \sqrt{\frac{2}{a}}\ \lambda_n T_1(y,t) = \frac{1}{\alpha}\frac{\partial \overline{T}}{\partial t} \tag{7.36}$$

Equation (7.36) is still a partial differential equation. We now remove the space variable y from Eq. (7.36). The range of the variable y is $(0,\infty)$. In this semi-infinite region, the integral transform $T(\lambda_n, \beta, t)$ of the function $\overline{T}(\lambda_n, y, t)$ with respect to the variable y can be defined as

$$\overline{\overline{T}}(\lambda_n, \beta, t) = \int_0^\infty K(\beta, y)\overline{T}(\lambda_n, y, t)\, dy \tag{7.37a}$$

with the inversion

$$\bar{T}(\lambda_n,y,t) = \int_0^\infty K(\beta,y)\bar{\bar{T}}(\lambda_n,\beta,t)d\beta \tag{7.37b}$$

where the kernel $K(\beta,y)$ is taken to be the characteristic functions of the following characteristic-value problem:

$$\frac{d^2\phi}{dy^2} + \beta^2\phi = 0 \tag{7.38a}$$

$$\phi(0) = 0 \quad \text{and} \quad |\phi(\infty)| \leq 1 \tag{7.38b,c}$$

Note that, in writing the auxiliary characteristic-value problem (7.38), the homogeneous boundary condition at $y = 0$ is taken to be of the same type as the boundary condition (7.30e) at $y = 0$ of the original problem; that is, they are both of the first kind. In fact, the kernel $K(\beta,y)$ can be taken from Table 7.2, and it is

$$K(\beta,y) = \sqrt{\frac{2}{\pi}} \sin \beta y$$

We also note that the kernel $K(\beta,y)$ satisfies the problem (7.38) for all values of β from zero to infinity.

We now take the transform of Eq. (7.36) with respect to variable y:

$$\int_0^\infty K(\beta,y) \frac{\partial^2\bar{T}}{\partial y^2} dy - \lambda_n^2 \int_0^\infty K(\beta,y)\bar{T}(\lambda_n,y,t)dy$$

$$+ (-1)^n\sqrt{\frac{2}{a}} \lambda_n \int_0^\infty K(\beta,y)T_1(y,t)dy$$

$$= \frac{1}{\alpha} \int_0^\infty K(\beta,y) \frac{\partial\bar{T}}{\partial t} dy \tag{7.39a}$$

which can also be written as

$$\int_0^\infty K(\beta,y) \frac{\partial^2\bar{T}}{\partial y^2} dy - \lambda_n^2\bar{\bar{T}}(\lambda_n,\beta,t) + (-1)^n\sqrt{\frac{2}{a}} \lambda_n\bar{T}_1(\beta,t) = \frac{1}{\alpha}\frac{d\bar{\bar{T}}}{dt} \tag{7.39b}$$

where we have defined

$$\bar{T}_1(\beta,t) = \int_0^\infty K(\beta,y)T_1(y,t)dy \tag{7.39c}$$

The integral on the left-hand side of Eq. (7.39b) can be evaluated by integrating it by parts twice:

$$\int_0^\infty K(\beta,y)\,\frac{\partial^2 \bar{T}}{\partial y^2}\,dy = -\beta^2 \bar{\bar{T}}(\lambda_n,\beta,t) - \sqrt{\frac{2}{\beta}}\;\beta\,\bar{T}_2(\lambda_n,t) \tag{7.40}$$

In writing Eq. (7.40) we assumed that the temperature and its first derivative with respect to y vanish as $y \to \infty$. Substitution of Eq. (7.40) into Eq. (7.39b) gives

$$\frac{d\bar{\bar{T}}}{dt} + \alpha(\beta^2 + \lambda_n^2)\bar{\bar{T}}(\lambda_n,\beta,t) = F(\lambda_n,\beta,t) \tag{7.41a}$$

where

$$F(\lambda_n,\beta,t) = -\alpha\left[\sqrt{\frac{2}{\beta}}\;\beta\,\bar{T}_2(\lambda_n,t) - (-1)^n\sqrt{\frac{2}{a}}\;\lambda_n\,\bar{T}_1(\beta,t)\right] \tag{7.41b}$$

The first-order differential equation (7.41a) for the double transform $\bar{\bar{T}}(\lambda_n,\beta,t)$ can be solved easily and the solution is

$$\bar{\bar{T}}(\lambda_n,\beta,t) = e^{-\alpha(\beta^2+\lambda_n^2)t}\left[\bar{\bar{T}}_i(\lambda_n,\beta) + \int_0^t e^{\alpha(\beta^2+\lambda_n^2)t'}F(\lambda_n,\beta,t')\,dt'\right] \tag{7.42a}$$

where

$$\bar{\bar{T}}_i(\lambda_n,\beta) = \int_0^a K(\lambda_n,x)\left(\int_0^\infty K(\beta,y)T_i(x,y)\,dy\right)dx \tag{7.42b}$$

Inverting this solution twice, first with the inversion formula (7.37b) and then with (7.31b), we get the solution for $T(x,y,t)$:

$$T(x,y,t) = \sum_{n=1}^\infty K(\lambda_n,x)\left\{\int_0^\infty K(\beta,y)\,e^{-\alpha(\beta^2+\lambda_n^2)t}\cdot\left[\bar{\bar{T}}_i(\lambda_n,\beta) + \int_0^t e^{\alpha(\beta^2+\lambda_n^2)t'}F(\lambda_n,\beta,t')\,dt'\right]d\beta\right\} \tag{7.43}$$

where all the terms have been defined previously.

This example demonstrates the procedure to solve a two-dimensional unsteady problem with the application of Fourier transforms. The method of solution of three-dimensional problems is exactly the same: By repeated application of Fourier transforms three times the problem is reduced to an ordinary differential equation for the transform of the temperature distribution with an initial condition. Solution of this ordinary differential equation yields the transform of the temperature distribution which can easily be inverted through the use of inversion relations.

7.6 STEADY TWO- AND THREE-DIMENSIONAL PROBLEMS IN RECTANGULAR COORDINATES

Steady two- and three-dimensional heat conduction problems in rectangular coordinates can also be solved by applying Fourier transforms. The procedure is the same as in solving unsteady problems; the partial derivatives with respect to space variables are removed by repeated application of Fourier transforms and thus the partial differential equation is reduced to an algebraic equation which is then inverted to find the temperature distribution. We may not prefer, however, to remove all the partial derivatives but leave one space variable and thus reduce the original partial differential equation to an ordinary differential equation. Solution of this differential equation gives the transform of the temperature distribution, which is then inverted to find the temperature distribution. We now solve one example problem in both ways to explain the method of solution.

Let us re-solve the problem discussed in Section 5.2 by applying Fourier transforms. The formulation of the problem is

$$\frac{\partial^2 T}{\partial x^2} + \frac{\partial^2 T}{\partial y^2} = 0 \tag{7.44a}$$

$$T(0,y) = T(a,y) = 0 \tag{7.44b,c}$$

$$T(x,0) = 0 , \qquad T(x,b) = f(x) \tag{7.44d,e}$$

In the finite interval (0,a), the Fourier transform of the temperature distribution $T(x,y)$ with respect to variable x can be defined as

$$\overline{T}(\lambda_n,y) = \int_0^a K(\lambda_n,x)T(x,y)dx \tag{7.45a}$$

with the inversion formula

$$T(x,y) = \sum_{n=1}^{\infty} K(\lambda_n,x)\overline{T}(\lambda_n,y) \tag{7.45b}$$

By examining the boundary conditions of the problem at $x = 0$ and $x = a$ we take the kernel from Table 7.1 as

$$K(\lambda_n,x) = \sqrt{\frac{2}{a}}\ \sin \lambda_n x \tag{7.45c}$$

with

$$\lambda_n = \frac{n\pi}{a} , \qquad n = 1,2,3,... \tag{7.45d}$$

We now obtain the transform of the partial differential equation (7.44a) by means of Eq. (7.45a):

$$\int_0^a K(\lambda_n,x)\,\frac{\partial^2 T}{\partial x^2}\,dx + \int_0^a K(\lambda_n,x)\,\frac{\partial^2 T}{\partial y^2}\,dx = 0 \tag{7.46a}$$

which can be written as

$$\int_0^a K(\lambda_n,x)\,\frac{\partial^2 T}{\partial x^2}\,dx + \frac{d^2\overline{T}}{dy^2} = 0 \tag{7.46b}$$

Integrating by parts twice, it can easily be shown that the integral on the left-hand side of Eq. (7.46b) is given by

$$\int_0^a K(\lambda_n,x)\,\frac{\partial^2 T}{\partial x^2}\,dx = -\lambda_n^2\overline{T}(\lambda_n,y) \tag{7.47}$$

Substitution of Eq. (7.47) into Eq. (7.46b) yields

$$\frac{d^2\overline{T}}{dy^2} - \lambda_n^2\overline{T}(\lambda_n,y) = 0 \tag{7.48}$$

Equation (7.48) is a second-order ordinary differential equation, the general solution of which can be written as

$$\overline{T}(\lambda_n,y) = A_n \sinh \lambda_n y + B_n \cosh \lambda_n y \tag{7.49}$$

The transforms of the boundary conditions at $y = 0$ and at $y = b$ yield

$$\overline{T}(\lambda_n,0) = \int_0^a K(\lambda_n,x)T(x,0)\,dx = 0 \tag{7.50a}$$

and

$$\overline{T}(\lambda_n,b) = \int_0^a K(\lambda_n,x)T(x,b)\,dx = \int_0^a K(\lambda_n,x)f(x)\,dx = \overline{f}(\lambda_n) \tag{7.50b}$$

Application of the condition (7.50a) gives $B_n \equiv 0$ in Eq. (7.49) and the condition (7.50b) yields

$$A_n = \frac{\overline{f}(\lambda_n)}{\sinh \lambda_n b}$$

Therefore, the transform of the temperature distribution is given by

$$\overline{T}(\lambda_n,y) = \overline{f}(\lambda_n)\,\frac{\sinh \lambda_n y}{\sinh \lambda_n b} \tag{7.51}$$

Equation (7.51) can now be inverted to find the temperature distribution:

$$T(x,y) = \sum_{n=1}^{\infty} K(\lambda_n,x)\overline{f}(\lambda_n)\,\frac{\sinh \lambda_n y}{\sinh \lambda_n b} \tag{7.52a}$$

which can also be written as

$$T(x,y) = \frac{2}{a}\sum_{n=1}^{\infty}\left(\int_0^a f(x')\sin\frac{n\pi}{a}x'\,dx'\right)\frac{\sinh\frac{n\pi}{a}y}{\sinh\frac{n\pi}{a}b}\sin\frac{n\pi}{a}x \tag{7.52b}$$

This result is the same as Eq. (5.16).

As an alternative approach, we define the Fourier transform of $\bar{T}(\lambda_n,y)$ with respect to variable y in the finite interval (0,b) as

$$\bar{\bar{T}}(\lambda_n,\beta_m) = \int_0^b K(\beta_m,y)\bar{T}(\lambda_n,y)\,dy \tag{7.53a}$$

with the inversion

$$\bar{T}(\lambda_n,y) = \sum_{m=1}^{\infty} K(\beta_m,y)\bar{\bar{T}}(\lambda_n,\beta_m) \tag{7.53b}$$

where the kernel $K(\beta_m,y)$ is given by

$$K(\beta_m,y) = \sqrt{\frac{2}{b}}\,\sin\beta_m y \tag{7.54a}$$

with the characteristic values

$$\beta_m = \frac{m\pi}{b}, \qquad m = 1,2,3,\ldots \tag{7.54b}$$

In taking the kernel $K(\beta_m,y)$ and the characteristic values β_m from Table 7.1, we have made use of the boundary conditions at y = 0 and y = b of the original problem.

We now obtain the transform of Eq. (7.48) with respect to variable y as follows:

$$\int_0^b K(\beta_m,y)\frac{d^2\bar{T}}{dy^2}\,dy - \lambda_n^2\int_0^b K(\beta_m,y)\bar{T}(\lambda_n,y)\,dy = 0 \tag{7.55a}$$

which can be written as

$$\int_0^b K(\beta_m,y)\frac{d^2\bar{T}}{dy^2}\,dy - \lambda_n^2\bar{\bar{T}}(\lambda_n,\beta_m) = 0 \tag{7.55b}$$

The integral in Eq. (7.55b) can be shown to be given by

$$\int_0^b K(\beta_m,y)\frac{d^2\bar{T}}{dy^2}\,dy = -\beta_m^2\bar{\bar{T}}(\lambda_n,\beta_m) + (-1)^{m+1}\sqrt{\frac{2}{b}}\,\beta_m\,\bar{f}(\lambda_n) \tag{7.56}$$

Substitution of Eq. (7.56) into Eq. (7.55b) yields the following algebraic equation:

$$(\lambda_n^2 + \beta_m^2)\bar{\bar{T}}(\lambda_n,\beta_m) = (-1)^{m+1}\sqrt{\frac{2}{b}}\ \beta_m\ \bar{f}(\lambda_n) \tag{7.57a}$$

or

$$\bar{\bar{T}}(\lambda_n,\beta_m) = (-1)^{m+1}\sqrt{\frac{2}{b}}\,\frac{\beta_m\bar{f}(\lambda_n)}{\lambda_n^2 + \beta_m^2} \tag{7.57b}$$

Inversion of Eq. (7.57b), through Eq. (7.53b), yields

$$\bar{T}(\lambda_n,y) = \sqrt{\frac{2}{b}}\sum_{m=1}^{\infty}(-1)^{m+1}K(\beta_m,y)\frac{\beta_m\bar{f}(\lambda_n)}{\lambda_n^2 + \beta_m^2} \tag{7.58}$$

and the inversion of Eq. (7.58), through Eq. (7.53b), gives the temperature distribution,

$$T(x,y) = \sqrt{\frac{2}{b}}\sum_{n=1}^{\infty}\sum_{m=1}^{\infty}(-1)^{m+1}K(\lambda_n,y)K(\beta_m,y)\,\frac{\beta_m\bar{f}(\lambda_n)}{\lambda_n^2 + \beta_m^2} \tag{7.59}$$

which can also be written as

$$T(x,y) = \frac{4}{ab}\sum_{n=1}^{\infty}\sum_{m=1}^{\infty}(-1)^{m+1}\int_0^a f(x')\sin\frac{n\pi}{a}x'\,dx'$$

$$\cdot\frac{\frac{m\pi}{b}\sin\frac{n\pi}{a}x\sin\frac{m\pi}{b}y}{\left(\frac{n\pi}{a}\right)^2 + \left(\frac{m\pi}{b}\right)^2} \tag{7.60}$$

We have thus obtained another expression for the same temperature distribution. Now the question arises: Are these two expressions the same? The answer is yes, because it can easily be shown that (see Problem 7.3)

$$\frac{\sinh\frac{n\pi}{a}y}{\sinh\frac{n\pi}{a}b} = \frac{2}{b}\sum_{m=1}^{\infty}(-1)^{m+1}\,\frac{\frac{m\pi}{b}\sin\frac{m\pi}{b}y}{\left(\frac{n\pi}{a}\right)^2 + \left(\frac{m\pi}{b}\right)^2} \tag{7.61}$$

One disadvantage of the second approach might be the presence of one more infinite series in the result.

Three-dimensional steady problems can be solved in exactly the same way as two-dimensional problems.

7.7 HANKEL TRANSFORMS

Integral transforms whose kernels are Bessel functions are called Hankel transforms and they are obtained from the expansion of an arbitrary function into an infinite series of Bessel functions, which arise in the problems in cylindrical coordinates. They are sometimes referred to as Bessel transforms. There is a great variety of these transforms, first because of the variety of Bessel functions that are the solutions of Bessel's differential equation

$$r^2 \frac{d^2R}{dr^2} + r\frac{dR}{dr} + (\lambda^2 r^2 - \nu^2)R = 0$$

containing two parameters λ and ν. Furthermore, the Hankel transforms may be developed over either a finite interval or a semi-infinite region with various boundary conditions, or even over the infinite region. Mathematical theory of Hankel transforms can be found in [3] and [5], and the applications of these transforms to heat conduction problems are given in [4].

In this text, we restrict our attention to finite Hankel transforms. The use of finite Hankel transforms was first suggested by Sneddon [6] in 1946, and later by Eringen [1] in 1954. These transforms were used by Özışık [4] in the solution of heat conduction problems posed in cylindrical coordinates.

The finite Hankel transform of an arbitrary function f(r) in the region (a,b) is defined as

$$\bar{f}(\lambda_n) = \int_0^b K_\nu(\lambda_n, r) f(r) r\, dr \tag{7.62a}$$

with the inversion

$$f(r) = \sum_{n=1}^{\infty} K_\nu(\lambda_n, r)\, \bar{f}(\lambda_n) \tag{7.62b}$$

where the kernel $K_\nu(\lambda_n, r)$ is the normalized characteristic function of the following characteristic-value problem:

$$r^2 \frac{d^2R}{dr^2} + r\frac{dR}{dr} + (\lambda^2 r^2 - \nu^2)R = 0 \tag{7.63a}$$

$$\alpha_1 R(a) + \beta_1 \frac{dR(a)}{dr} = 0 \tag{7.63b}$$

$$\alpha_2 R(b) + \beta_1 \frac{dR(b)}{dr} = 0 \tag{7.63c}$$

The kernel $K_\nu(\lambda_n, r)$ and the characteristic values have been evaluated and listed by Ozisik [4] for the nine different combinations of the boundary conditions at r = a and r = b.

Table 7.3 Kernels and characteristic-values for use in finite Hankel Transforms in the finite region.

$$\bar{f}(\lambda_n) = \int_0^b K_\nu(\lambda_n, r) f(r) r \, dr \qquad r^2 \frac{d^2R}{dr^2} + r\frac{dR}{dr} + (\lambda^2 r^2 - \nu^2)R = 0$$

$$R(0) = \text{finite}$$

$$f(r) = \sum_{n=1}^{\infty} K_\nu(\lambda_n, r)\, \bar{f}(\lambda_n) \qquad \alpha R(b) + \beta \frac{dR(b)}{dr} = 0$$

B.C. at r=b	Kernel* $K_\nu(\lambda_n,r)$	Characteristic Values λ_n are the positive roots of**
Third kind ($\alpha \neq 0$, $\beta \neq 0$)	$\frac{\sqrt{2}}{b} \frac{1}{\left[1 + \frac{1}{\lambda_n^2}\left(H^2 - \frac{\nu^2}{b^2}\right)\right]^{1/2}} \frac{J_\nu(\lambda_n r)}{J_\nu(\lambda_n b)}$	$HJ_\nu(\lambda b) + \frac{dJ_\nu(\lambda b)}{dr} = 0$
Second kind ($\alpha=0$)	$\frac{\sqrt{2}}{b} \frac{1}{\left[1 - \left(\frac{\nu}{\lambda_n b}\right)^2\right]^{1/2}} \frac{J_\nu(\lambda_n r)}{J_\nu(\lambda_n b)}$	$\frac{dJ_\nu(\lambda b)}{dr} = 0$**
First kind ($\beta=0$)	$\frac{\sqrt{2}}{b} \frac{J_\nu(\lambda_n r)}{J_{\nu+1}(\lambda_n b)}$	$J_\nu(\lambda b) = 0$

*$H = \alpha/\beta$.

**When $\nu=0$, $\lambda_0=0$ is also characteristic value for this case.

When the region of the transform is (0,b), we saw in Section 4.9 that the kernel $K_\nu(\lambda_n,r)$ is the normalized characteristic function of the following characteristic-value problem:

$$r^2 \frac{d^2R}{dr^2} + r\frac{dR}{dr} + (\lambda^2 r^2 - \nu^2)R = 0 \tag{7.64a}$$

$$R(0) = \text{finite} \tag{7.64b}$$

$$\alpha R(b) + \beta \frac{dR(b)}{dr} = 0 \tag{7.64c}$$

We have already evaluated the kernels $K_\nu(\lambda_n,r)$ and the characteristic values λ_n in Section 4.8 for each special case of the boundary condition at r=0. Table 7.3 summarizes the results obtained in Chapter 4.

Theory of Hankel transforms for the half-space can be found in [3]. Applications to heat conduction problems are given in [4].

7.8 PROBLEMS IN CYLINDRICAL COORDINATES

Heat conduction problems posed in cylindrical coordinates, in general, will involve the following terms

$$\frac{\partial^2 T}{\partial r^2} + \frac{1}{r}\frac{\partial T}{\partial r} + \frac{1}{r^2}\frac{\partial^2 T}{\partial \phi^2} + \frac{\partial^2 T}{\partial z^2}$$

in the differential equation. The partial derivative with respect to variable z can be eliminated from the problem by applying Fourier transforms in the z-direction. The partial derivative with respect to variable ϕ can also be eliminated by applying Fourier transforms in the ϕ-direction, provided that the range of ϕ is $(0,\phi_0)$ where $\phi_0 < 2\pi$. If the range of ϕ is $(0,2\pi)$, then development of a new transform is necessary to remove the partial derivative with respect to variable ϕ [4]. The partial derivatives with respect to variable r are removed by the application of Hankel transforms. In case the problem involves both variables r and ϕ, the partial derivative with respect to variable ϕ should be removed first because of the coefficient $1/r^2$ of this derivative.

As an example, let us consider a half cylinder of semi-infinite length, $0 \leq r \leq b$, $0 \leq \phi \leq \pi$, and $0 \leq z \leq \infty$. Internal energy is generated in this cylinder at a rate of $\dot{q}(r,\phi)$ per unit volume. The surfaces at $\phi = 0$, $\phi = \pi$ and $z = 0$ are at zero temperature while the surface at $r = b$ is kept at temperature $T_w(\phi)$. We wish to find the steady temperature distribution in the cylinder. Assuming constant thermo-physical properties, the formulation of the problem in terms of $T = T(r,\phi,z)$ can be given as follows (Fig. 7.3):

$$\frac{\partial^2 T}{\partial r^2} + \frac{1}{r}\frac{\partial T}{\partial r} + \frac{1}{r^2}\frac{\partial^2 T}{\partial \phi^2} + \frac{\partial^2 T}{\partial z^2} + \frac{\dot{q}(r,\phi)}{k} = 0 \qquad (7.65a)$$

$$T(0,\phi,z) \text{ is finite} \qquad (7.65b)$$

$$T(r,0,z) = T(r,\pi,z) = T(r,\phi,0) = 0 \qquad (7.65c,d,e)$$

$$T(r,\phi,\infty) \rightarrow \text{finite} \qquad (7.65f)$$

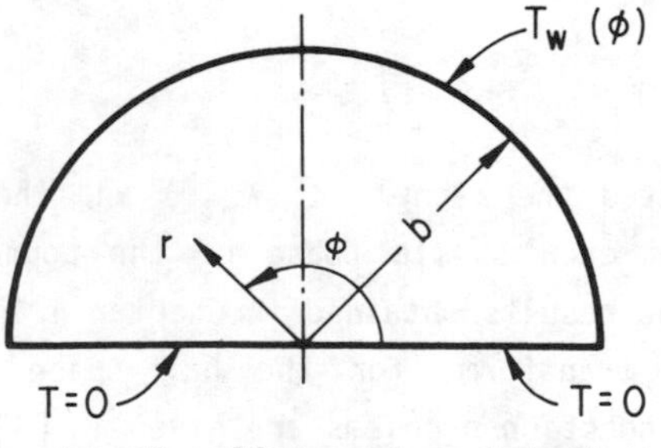

Fig. 7.3 A half cylinder of semi-infinite length.

$$T(b,\phi, z) = T_w(\phi) \tag{7.65g}$$

The partial derivative with respect to ϕ-variable can be eliminated by Fourier transforms. The range of variable ϕ is $(0,\pi)$, and in this finite interval, the finite Fourier transform $\bar{T}(r,n,z)$ of the temperature function $T(r,\phi,z)$ with respect to the variable ϕ can be defined as

$$\bar{T}(r,n,z) = \int_0^{\pi} K(n,\phi)T(r,\phi,z)d\phi \tag{7.66a}$$

with the inversion

$$T(r,\phi,z) = \sum_{n=1}^{\infty} K(n,\phi)\bar{T}(r,n,z) \tag{7.66b}$$

where the kernel $K(n,\phi)$ is the normalized characteristic function of the following characteristic-value problem:

$$\frac{d^2\psi}{d\phi^2} + n^2\psi = 0 \tag{7.67a}$$

$$\psi(0) = \psi(\pi) = 0 \tag{7.67b}$$

The kernel $K(n,\phi)$ can, however, be taken from Table 7.1:

$$K(n,\phi) = \sqrt{\frac{2}{\pi}} \sin n\phi , \qquad n = 1,2,3,...$$

The transform of the differential equation (7.65a) with respect to variable ϕ, through the use of Eq. (7.66a), yields

$$\frac{\partial^2\bar{T}}{\partial r^2} + \frac{1}{r}\frac{\partial\bar{T}}{\partial r} - \frac{n^2}{r^2}\bar{T}(r,n,z) + \frac{\partial^2\bar{T}}{\partial z^2} + \frac{\bar{\dot{q}}(r,n)}{k} = 0 \tag{7.68}$$

where we have defined

$$\bar{\dot{q}}(r,n) = \int_0^{\pi} K(n,\phi)\ \dot{q}(r,\phi)\ d\phi \tag{7.69}$$

Equation (7.68) involves the differential operator

$$\frac{\partial^2}{\partial r^2} + \frac{1}{r}\frac{\partial}{\partial r} - \frac{n^2}{r^2}$$

In order to remove this differential operator with respect to variable r, we define the finite Hankel transform $\bar{T}(\lambda_m,n,z)$ of the function $\bar{T}(r,n,z)$ in the finite interval $(0,b)$ as

$$\bar{\bar{T}}(\lambda_m,n,z) = \int_0^{b} K_n(\lambda_m,r)\bar{T}(r,n,z)r\ dr \tag{7.70a}$$

with the inversion

$$\bar{T}(r,n,z) = \sum_{m=1}^{\infty} K_n(\lambda_m,r)\bar{\bar{T}}(\lambda_m,n,z) \tag{7.70b}$$

where the kernel $K_n(\lambda_m,r)$ is the normalized characteristic function of the following characteristic-value problem:

$$r^2 \frac{d^2R}{dr^2} + r\frac{dR}{dr} + (\lambda^2 r^2 - n^2)R = 0 \tag{7.71a}$$

$$R(0) = \text{finite} \tag{7.71b}$$

$$R(b) = 0 \tag{7.71c}$$

The kernel $K_n(\lambda_m,r)$ from Table 7.3 is

$$K_n(\lambda_m,r) = \frac{\sqrt{2}}{b}\frac{J_n(\lambda_m r)}{J_{n+1}(\lambda_m b)} \tag{7.72a}$$

where the characteristic values are positive roots of

$$J_n(\lambda b) = 0 \tag{7.72b}$$

Now, the transform of Eq.(7.68) with respect to variable r, through the use of transform (7.70a), yields

$$\frac{d^2\bar{\bar{T}}}{dz^2} - \lambda_m^2\bar{\bar{T}}(\lambda_m,n,z) = b\,\frac{dK_n(\lambda_m,b)}{dr}\,\bar{T}_w(n) - \frac{1}{k}\,\bar{\bar{\dot{q}}}(\lambda_m,n) \tag{7.73}$$

where we have defined

$$\bar{T}_w(n) = \int_0^{\pi} K(n,\phi)T_w(\phi)d\phi \tag{7.74a}$$

and

$$\bar{\bar{\dot{q}}}(\lambda_m,n) = \int_0^b K_n(\lambda_m,r)\bar{\dot{q}}(r,n)r\,dr \tag{7.74b}$$

Equation (7.73) can be transformed with respect to variable z in the semi-infinite interval (0,∞) to reduce it to an algebraic equation. However, we shall prefer to solve this ordinary differential equation. The solution can be written as

$$\bar{\bar{T}}(\lambda_m,n,z) = A_n^m e^{-\lambda_m z} + B_n^m e^{\lambda_m z} + \bar{\bar{T}}_p(\lambda_m,n) \tag{7.75}$$

where the particular solution can be shown to be

$$\bar{\bar{T}}_p(\lambda_m,n) = -\frac{1}{\lambda_m^2}\left[b\,\frac{d\,K_n(\lambda_m,b)}{dr}\,\bar{T}_w(n) - \frac{1}{k}\,\bar{\bar{\dot{q}}}(\lambda_m,n)\right] \tag{7.76}$$

Since $\lim_{z\to\infty} T(r,\phi,z)$ is finite,

$$\lim_{z\to\infty} \bar{\bar{T}}(\lambda_m,n,z) \to \text{finite}$$

which yields $B_n^m \equiv 0$. On the other hand, since $T(r,\phi,0) = 0$,

$$\bar{\bar{T}}(\lambda_m,n,0) = 0$$

which gives

$$A_n^m = -\bar{\bar{T}}_p(\lambda_m,n)$$

The solution for $\bar{\bar{T}}(\lambda_m,n,z)$ now becomes

$$\bar{\bar{T}}(\lambda_m,n,z) = \bar{\bar{T}}_p(\lambda_m,n)(1 - e^{-\lambda_m z}) \tag{7.77}$$

When this double transform is inverted, through the use of inversion relations (7.70b) and (7.66b), we obtain the temperature distribution,

$$T(r,\phi,z) = \sum_{n=1}^{\infty}\sum_{m=1}^{\infty} (1 - e^{-\lambda_m z})K(n,\phi)K_n(\lambda_m,r)\bar{\bar{T}}_p(\lambda_m,n) \tag{7.78a}$$

which can also be written as

$$T(r,\phi,z) = \frac{4}{\pi b}\sum_{n=1}^{\infty}\sum_{m=1}^{\infty} \frac{[1 - e^{-\lambda_m z}]}{\lambda_m^2}\sin n\phi\,\frac{J_n(\lambda_m r)}{J_{n+1}(\lambda_m b)}$$

$$\cdot\left[\lambda_m \int_0^{\pi} \sin n\phi'\, T_w(\phi')d\phi'\right.$$

$$\left. + \frac{1}{kb\,J_{n+1}(\lambda_m b)} \int_0^b \int_0^{\pi} J_n(\lambda_m r')\sin n\phi'\,\dot{q}(r',\phi')r'\,dr'\,d\phi'\right]$$

...(7.78b)

The method of solution of unsteady heat conduction problems in cylindrical coordinates follows the same procedure as in rectangular coordinates.

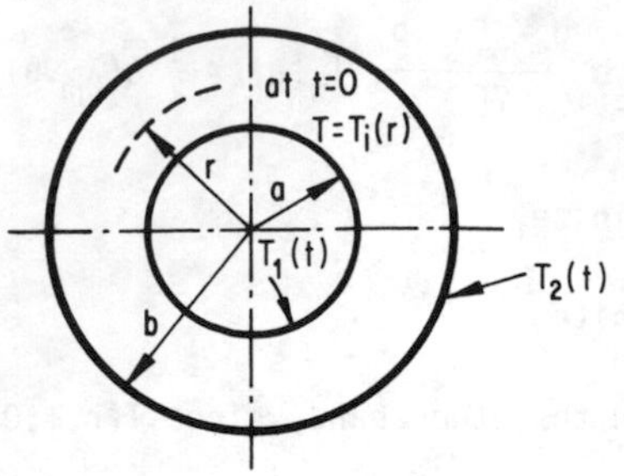

Fig. 7.4 Hollow sphere.

7.9 PROBLEMS IN SPHERICAL COORDINATES

As discussed already in Chapter 6, one-dimensional linear heat conduction problems posed in spherical coordinates can be transformed into rectangular coordinates by defining a new temperature function as $\theta = r.T$. Fourier transforms can then be used to solve the problem. The following example explains the method of solution.

Consider the hollow sphere shown in Fig. 7.4 which is initially at temperature $T_i(r)$. The surfaces at $r = a$ and $r = b$ are kept at temperatures $T_1(t)$ and $T_2(t)$, respectively, for times $t > 0$. We wish to find the unsteady-state temperature distribution in this spherical shell. Assuming constant thermo-physical properties, the formulation of the problem in terms of $T(r,t)$ can be given by

$$\frac{\partial^2 T}{\partial r^2} + \frac{2}{r}\frac{\partial T}{\partial r} = \frac{1}{\alpha}\frac{\partial T}{\partial t} \tag{7.79a}$$

$$T(r,0) = T_i(r) \tag{7.79b}$$

$$T(a,t) = T_1(t) \qquad \text{and} \qquad T(b,t) = T_2(t) \tag{7.79c,d}$$

We now define a new temperature function $\theta(r,t)$ as

$$\theta(r,t) = r.T(r,t)$$

Then the formulation of the problem in terms of $\theta(r,t)$ becomes

$$\frac{\partial^2 \theta}{\partial r^2} = \frac{1}{\alpha}\frac{\partial \theta}{\partial t} \tag{7.80a}$$

$$\theta(r,0) = r\,T_i(r) \tag{7.80b}$$

$$\theta(a,t) = a\,T_1(t) \qquad \text{and} \qquad \theta(b,t) = b\,T_2(t) \tag{7.80c,d}$$

Now a change of variable $r = x + a$ yields

$$\frac{\partial^2\theta}{\partial x^2} = \frac{1}{\alpha}\frac{\partial\theta}{\partial t} \tag{7.81a}$$

$$\theta(x,0) = (x + a)T_i(x + a) \tag{7.81b}$$

$$\theta(0,t) = a\ T_1(t) \quad \text{and} \quad \theta(b-a,t) = b\ T_2(t) \tag{7.81c,d}$$

which can be solved by using Fourier transforms. We note that the range of variable x is $(0,b-a)$ and in this finite interval the Fourier transform $\overline{\theta}(\lambda_n,t)$ of the function $\theta(x,t)$ can be defined as

$$\overline{\theta}(\lambda_n,t) = \int_0^{b-a} \theta(x,t)K(\lambda_n,x)dx \tag{7.82a}$$

with the inversion

$$\theta(x,t) = \sum_{n=1}^{\infty} \overline{\theta}(\lambda_n,t)K(\lambda_n,x) \tag{7.82b}$$

where the kernel $K(\lambda_n,x)$ is given by

$$K(\lambda_n,x) = \sqrt{\frac{2}{b-a}}\ \sin \lambda_n x \tag{7.82c}$$

with the characteristic values

$$\lambda_n = \frac{n\pi}{b-a}, \qquad n = 1,2,3,\ldots \tag{7.82d}$$

The transform of Eq. (7.81a), through the use of Eq. (7.82a), yields

$$\frac{d\overline{\theta}}{dt} + \alpha\lambda_n^2\ \overline{\theta}(\lambda_n,t) = \alpha\sqrt{\frac{2}{b-a}}\ [(-1)^n\ a\ T_1(t) - b\ T_2(t)] \tag{7.83}$$

the solution of which can be written as

$$\overline{\theta}(\lambda_n,t) = e^{-\lambda_n^2\alpha t}\left[\overline{\theta}(\lambda_n,0) + \alpha\sqrt{\frac{2}{b-a}}\right.$$

$$\left.\cdot\int_0^t [(-1)^n\ a\ T_1(t') - b\ T_2(t')]e^{\alpha\lambda_n^2 t'}\ dt'\right] \tag{7.84}$$

where

$$\overline{\theta}(\lambda_n,0) = \int_0^{b-a} (x + a)T_i(x + a)K(\lambda_n,x)dx \tag{7.85}$$

Equation (7.84) can now be inverted using Eq. (7.82b) to obtain $\theta(x,t)$:

$$\theta(x,t) = \frac{2}{b-a}\sum_{n=1}^{\infty} e^{-\alpha\lambda_n^2 t} \sin \lambda_n x \Big[\int_0^{b-a} (x' + a)\, T_i(x' + a) \sin \lambda_n x\, dx'$$

$$+ \alpha \int_0^t [(-1)^n a\, T_1(t') - b\, T_2(t')] e^{\alpha\lambda_n^2 t'} dt' \Big] \tag{7.86a}$$

which can also be written as

$$\theta(r,t) = \frac{2}{b-a}\sum_{n=1}^{\infty} e^{-\alpha\lambda_n^2 t} \sin \lambda_n (r-a)$$

$$\cdot \Big[\int_0^b r'\, T_i(r') \sin \lambda_n (r'-a) dr'$$

$$+ \alpha \int_0^t [(-1)^n a\, T_1(t') - b\, T_2(t')] e^{\alpha\lambda_n^2 t'} dt' \Big] \tag{7.86b}$$

The temperature distribution T(r,t) then becomes

$$T(r,t) = \frac{2}{r(b-a)}\sum_{n=1}^{\infty} e^{-\alpha\lambda_n^2 t} \sin \lambda_n (r-a)$$

$$\cdot \Big[\int_a^b r'\, T_i(r') \sin \lambda_n (r'-a) dr'$$

$$+ \alpha \int_0^t [(-1)^n a\, T_1(t') - b\, T_2(t')] e^{\alpha\lambda_n^2 t'} dt' \Big] \tag{7.87}$$

In order to solve two- and three-dimensional heat conduction problems in spherical coordinates by using integral transforms, we need to develop new transforms to remove the partial derivatives with respect to variable r. These transforms can be obtained from the expansion of an arbitrary function f(r) into an infinite series of Legendre polynomials. These transforms are called *Legendre transforms*. They had been first discussed by Tranter [7] and Churchill [8] and later applied to the solutions of linear heat conduction problems in spherical coordinates by Özışık [4].

7.10 OBSERVATIONS ON THE METHOD

In all the cases we have considered in this chapter, the problem of determining the solution of a heat conduction problem has been reduced by means of an integral transform to a much simpler problem consisting of an ordinary differential equation together with an initial condition or two

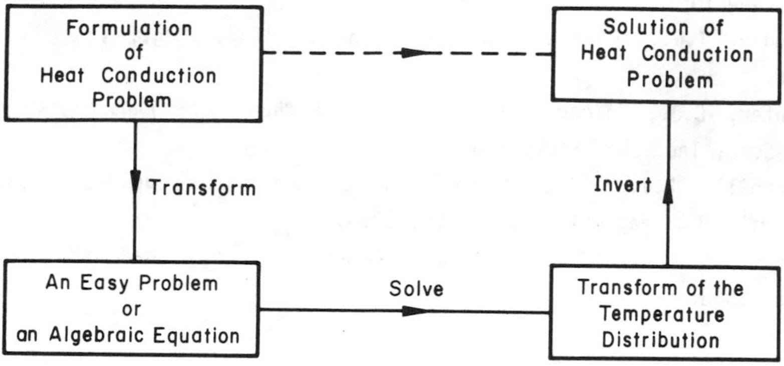

Fig. 7.5 From the formulation to the solution.

boundary conditions. We have also observed that steady problems, if preferred, can be reduced to an algebraic equation for the transform of the temperature distribution. Once the solution of this simple problem or the algebraic equation is found, the inversion relations, which are given at the onset of the problem, are used to find the solution of the original problem. To explain the whole philosophy of the use of the integral transform methods, the process can be described diagramatically as shown in Fig. 7.5. In the next chapter we shall see that the same procedure is followed in the application of Laplace transforms for the solution of unsteady-state heat conduction problems. The path in Fig. 7.5 from the formulation to the solution is shown by a broken line because it can only seldom be realized and the usual path is along the full line.

REFERENCES

1. Eringen, A.C., "The Finite Sturm-Liouville Transform", *Quart. J. Math.*, Oxford (2), Vol. 5, p. 120, 1954.
2. Churchill, R.V., "Generalized Finite Fourier Cosine Transforms", *Mich. Math. Journal*, Vol. 3, p. 85, 1955.
3. Sneddon, I.N., *The Use of Integral Transforms*, McGraw-Hill Book Co., New York, 1972.
4. Özışık, M.N., *Boundary Value Problems of Heat Conduction*, International Textbook Co., Scranton, Pennsylvania, 1968.
5. Churchill, R.V., *Operational Mathematics*, 3rd ed., McGraw-Hill Book

Co., New York, 1972.
6. Sneddon, I.N., "Finite Hankel Transforms", *Phil. Mag.*, Vol. 37, p. 17, 1946.
7. Tranter, C.J., *Integral Transforms in Mathematical Physics*, John Wiley and Sons, Inc., New York, 1962.
8. Churchill, R.V., "The Operational Calculus of Legendre Transforms", *J. Math. Physics*, Vol. 33, p. 165, 1954.
9. Luikov, A.V., *Analytical Heat Diffusion Theory*, Academic Press, New York, 1968.

PROBLEMS

7.1 Given the Sturm-Liouville system

$$\frac{d^2y}{dx^2} + \lambda^2 y = 0$$

$$y(0) = 0$$

$$\alpha_2 y(L) + \beta_2 \frac{dy(L)}{dx} = 0$$

where α_2, β_2 and L are given non-zero constants. Find (a) the characteristi values and the characteristic functions of the system. (b) Expand an arbitrary function f(x) into a series of these characteristic functions in the interval (0,L) and determine the expansion coefficients. (c) Rewrite this expansion in two parts as an integral transform of the function f(x) and the corresponding inversion formula.

7.2 If an arbitrary function f(x) and df/dx are piecewise continuous over each bounded interval of the half line $x > 0$, and f(x) is absolutely integrable from zero to ∞, then show that (a) for $x > 0$, function f(x) can be represented by the *Fourier cosine integral* formula.

$$f(x) = \frac{2}{\pi}\int_0^\infty \cos\beta x \int_0^\infty f(x')\cos\beta x'\,dx'\,d\beta, \qquad x > 0$$

(b) when f(x) is even, the above representation is valid for all values of x, (c) the Fourier cosine integral representation of f(x) can be rearranged as the *Fourier cosine transform* in the interval $(0,\infty)$ as

$$\bar{f}(\beta) = \sqrt{\frac{2}{\pi}}\int_0^\infty f(x)\cos\beta x\,dx$$

with the inversion

$$f(x) = \sqrt{\frac{2}{\pi}} \int_0^\infty \bar{f}(\beta) \cos \beta x \, d\beta$$

(d) the kernel of the Fourier cosine transform satistifes the following characteristic value problem:

$$\frac{d^2y}{dx^2} + \beta^2 y(x) = 0$$

$$\frac{dy(0)}{dx} = 0 \qquad \text{and} \qquad |y(\infty)| \leq 1$$

for all values of β from zero to ∞.

7.3 Show that in the interval (0,b) the following is a valid expression:

$$\frac{\sinh \frac{n\pi}{a} y}{\sinh \frac{n\pi}{a} b} = \frac{2}{b} \sum_{m=1}^{\infty} (-1)^{m+1} \frac{\frac{m\pi}{b} \sin \frac{m\pi}{b} y}{(\frac{n\pi}{a})^2 + (\frac{m\pi}{b})^2}, \qquad 0 < y < b$$

7.4 Consider a plane wall of thickness L, which is initially at a non-uniform temperature $T_i(x)$. Assume that for times $t > 0$ heat is generated in this wall at a rate of $\dot{q}(x,t)$ per unit volume and heat is dissipated by convection from the boundaries at $x = 0$ and $x = L$ into a surrounding medium whose temperature $T_\infty(t)$ varies with time. The properties k and α, and the surface heat transfer coefficients h_1 at the surface at $x = 0$ and h_2 at the surface at $x = L$ are constants. Find a relation for the unsteady temperature distribution in the wall.

7.5 Find an expression for the steady-state temperature distribution in the semi-infinite rectangular strip shown in Fig. 7.6.

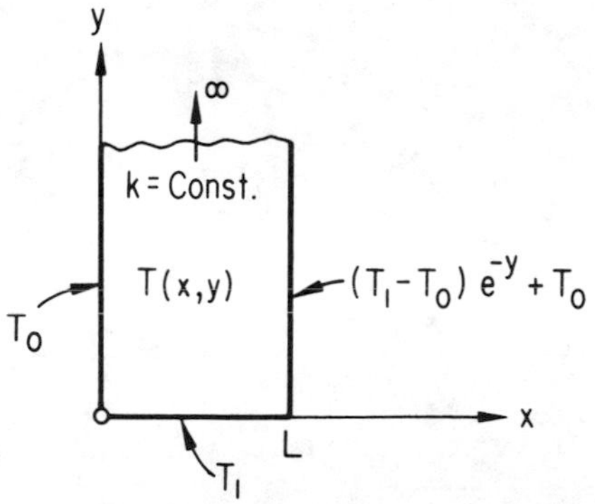

Fig. 7.6 Semi-infinite rectangular strip in Problem 7.5.

7.6 Consider a long solid cylinder of radius b, which is initially at a non-uniform temperature $T_i(r)$. Assume that for times $t \geq 0$ heat is generated in this cylinder at a rate of $\dot{q}(r,t)$ per unit volume, and heat is dissipated by convection from the boundary surface at $r = b$ into a medium whose temperature $T_\infty(t)$ varies with time. The properties k and α, and the surface heat transfer coefficient h are constants. Determine an expression for the unsteady temperature distribution in the cylinder for times $t > 0$.

7.7 Consider a long solid cylinder enclosed by the surface $r = b$, and by the planes $\phi = 0$ and $\phi = \pi/2$. This cylinder is initially at temperature $T_i(r,\phi)$. For times $t \geq 0$, the boundary surfaces are maintained at temperatures $T_1(r,t)$ (at $\phi = 0$), $T_2(r,t)$ (at $\phi = \pi/2$), and $T_3(\phi,t)$ (at $r = b$). Find an expression for the unsteady temperature distribution in the cylinder for times $t > 0$.

7.8 Find an expression for the steady-state temperature distribution in the solid cylinder of Problem 7.7.

7.9 A solid sphere of radius r_0 is initially at a non-uniform temperature $T_i(r)$. For times $t \geq 0$, heat is generated in the sphere at a rate $\dot{q}(r,t)$ per unit volume and it loses heat by convection from the surface at $r = r_0$ to a medium whose temperature $T_\infty(t)$ varies with time. The properties k and α of the sphere, and the heat transfer coefficient h are constants. Obtain an expression for the unsteady temperature distribution in the sphere.

8

Solutions with Laplace Transforms

8.1 INTRODUCTION

In Chapter 7, we have seen how heat conduction problems can be solved by removing the space variables by means of integral transforms. In solving time-dependent problems, on the other hand, the time variable may be removed from the problem by Laplace transforms. The Laplace transform method is also an integral transform method. There are advantages and disadvantages of this method as compared to the integral transform methods introduced in Chapter 7. The Laplace transform method is particularly applicable to time-dependent problems and is, in fact, attractive for one-dimensional problems. When two or three-dimensional time-dependent problems are solved with Laplace transforms, the resulting simple problem again involves partial derivatives with respect to space variables. With Laplace transforms, inversion may not be as easy as with integral transforms. On the other hand, Laplace transforms may yield closed form solutions of the problems where integral transforms give solutions in terms of series expansions.

In this chapter, we first examine the basic properties of Laplace transforms and then discuss the method of solution of time-dependent heat conduction problems by Laplace transforms in terms of examples.

8.2 DEFINITION OF THE LAPLACE TRANSFORM

Let $f(t)$ be a function of t defined for $t > 0$. If $f(t)$ is multiplied by e^{-pt} and integrated with respect to t from zero to infinity, a new function of the parameter p is obtained; that is,

$$\overline{f}(p) = \mathcal{L}\{f(t)\} = \int_0^\infty e^{-pt} f(t)\,dt \tag{8.1}$$

This new function, denoted by $\overline{f}(p)$ or $\mathcal{L}\{f(t)\}$, is called the *Laplace transform* of the function f(t). The Laplace transform of f(t) is said to exist if the integral in Eq. (8.1) converges for some value of the parameter p; otherwise Laplace transform of f(t) does not exist.

Simple examples can be given to illustrate the calculation of Laplace transforms. For instance, let f(t) = 1 when t > 0, then

$$\mathcal{L}\{1\} = \int_0^\infty e^{-pt}dt = \frac{1}{p}, \qquad (p > 0) \tag{8.2}$$

Another example is the transform of f(t) = t when t > 0;

$$\mathcal{L}\{t\} = \int_0^\infty t\, e^{-pt}dt = \frac{1}{p^2}, \qquad (p > 0) \tag{8.3}$$

The Laplace transform of $f(t) = e^{\alpha t}$ when t > 0, where α is a constant, is

$$\mathcal{L}\{e^{\alpha t}\} = \int_0^\infty e^{-pt}e^{\alpha t}dt = \frac{1}{p-\alpha}, \qquad (p > \alpha) \tag{8.4}$$

Let us now consider another example where

$$f(t) = \begin{cases} 1 & \text{for} \quad 0 < t < 2 \\ t & \text{for} \quad t > 2 \end{cases}$$

We find that

$$\mathcal{L}\{f(t)\} = \int_0^\infty f(t)e^{-pt}dt = \int_0^2 e^{-pt}dt + \int_2^\infty t\, e^{-pt}dt$$

$$= \frac{1}{p} + e^{-2p}\left(\frac{1}{p} + \frac{1}{p^2}\right), \qquad (p > 0) \tag{8.5}$$

With the help of elementary methods of integration, Laplace transforms of many other functions can easily be found. For example,

$$\mathcal{L}\{\sin \omega t\} = \int_0^\infty \sin \omega t\, e^{-pt}dt = \frac{\omega}{p^2 + \omega^2}, \qquad (p > 0) \tag{8.6}$$

and

$$\mathcal{L}\{\cos \omega t\} = \int_0^\infty \cos \omega t\, e^{-pt}dt = \frac{p}{p^2 + \omega^2}, \qquad (p > 0) \tag{8.7}$$

It follows from elementary properties of integrals that the Laplace transformation is *linear*. That is, if C_1 and C_2 are two constants while $f_1(t)$ and $f_2(t)$ are functions of t with Laplace transforms $f_1(p)$ and $f_2(p)$, then

$$\mathcal{L}\{C_1 f_1(t) + C_2 f_2(t)\} = C_1 \mathcal{L}\{f_1(t)\} + C_2 \mathcal{L}\{f_2(t)\}$$
$$= C_1 \bar{f}_1(p) + C_2 \bar{f}_2(p) \tag{8.8}$$

Thus the Laplace transform of the sum of two functions is the sum of Laplace transforms of each individual function.

The Laplace transform of f(t) exists; that is, the integral of Eq. (8.1) converges, if the following conditions are satisfied:

a) The function f(t) is *piecewise continuous* in any interval $t_0 \leq t \leq t_2$, where $t_0 > 0$.
b) $\lim t^n |f(t)| \to 0$ for some constant n such that $0 < n < 1$.
c) The function f(t) is of *exponential order* γ as $t \to \infty$. That is, two real constants $M > 0$ and $\gamma > 0$ should exist such that

$$|e^{-\gamma t} f(t)| < M \qquad \text{or} \qquad |f(t)| < M\, e^{\gamma t}$$

for all t greater than some finite number t_1. Intuitively, the absolute value of the functions of exponential order cannot grow more rapidly than $M\, e^{\gamma t}$ as t increases. In applications, however, this is not a restriction since M and γ can be as large as desired. The bounded functions, such as $\sin \omega t$ or $\cos \omega t$, are of exponential order.

8.3 AN INTRODUCTORY EXAMPLE

In this section, we solve a simple heat conduction problem by using the definition of the Laplace transform and thus introduce the method of solution: Consider a solid body initially at a uniform temperature T_i. At time t = 0 this solid body is exposed to a fluid at temperature T_∞ which varies periodically with time as

$$T_\infty(t) = T_0 + (\Delta T)_0 \sin \omega t \tag{8.9}$$

We assume that the thermal conductivity of the solid, k, surface heat transfer coefficient, h, and the dimensions of the body are such that $Bi = h(V/A)/k \ll 1$. So that the heat transferred to the solid body is distributed instantaneously and uniformly throughout the body, which results in a uniform temperature in the system at any time. The first law of thermodynamics as applied to the system shown in Fig. 8.1 yields

$$\frac{d\theta}{dt} + m\theta(t) = m \sin \omega t \tag{8.10a}$$

where

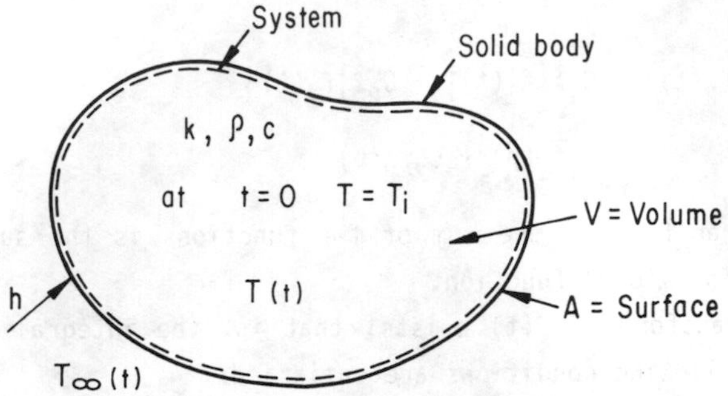

Fig. 8.1 Lumped-heat capacity system.

$$\theta(t) = \frac{T(t) - T_o}{(\Delta T)_o}, \qquad m = \frac{hA}{\rho c V}$$

ρ is the density and c is the specific heat of the solid.

The initial condition of the problem is

$$\theta(0) = \theta_i = \frac{T_i - T_o}{(\Delta T)_o} \tag{8.10b}$$

In order to find the temperature in the solid at any time t by using the definition of the Laplace transform introduced in the previous section, we first multiply both sides of Eq. (8.10a) by e^{-pt} and then integrate the resulting expression with respect to t from zero to ∞:

$$\int_0^\infty e^{-pt} \frac{d\theta}{dt} dt + m \int_0^\infty e^{-pt} \theta(t) dt = m \int_0^\infty e^{-pt} \sin \omega t \, dt \tag{8.11}$$

The first integral on the left-hand side can be integrated by integrating by parts:

$$\int_0^\infty e^{-pt} \frac{d\theta}{dt} dt = \theta(t) e^{-pt} \Big|_0^\infty - \int_0^\infty \theta(t)(-pe^{-pt}) dt$$

$$= -\theta_i + p \int_0^\infty \theta(t) e^{-pt} dt \tag{8.12a}$$

The integral on the right-hand side of Eq. (8.11) is the Laplace transform of the function sin ωt and from Eq. (8.6) we have

$$\int_0^\infty e^{-pt} \sin \omega t \, dt = \frac{\omega}{p^2 + \omega^2}, \qquad (p > 0) \tag{8.12b}$$

Now, inserting these results into Eq. (8.11) we get

$$\int_0^\infty \theta(t)e^{-pt}dt = \frac{m\omega}{(p^2 + \omega^2)(p + m)} + \frac{\theta_i}{p + m} \tag{8.13a}$$

Thus, the problem is to find a function θ(t) whose Laplace transform is the right-hand side of Eq. (8.13a). In order to facilitate the procedure of finding this function we rearrange the above equation as (see Section 8.6)

$$\int_0^\infty \theta(t)e^{-pt}dt = \frac{m}{m^2 + \omega^2}\left[\frac{m\omega}{p^2 + \omega^2} - \frac{\omega p}{p^2 + \omega^2}\right] + \left[\frac{m\omega}{m^2 + \omega^2} + \theta_i\right]\frac{1}{p + m} \tag{8.13b}$$

A close examination of the right-hand side of Eq. (8.13b), with the help of Eqs. (8.4), (8.6) and (8.7), reveals that the function θ(t) is

$$\theta(t) = \frac{m}{m^2 + \omega^2}\left[m \sin \omega t - \omega \cos \omega t\right] + \left[\frac{m\omega}{m^2 + \omega^2} + \theta_i\right] e^{-mt} \tag{8.14a}$$

This result can also be written as

$$\theta(t) = \frac{m\omega}{\sqrt{m^2 + \omega^2}} \sin (\omega t - \phi) + \left(\frac{m\omega}{m^2 + \omega^2} + \theta_i\right) e^{-mt} \tag{8.14b}$$

where

$$\phi = \tan^{-1} \frac{\omega}{m} \tag{8.14c}$$

We have solved a simple lumped-heat capacity system problem by using the definition of the Laplace transform and thus introduced the method of solution. Actually, most of the steps in this procedure can be eliminated if certain properties of Laplace transforms are used during the solution process. Therefore, in the following section we shall study some of the important properties of Laplace transforms.

8.4 SOME IMPORTANT PROPERTIES OF LAPLACE TRANSFORMS

Some of the important properties of Laplace transforms that we use in solving heat conduction problems are as follows:

a) *First shifting property:*

$$\text{If } \bar{f}(p) = \mathfrak{L}\{f(t)\} \qquad \text{then } \mathfrak{L}\{e^{at}f(t)\} = \bar{f}(p - a) \tag{8.15}$$

b) *Second shifting property:*

If $\bar{f}(p) = \mathcal{L}\{f(t)\}$ and $g(t) = \begin{cases} f(t-a), & t > a \\ 0, & t < a \end{cases}$

then

$$\mathcal{L}\{g(t)\} = e^{-ap}\bar{f}(p) \tag{8.16}$$

c) *Change of scale property:*

If $\bar{f}(p) = \mathcal{L}\{f(t)\}$ then $\mathcal{L}\{f(at)\} = \frac{1}{a}\bar{f}(\frac{p}{a})$ (8.17)

d) *Laplace transform of derivatives:* Let $f(t)$ be continuous with a piecewise continuous derivative df/dt over every interval $0 \leq t \leq t_1$. Also let $f(t)$ be of exponential order for $t > t_1$, then

$$\mathcal{L}\{\frac{df}{dt}\} = p\,\bar{f}(p) - f(0) \tag{8.18a}$$

If $f(t)$ fails to be continuous at $t = 0$ but $\lim_{t\to 0^+} f(t) = f(0^+)$ exists then

$$\mathcal{L}\{\frac{df}{dt}\} = p\,\bar{f}(p) - f(0^+) \tag{8.18b}$$

Here it should be noted that $f(0^+)$ is not equal to $f(0)$ which may or may not exist. If $f(t)$ fails to be continuous at $t = a$ then

$$\mathcal{L}\{\frac{df}{dt}\} = p\,\bar{f}(p) - f(0) - e^{-ap}[f(a^+) - f(a^-)] \tag{8.18c}$$

where $f(a^+) - f(a^-)$ is the jump at the discontinuity at $t = a$. For more than one discontinuity, appropriate modifications can be made.

If $f(t)$, $dt/dt, \ldots, d^{n-1}f/dt^{n-1}$ are continuous in the interval $0 \leq t \leq t_1$ and are of exponential order for $t > t_1$ while d^nf/dt^n is piecewise continuous for $0 \leq t \leq t_1$, then

$$\mathcal{L}\{\frac{d^nf}{dt^n}\} = p^n\bar{f}(p) - p^{n-1}f(0) - p^{n-2}\frac{df(0)}{dt} - \ldots$$

$$- p\frac{d^{n-2}f(0)}{dt^{n-2}} - \frac{d^{n-1}f(0)}{dt^{n-1}} \tag{8.18d}$$

If $f(t)$, $df/dt, \ldots, d^{n-1}f/dt^{n-1}$ have discontinuities, appropriate modification in Eq. (8.18d) can be made as in Eq. (8.18b) or Eq. (8.18c).

e) *Laplace transform of integrals:*

If $\overline{f}(p) = \mathcal{L}\{f(t)\}$ then $\mathcal{L}\left\{\int_0^t f(t')dt'\right\} = \frac{1}{p}\overline{f}(p)$ (8.19)

f) *Multiplication by t^n:*

If $\overline{f}(p) = \mathcal{L}\{f(t)\}$ then $\mathcal{L}\{t^n f(t)\} = (-1)^n \frac{d^n}{dp^n}\overline{f}(p)$ (8.20)

g) *Division by t:*

If $\overline{f}(p) = \mathcal{L}\{f(t)\}$ then $\mathcal{L}\{\frac{f(t)}{t}\} = \int_p^\infty \overline{f}(p')dp'$ (8.21)

provided that $\lim_{t\to 0} \frac{f(t)}{t}$ exists.

8.5 THE INVERSE LAPLACE TRANSFORM

In solving heat conduction problems by Laplace transforms, as seen in Section 8.3, we encounter the problem of determining the inverse of a function which has a given transform. The notation $\mathcal{L}^{-1}\{\overline{f}(p)\}$ is conveniently used for a function whose Laplace transform is $\overline{f}(p)$. That is, if

$$\mathcal{L}\{f(t)\} = \overline{f}(p)$$

then

$$f(t) = \mathcal{L}^{-1}\{\overline{f}(p)\}$$

Therefore, in order to determine the inverse transform of $\overline{f}(p)$, it is necessary to find a function $f(t)$ which satisfies the equation

$$\int_0^\infty e^{-pt}f(t)dt = \overline{f}(p) \tag{8.22}$$

Since the unknown function $f(t)$ appears under an integral sign, Eq. (8.22) is an integral equation. If we restrict ourselves to functions $f(t)$ which are piecewise continuous in every finite interval $0 \le t \le t_1$ and of exponential order for $t > t_1$, then if Eq. (8.22) has a solution that solution is unique [1]. Thus, for example, Eq. (8.4) shows that the solution of

$$\int_0^\infty f(t)e^{-pt}dt = \frac{1}{p-a} \tag{8.23a}$$

is $f(t) = e^{at}$; that is,

$$\mathcal{L}^{-1}\left\{\frac{1}{p-a}\right\} = e^{at} \tag{8.23b}$$

Although there are methods for the direct determination of the inverse transform of a given function of p, the most obvious way of finding the inverse transform is to read the result from a table of Laplace transforms. Extensive tables of functions and corresponding transforms are available in the literature and their use is sufficient for our purpose [1-5]. A short table of Laplace transforms is given here in Table 8.1. More typically, the transform we are looking for may not be available in the table of transforms. In such cases, the methods discussed in the following two sections are very useful in obtaining both direct and inverse transforms.

8.5.1 Method of Partial Fractions

The method of partial fractions is a useful tool to employ in conjunction with the table of transforms to determine both direct and inverse transforms. This method is used to split complicated fractions, whose inverse transforms cannot be obtained directly from the Laplace transform tables, into simple ones which can be found in the tables. In order to illustrate how the method of partial fractions can be used to obtain inverse transforms of quotients of polynomials in p, consider

$$\mathcal{L}^{-1}\left\{\frac{1}{(p+1)(p^2+1)}\right\}$$

To find the inverse transform, the fraction can be written as the sum of several fractions whose least common denominator is the denominator in this fraction. Thus we first assume an expansion of the form

$$\frac{1}{(p+1)(p^2+1)} = \frac{A}{p+1} + \frac{Bp+C}{p^2+1}$$

After clearing fractions, we require this equation to be an identity and obtain $A = -B = C = 1/2$. Hence

$$\frac{1}{(p+1)(p^2+1)} = \frac{1}{2}\frac{1}{p+1} - \frac{1}{2}\frac{p-1}{p^2+1}$$

and the use of Table 8.1 gives the inverse transform as

$$\mathcal{L}^{-1}\left\{\frac{1}{(p+1)(p^2+1)}\right\} = \frac{1}{2}[e^{-t} + \sin t - \cos t]$$

If, in general, the transform is in the form of a rational function $P(p)/Q(p)$ where $P(p)$ and $Q(p)$ are polynomials as in the above example, then it can be rewritten as the sum of rational functions, called partial fractions, having the form

Table 8.1 Laplace Transforms.

No.	$\bar{f}(p)$	$f(t)$ $(t > 0)$
1	$1/p$	1
2	$1/p^2$	t
3	$1/p^n$, $n = 1,2,3,..$	$\frac{t^{n-1}}{(n-1)!}$, $0!=1$
4	$\frac{1}{\sqrt{p}}$	$\frac{1}{\sqrt{\pi t}}$
5	$\frac{1}{p - a}$	e^{at}
6	$\frac{1}{(p - a)^n}$, $n = 1,2,3,..$	$\frac{1}{(n-1)!} t^{n-1} e^{at}$, $0! = 1$
7	$\frac{a}{p^2 + a^2}$	$\sin at$
8	$\frac{p}{p^2 + a^2}$	$\cos at$
9	$\frac{a}{p^2 - a^2}$	$\sinh at$
10	$\frac{p}{p^2 - a^2}$	$\cosh at$
11	$\frac{2ap}{(p^2 + a^2)^2}$	$t \sin at$
12	$\frac{p^2 - a^2}{(p^2 + a^2)^2}$	$t \cos at$
13	$\frac{2ap}{(p^2 - a^2)^2}$	$t \sinh at$
14	$\frac{p^2 + a^2}{(p^2 - a^2)}$	$t \cosh at$
15	$\frac{2a^3}{(p^2 + a^2)^2}$	$\sin at - at \cos at$

Table 8.1 (Continued)

No.	$\bar{f}(p)$	$f(t)$ $(t > 0)$
16	$\dfrac{2a^3}{(p^2 + a^2)^2}$	at coshat - sinhat
17	$\dfrac{4a^3}{p^4 + 4a^4}$	sinat coshat - cosat sinhat
18	$\dfrac{2a^2p}{p^4 + 4a^4}$	sinat sinhat
19	$\dfrac{1}{\sqrt{p^2 + a^2}}$	$J_0(at)$
20	$\dfrac{1}{\sqrt{p^2 - a^2}}$	$I_0(at)$
21	$\dfrac{a}{(p^2 + a^2)^{3/2}}$	$t\ J_1(at)$
22	$\dfrac{p}{(p^2 + a^2)^{3/2}}$	$t\ J_0(at)$
23	$\dfrac{a}{(p^2 + a^2)^{3/2}}$	$t\ I_1(at)$
24	$\dfrac{p}{(p^2 + a^2)^{3/2}}$	$t\ I_0(at)$
25	$e^{-x\sqrt{p/a}}$	$\dfrac{x}{2(\pi a t^3)^{1/2}}\ e^{-x^2/4at}$
26	$\dfrac{e^{-x\sqrt{p/a}}}{\sqrt{p/a}}$	$\left(\dfrac{a}{\pi t}\right)^{1/2} e^{-x^2/4at}$
27	$\dfrac{e^{-x\sqrt{p/a}}}{p}$	$\mathrm{erfc}\left[\dfrac{x}{2(at)^{1/2}}\right]$
28	$\dfrac{e^{-x\sqrt{p/a}}}{p\sqrt{p/a}}$	$2\left(\dfrac{at}{\pi}\right)^{1/2} e^{-x^2/4at} - x\ \mathrm{erfc}\left[\dfrac{x}{2(at)^{1/2}}\right]$

Table 8.1 (Continued)

No.	$\bar{f}(p)$	$f(t)$ $(t > 0)$
29	$\frac{e^{-x\sqrt{p/a}}}{p^2}$	$t + \frac{x^2}{2a} \operatorname{erfc}\left[\frac{x}{2(at)^{\frac{1}{2}}}\right] - x\left(\frac{t}{\pi a}\right)^{\frac{1}{2}} e^{-x^2/4at}$
30	$\frac{e^{-a/p}}{p^{3/2}}$	$\frac{\sin 2\sqrt{at}}{\sqrt{\pi a}}$
31	$\frac{e^{-a/p}}{\sqrt{p}}$	$\frac{\cos 2\sqrt{at}}{\sqrt{\pi t}}$
32	$-\frac{(\gamma + \ln p)}{p}$ γ=Euler's constant=.5772156...	$\ln t$
33	$\frac{\sinh px}{p \sinh pa}$	$\frac{x}{a} + \frac{2}{\pi} \sum_{n=1}^{\infty} \frac{(-1)^n}{n} \sin \frac{n\pi x}{a} \cos \frac{n\pi t}{a}$
34	$\frac{\sinh px}{p \cosh pa}$	$\frac{4}{\pi} \sum \frac{(-1)^n}{2n-1} \sin \frac{(2n-1)\pi x}{2a} \sin \frac{(2n-1)\pi t}{2a}$
35	$\frac{\cosh px}{p \sinh ps}$	$\frac{t}{a} + \frac{2}{\pi} \sum_{n=1}^{\infty} \frac{(-1)^n}{n} \cos \frac{n\pi x}{a} \sin \frac{n\pi t}{a}$
36	$\frac{\cosh px}{p \cosh pa}$	$1 + \frac{4}{\pi} \sum_{n=1}^{\infty} \frac{(-1)^n}{2n-1} \cos \frac{(2n-1)\pi x}{2a} \cos \frac{(2n-1)\pi t}{2a}$
37	$\frac{\sinh x\sqrt{p}}{\sinh a\sqrt{p}}$	$\frac{2\pi}{a^2} \sum_{n=1}^{\infty} (-1)^n n\, e^{-n^2\pi^2 t/a^2} \sin \frac{n\pi x}{a}$
38	$\frac{\cosh x\sqrt{p}}{\cosh a\sqrt{p}}$	$\frac{\pi}{a^2} \sum_{n=1}^{\infty} (-1)^{n-1}(2n-1) e^{-(2n-1)^2\pi^2 t/4a^2} \cos \frac{(2n-1)\pi x}{2a}$
39	$\frac{\sinh x\sqrt{p}}{p \sinh a\sqrt{p}}$	$\frac{x}{a} + \frac{2}{\pi} \sum_{n=1}^{\infty} \frac{(-1)^n}{n} e^{-n^2\pi^2 t/a^2} \sin \frac{n\pi x}{a}$
40	$\frac{\cosh x\sqrt{p}}{p \cosh a\sqrt{p}}$	$1 + \frac{4}{\pi} \sum_{n=1}^{\infty} \frac{(-1)^n}{2n-1} e^{-(2n-1)^2\pi^2 t/4a^2} \cos \frac{(2n-1)\pi x}{2a}$

$$\frac{A}{(ap + b)^r}, \qquad \frac{Ap + B}{(ap^2 + bp + c)^r}$$

where $r = 1,2,3,\ldots$. By finding the inverse Laplace transform of each partial fraction, we can determine $\mathcal{L}^{-1}\{P(p)/Q(p)\}$.

As a second example, let us consider

$$\frac{2p - 5}{(3p - 4)(2p + 1)^3} = \frac{A}{3p - 4} + \frac{B}{2p + 1} + \frac{C}{(2p + 1)^2} + \frac{D}{(2p + 1)^3}$$

The constants A, B, C, and D can be found by clearing fractions and equating the coefficients of like powers of p on both sides of the resulting equation.

If the degree of the polynomial P(p) is less than that of the polynomial Q(p) and if Q(p) has n distinct real zeros P_k, $k = 1,2,\ldots,n$, then

$$\mathcal{L}^{-1}\left\{\frac{P(p)}{Q(p)}\right\} = \sum_{k=1}^{n} \frac{P(P_k)}{\frac{dQ(P_k)}{dp}} e^{P_k t} \tag{8.24}$$

This is often called *Heaviside's expansion theorem.*

8.5.2 Convolution Theorem

Several properties and theorems regarding Laplace transforms other than the above ones could also be mentioned. The most useful one of these is the convolution theorem. It frequently happens that although a given $\overline{f}(p)$ is not the transform of a known function, it can be expressed as the product of two functions, each of which is the transform of a known function; that is,

$$\overline{f}(p) = \overline{g}(p)\overline{h}(p)$$

where $\overline{g}(p)$ and $\overline{h}(p)$ are the transforms of the known functions g(t) and h(t), respectively. In this case

$$\mathcal{L}^{-1}\{\overline{f}(p)\} = \mathcal{L}^{-1}\{\overline{g}(p)\overline{h}(p)\} = \int_{t'=0}^{t} g(t-t')h(t')dt' \tag{8.25}$$

This integral, sometimes denoted by g*h, is called the convolution of g(t) and h(t). It has the following property:

$$g*h = h*g$$

Although it is easy to prove this theorem we will not give the proof here. Interested readers may refer to one of the references [1,2,3] for the proof.

In order to illustrate the use of the convolution theorem, let us consider the same example we discussed in Section 8.6, where the inverse of the following expression was desired:

$$\bar{f}(p) = \frac{1}{(p + 1)(p^2 + 1)}$$

This expression can be treated as a product of

$$\bar{g}(p) = \frac{1}{p + 1} \quad \text{and} \quad \bar{h}(p) = \frac{1}{p^2 + 1}$$

The inverses of these are given in Table 8.1 by

$$g(t) = e^{-t} \quad \text{and} \quad h(t) = \cos t$$

Thus, the convolution theorem gives

$$f(t) = \int_0^t e^{-(t-t')} \cos t' \, dt' = e^{-t} \int_0^t e^{t'} \cos t' \, dt'$$

The integral may be evaluated using integral tables to give

$$f(t) = \frac{1}{2}[e^{-t} + \sin t - \cos t]$$

This is the same result we obtained previously by the method of partial fractions.

8.6 LAPLACE TRANSFORMS AND HEAT CONDUCTION PROBLEMS

The Laplace transform method is particularly convenient and efficient to solve one-dimensional time-dependent heat conduction problems. The procedure is very much the same as followed in the solution of the problem discussed in Section 8.3. Briefly, the procedure for solving one-dimensional time-dependent problems can be outlined as shown in Fig. 8.2. This procedure will be illustrated in the following sections in terms of several representative examples.

8.7 PLANE WALL

Consider a plane wall of thickness 2L kept initially at a uniform temperature T_i. Assume that the temperatures of the surfaces are changed to T_w at $t = 0$ and kept constant at this value during the whole heat transfer process (heating and cooling). We wish to find the unsteady temperature distribution in the wall for times $t > 0$.

If we place the origin of the x-coordinate at the center of the wall as shown in Fig. 8.3 then the problem can be formulated in terms of $T = T(x,t)$ as follows:

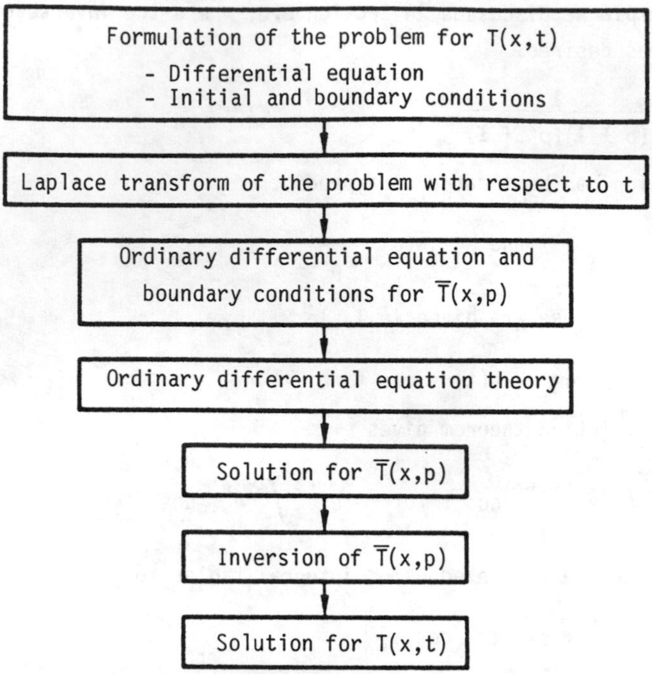

Fig. 8.2 Procedure for solving one-dimensional time-dependent problems.

$$\frac{\partial^2 T}{\partial x^2} = \frac{1}{\alpha}\frac{\partial T}{\partial t} \tag{8.26a}$$

$$T(x,0) = T_i \tag{8.26b}$$

$$\frac{\partial T(0,t)}{\partial x} = 0 \qquad \text{and} \qquad T(L,t) = T_w \tag{8.26c,d}$$

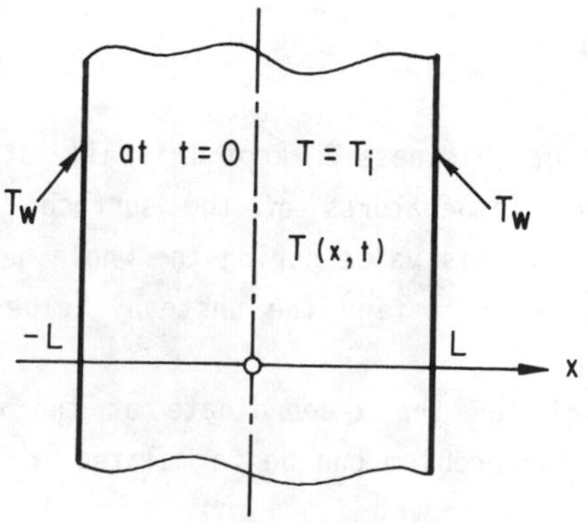

Fig. 8.3 Plane wall of thickness 2L

Let us first find the Laplace transform of Eq. (8.26a); that is,

$$\mathcal{L}\left\{\frac{\partial^2 T}{\partial x^2}\right\} = \frac{1}{\alpha}\,\mathcal{L}\left\{\frac{\partial T}{\partial t}\right\} \tag{8.27}$$

The use of the properties of Laplace transforms yields

$$\mathcal{L}\left\{\frac{\partial^2 T}{\partial x^2}\right\} = \frac{d^2\overline{T}}{dx^2} \tag{8.28a}$$

and

$$\mathcal{L}\left\{\frac{\partial T}{\partial t}\right\} = p\overline{T}(x,p) - T_i \tag{8.28b}$$

where

$$\overline{T} = \overline{T}(x,p) = \mathcal{L}\{T(x,t)\} = \int_0^\infty e^{-pt}T(x,t)dt \tag{8.28c}$$

Thus, Eq. (8.27) can be written as

$$\frac{d^2\overline{T}}{dx^2} - \frac{p}{\alpha}\overline{T} = -\frac{T_i}{\alpha} \tag{8.29}$$

The Laplace transform of the boundary conditions (8.26c,d) are

$$\frac{d\overline{T}(0,p)}{dx} = 0\ , \qquad \overline{T}(L,p) = \frac{T_w}{p} \tag{8.30a,b}$$

Equation (8.29) is an ordinary differential equation for $\overline{T}(x,p)$ which is considered to be a function of x with the constant parameter p carried along. The solution of this equation may be written as

$$\overline{T}(x,p) = C_1 \cosh mx + C_2 \sinh mx + \frac{T_i}{p} \tag{8.31}$$

where $m^2 = p/\alpha$, and C_1 and C_2 are the constants of integration to be determined from the boundary conditions. Application of the boundary condition at $x = 0$ yields $C_2 = 0$, and the boundary condition at $x = L$ gives

$$C_1 = -\frac{T_i - T_w}{p\cosh mL} \tag{8.32}$$

Thus, the solution for $\overline{T}(x,p)$ may now be written as

$$\overline{T}(x,p) = \frac{T_i}{p} - (T_i - T_w)\frac{\cosh mx}{p\cosh mL} \tag{8.33}$$

The final step now is to invert $\overline{T}(x,p)$ to obtain the desired solution for

T(x,t); that is,

$$T(x,t) = T_i \mathcal{L}^{-1}\{\frac{1}{p}\} - (T_i - T_w)\, \mathcal{L}^{-1}\left\{\frac{\cosh mx}{p \cosh mL}\right\} \tag{8.34}$$

Now, the table of transforms, Table 8.1, should be checked for the transforms in Eq. (8.34). Both transforms are available in the table. We, therefore, have, from the transform No. 2,

$$\mathcal{L}^{-1}\{\frac{1}{p}\} = 1 \tag{8.35}$$

and, since $m = \sqrt{(p/\alpha)}$, from transform No. 40,

$$\mathcal{L}^{-1}\left\{\frac{\cosh mx}{p \cosh mL}\right\} = 1 + \frac{4}{\pi} \sum_{n=1}^{\infty} \frac{(-1)^n}{2n - 1} \exp\left[- \frac{(2n - 1)^2\pi^2\alpha}{4L^2} t\right] \cdot \cos \frac{(2n - 1)\pi x}{2L} \tag{8.36}$$

Thus, the solution for T(x,t) can be written as

$$T(x,t) = T_i - (T_i - T_w)\left\{1 + \frac{4}{\pi} \sum_{k=1}^{\infty} \frac{(-1)^k}{(2k - 1)} \exp\left[- \frac{(2k - 1)^2\pi^2\alpha}{4L^2} t\right] \cdot \cos\left[\frac{(2k - 1)\pi}{2} \frac{x}{L}\right]\right\} \tag{8.37}$$

This is the same solution as given by Eq. (6.35) which was obtained by the method of separation of variables.

An alternative expression for T(x,t) can be obtained by following a different approach to invert the second term in Eq. (8.34). Note that,

$$\frac{\cosh mx}{\cosh mL} = \frac{e^{mx} + e^{-mx}}{e^{mL} + e^{-mL}} = e^{-mL}(e^{mx} + e^{-mx}) \frac{1}{1 + e^{-2mL}}$$

$$= \left[e^{-m(L+x)} + e^{-m(L+x)}\right] \sum_{k=0}^{\infty} (-1)^k e^{-2kmL}$$

$$= \sum_{k=0}^{\infty} (-1)^k \left\{\exp[-m(nL - x)] + \exp[-m(nL + x)]\right\} \tag{8.38}$$

where n = 2k + 1. From Table 8.1, the transform No. 27, we have

$$\frac{1}{p} e^{-a\sqrt{p}} = \mathcal{L}\left\{\operatorname{erfc}\left(\frac{a}{2\sqrt{t}}\right)\right\} \tag{8.39}$$

It then follows that

$$\mathcal{L}^{-1}\left\{\frac{\cosh mx}{p \cosh mL}\right\} = \sum_{k=0}^{\infty} (-1)^k \left\{\operatorname{erfc}\left[\frac{(2k+1)L - x}{2\sqrt{\alpha t}}\right] + \operatorname{erfc}\left[\frac{(2k+1)L + x}{2\sqrt{\alpha t}}\right]\right\} \tag{8.40}$$

Hence, the solution for T(x,t) can be written as

$$T(x,t) = T_i - (T_i - T_w) \sum_{k=0}^{\infty} (-1)^k \left\{\operatorname{erfc}\left[\frac{(2k+1)L - x}{2\sqrt{\alpha t}}\right] + \operatorname{erfc}\left[\frac{(2k+1)L + x}{2\sqrt{\alpha t}}\right]\right\} \tag{8.41}$$

We have thus obtained another expression for T(x,t) in the form of a series expansion in terms of complementary error functions. Since the value of the complementary error function decreases rapidly as the argument increases, the solution (8.41) will converge rapidly for small values of t, whereas the convergence of the solution (8.37) is fast for large values of t because of the negative argument of the exponential term.

8.8 SEMI-INFINITE SOLID

We now consider a semi-infinite solid as shown in Fig. 8.4, which is initially at a uniform temperature T_i. The surface temperature is changed to T_w at t = 0 and maintained at this constant value for times t > 0. Assuming that the thermo-physical properties are constant, let us find the unsteady temperature distribution T(x,t) in the solid.

The formulation of the problem in terms of $\theta = T(x,t) - T_i$ is given by

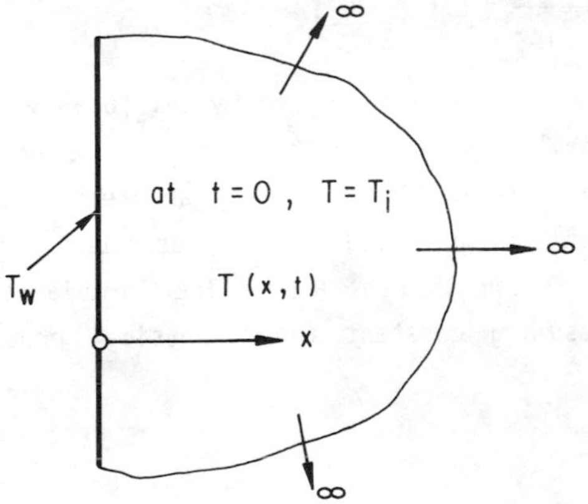

Fig. 8.4 Semi-infinite solid with an isothermal surface temperature.

$$\frac{\partial^2 \theta}{\partial x^2} = \frac{1}{\alpha}\frac{\partial \theta}{\partial t} \tag{8.42a}$$

$$\theta(x,0) = 0 \tag{8.42b}$$

$$\theta(0,t) = T_w - T_i \equiv \theta_w \tag{8.42c}$$

$$\lim_{x\to\infty} \theta(x,t) \to 0 \tag{8.42d}$$

The transformed differential equation and the boundary conditions are

$$\frac{d^2\overline{\theta}}{dx^2} - \frac{p}{\alpha}\overline{\theta} = 0 \tag{8.43a}$$

$$\overline{\theta}(0,p) = \frac{\theta_w}{p}, \qquad \lim_{x\to\infty} \overline{\theta}(x,p) \to 0 \tag{8.43b,c}$$

where $\overline{\theta}(x,p)$ is the Laplace transform of $\theta(x,t)$. The general solution of Eq. (8.43a) may be written as

$$\overline{\theta}(x,p) = C_1 e^{-mx} + C_2 e^{mx} \tag{8.44}$$

where C_1 and C_2 are the constants of integration and $m^2 = p/\alpha$. Application of the boundary condition (8.43c) yields $C_2 = 0$, and the boundary condition (8.43b) gives $C_1 = \theta_w/p$. Then it follows that

$$\frac{\overline{\theta}(x,p)}{\theta_w} = \frac{e^{-mx}}{p} = \frac{e^{-x\sqrt{p/\alpha}}}{p} \tag{8.45}$$

The inverse transform of Eq. (8.45) is found in the table of Laplace transforms, Table 8.1, in terms of the complementary error function. Thus, we obtain

$$\frac{\theta(x,t)}{\theta_w} = \frac{T(x,t) - T_i}{T_w - T_i} = \operatorname{erfc}\left(\frac{x}{2\sqrt{\alpha t}}\right) \tag{8.46}$$

which is identical with the solution given by Eq. (6.93).

As a second example, let us now obtain the unsteady temperature distribution in a semi-infinite solid, initially at zero temperature, when the surface temperature variation is specified as a prescribed function $f(t)$ of time for times $t > 0$ as shown in Fig. 8.5, The formulation of the problem in terms of $T(x,t)$, assuming constant thermo-physical properties, is given by

$$\frac{\partial^2 T}{\partial x^2} = \frac{1}{\alpha}\frac{\partial T}{\partial t} \tag{8.47a}$$

$$T(x,0) = 0 \tag{8.47b}$$

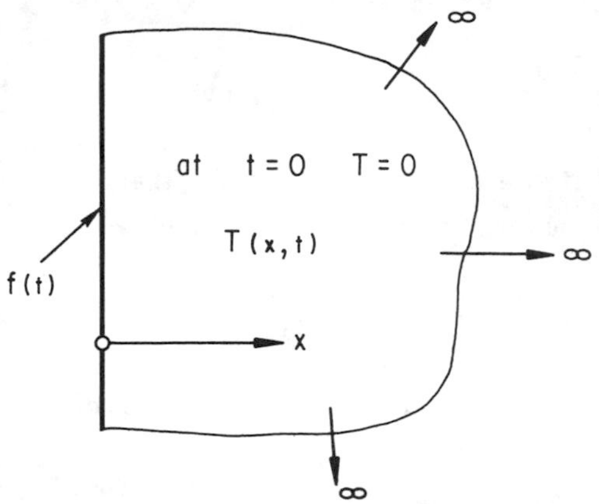

Fig. 8.5 Semi-infinite solid with a time-dependent surface temperature.

$$T(0,t) = f(t) \tag{8.47c}$$

$$\lim_{x\to\infty} T(x,t) \to 0 \tag{8.47d}$$

The transformed differential equation and boundary conditions in terms of the transform $\overline{T}(x,p)$ of $T(x,t)$ are

$$\frac{d^2\overline{T}}{dx^2} - \frac{p}{\alpha}\overline{T} = 0 \tag{8.48a}$$

$$\overline{T}(0,p) = \overline{f}(p) \qquad \text{and} \qquad \lim_{x\to\infty} \overline{T}(x,p) \to 0 \tag{8.48b,c}$$

where $\overline{f}(p)$ is the Laplace transform of $f(t)$; that is,

$$\overline{f}(p) = \mathcal{L}\{f(t)\} = \int_0^\infty e^{-pt} f(t)dt \tag{8.49}$$

The solution of Eq. (8.48a) is

$$\overline{T}(x,p) = C_1\, e^{-mx} + C_2\, e^{mx} \tag{8.50}$$

where C_1 and C_2 are the constants of integration and $m^2 = p/\alpha$. Application of Eq. (8.48c) gives $C_2 = 0$ and Eq. (8.48b) yields $C_1 = \overline{f}(p)$. Thus,

$$\overline{T}(x,p) = \overline{f}(p)e^{-mx} = \overline{f}(p)e^{-x\sqrt{p/\alpha}} \tag{8.51}$$

Since $\overline{T}(x,p)$ is expressed as a product of two functions of p, each of which can be inverted easily, the convolution theorem can be used to find $T(x,t)$. From Table 8.1, the transform No. 25, we have

$$\mathcal{L}^{-1}\{e^{-x\sqrt{p/\alpha}}\} = \frac{x}{2\sqrt{\pi\alpha t^3}} \exp\left(-\frac{x^2}{4\alpha t}\right)$$

Hence, by the convolution theorem we get

$$T(x,t) = \frac{x}{2\sqrt{\pi\alpha}} \int_0^t \frac{f(t - t')}{(t')^{3/2}} \exp\left(- \frac{x^2}{4\alpha t'}\right) dt' \tag{8.52}$$

If the surface temperature is constant; that is, if $f(t) = T_w$, then the temperature distribution in the solid is given by

$$T(x,t) = T_w \frac{x}{2\sqrt{\pi\alpha}} \int_0^t \frac{1}{(t')^{3/2}} \exp\left(- \frac{x^2}{4\alpha t'}\right) dt \tag{8.53a}$$

which can be written in terms of the complementary error function as

$$T(x,t) = T_w \operatorname{erfc}\left(\frac{x}{2\sqrt{\alpha t}}\right) \tag{8.53b}$$

This is the same result as the one given by Eq. (8.46) with $T_i = 0$.

8.9 SOLID CYLINDER

Let us consider a solid cylinder of constant thermo-physical properties and radius r_0 as shown in Fig. 8.6, which is initially at a uniform temperature T_i. Assume that the temperature of the surface is suddenly changed to T_w at time $t = 0$ and is maintained constant at this value for $t > 0$. We wish to find the unsteady temperature distribution in the solid cylinder during the heating (or cooling) process.

The formulation of the problem in terms of $\theta(r,t) = T(r,t) - T_i$ is given by

$$\frac{\partial^2\theta}{\partial r^2} + \frac{1}{r}\frac{\partial\theta}{\partial r} = \frac{1}{\alpha}\frac{\partial\theta}{\partial t} \tag{8.54a}$$

$$\theta(r,0) = 0 \tag{8.54b}$$

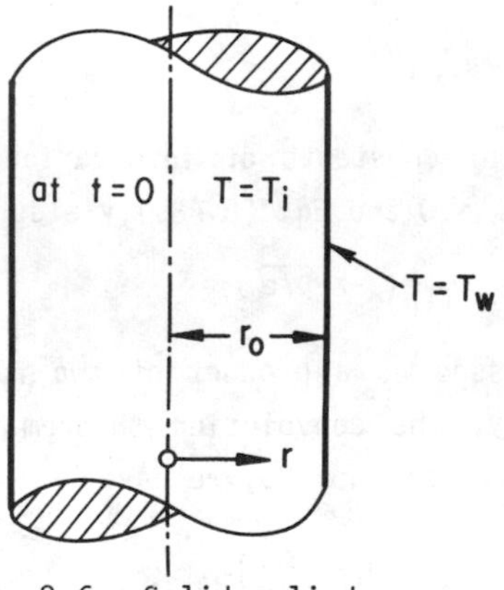

Fig. 8.6 Solid cylinder

$$\theta(0,t) = \text{finite} \tag{8.54c}$$

$$\theta(r_0,t) = T_w - T_i \equiv \theta_w \tag{8.54d}$$

If $\bar{\theta}(x,p)$ denotes the Laplace transform of $\theta(x,t)$, then it satisfies the following boundary-value problem:

$$\frac{d^2\bar{\theta}}{dr^2} + \frac{1}{r}\frac{d\bar{\theta}}{dr} - \frac{p}{\alpha}\bar{\theta} = 0 \tag{8.55a}$$

$$\bar{\theta}(0,p) = \text{finite} , \qquad \bar{\theta}(r_0,p) = \frac{\theta_w}{p} \tag{8.55b,c}$$

The solution of this problem is easily found to be

$$\frac{\bar{\theta}(x,p)}{\theta_w} = \frac{I_0(r\sqrt{p/\alpha})}{pI_0(r_0\sqrt{p/\alpha})} \tag{8.56}$$

This expression is not found in Table 8.1 (nor in most of the tables in the references). The inverse transform can, however, be found by the Heaviside's expansion theorem (8.24) as follows: Note that Eq. (8.56) can be written as

$$\frac{\bar{\theta}(x,p)}{\theta_w} = \frac{P(p)}{Q(p)} \tag{8.57a}$$

where the polynomials P(p) and Q(p) are

$$P(p) = I_0(r\sqrt{p/\alpha}) = \sum_{k=0}^{\infty} \frac{1}{(k!)^2}\left(\frac{r}{2}\right)^{2k}\left(\frac{p}{\alpha}\right)^{k} \tag{8.57b}$$

$$Q(p) = p\, I_0(r_0\sqrt{p/\alpha}) = \sum_{k=0}^{\infty} \frac{1}{(k!)^2}\left(\frac{r_0}{2}\right)^{2k}\left(\frac{p^{k+1}}{\alpha^k}\right) \tag{8.57c}$$

The Heaviside's expansion theorem, therefore, gives

$$\mathcal{L}^{-1}\left\{\frac{I_0(r\sqrt{p/\alpha})}{pI_0(r_0\sqrt{p/\alpha})}\right\} = \sum_{k=0}^{\infty} \frac{P(P_k)}{\frac{dQ(P_k)}{dp}}\, e^{P_k t} \tag{8.58}$$

where P_k's are the zeros of the polynomial Q(p); that is, they are the zeros of

$$p\, I_0(r_0\sqrt{p/\alpha}) = 0 \tag{8.59a}$$

which yields

$$P_0 = 0 \tag{8.59b}$$

and

$$I_0(r_0\sqrt{p_k/\alpha}) = 0\ , \qquad k = 1,2,3,... \tag{8.59c}$$

Equation (8.59c) is also equal to stating that

$$J_0 = (ir_0\sqrt{p_k/\alpha}) = 0\ , \qquad k = 1,2,3,... \tag{8.59d}$$

If λ_k's represent the zeros of the Bessel function $J_0(z)$, then

$$ir_0\sqrt{p_k/\alpha} = \lambda_k \qquad \text{or} \qquad p_k = -\frac{\alpha\lambda_k^2}{r_0^2}\ , \qquad k = 1,2,3,... \tag{8.59e}$$

The zeros of the Bessel function $J_0(z)$ are given in Appendix B and the first four zeros are

$$\lambda_1 = 2.4048,\ \lambda_2 = 5.5201,\ \lambda_3 = 8.6537,\ \lambda_4 = 11.7915$$

The functions $P(p_k)$ can now be found as

$$P(0) = 1$$

$$P(p_k) = I_0(-ir\ \lambda_k/r_0) = J_0\left(\lambda_k \frac{r}{r_0}\right)$$

Since

$$\frac{dQ}{dp} = I_0(r_0\sqrt{p/\alpha}\ ) + \frac{r_0}{2}\sqrt{\frac{p}{\alpha}}\, I_1(r_0\sqrt{p/\alpha}\ )$$

we obtain

$$\frac{dQ(0)}{dp} = 1$$

and

$$\frac{dQ(p_k)}{dp} = -\frac{i\lambda_k}{2} I_1(-i\lambda_k) = -\frac{\lambda_k}{2} J_1(\lambda_k), \qquad k = 1,2,...$$

Thus, the solution can be written as

$$\frac{\theta(x,t)}{\theta_w} = \frac{T(x,t) - T_i}{T_w - T_i} = 1 - 2\sum_{k=1}^{\infty} \frac{J_0(\lambda_k \frac{r}{r_0})}{\lambda_k J_1(\lambda_k)} \exp\left(-\frac{\alpha\lambda_k^2}{r_0^2} t\right) \tag{8.60}$$

which gives the temperature distribution in the cylinder for times $t > 0$.

8.10 SOLID SPHERE

Finally, consider a solid sphere of radius r_0, which is initially at a uniform temperature T_i. The temperature of the surface at $r = r_0$ is changed at $t = 0$ to T_w and maintained constant at this value for times $t > 0$. Let us now find the unsteady temperature distribution $T(r,t)$ in the sphere.

Assuming constant thermo-physical properties, the formulation of the problem in terms of $\theta(r,t) = T(r,t) - T_i$ is given by

$$\frac{\partial^2\theta}{\partial r^2} + \frac{2}{r}\frac{\partial\theta}{\partial r} = \frac{1}{\alpha}\frac{\partial\theta}{\partial t} \tag{8.61a}$$

$$\theta(r,0) = 0 \tag{8.61b}$$

$$\theta(0,t) = \text{finite}, \qquad \theta(r_0,t) = T_w - T_i \equiv \theta_w \tag{8.61c,d}$$

The Laplace transform of the differential equation (8.61a) and the boundary conditions (8.61c,d) are

$$\frac{d^2\bar\theta}{dr^2} + \frac{2}{r}\frac{d\bar\theta}{dr} - \frac{p}{\alpha}\bar\theta = 0 \tag{8.62a}$$

$$\bar\theta(0,p) = \text{finite}, \qquad \bar\theta(r_0,p) = \frac{\theta_w}{p} \tag{8.62b,c}$$

where $\bar\theta(r,p)$ is the Laplace transform of $\theta(r,t)$. The differential equation (8.62a) can be rewritten as

$$\frac{d^2}{dr^2}(r\bar\theta) - \frac{p}{\alpha}(r\bar\theta) = 0 \tag{8.63}$$

the solution of which gives

$$\bar\theta(r,p) = \frac{1}{r}[C_1 \cosh mr + C_2 \sinh mr] \tag{8.64}$$

where $m^2 = p/\alpha$. Since $\bar\theta(0,p) = \text{finite}$, $C_1 \equiv 0$. Furthermore, the boundary condition (8.62c) yields

$$C_2 = \theta_w \frac{r_0}{p \sinh mr_0} \tag{8.65}$$

Then, the solution for $\bar\theta(r,p)$ becomes

$$\frac{\bar\theta}{\theta_w} = \frac{r_0}{r}\frac{\sinh mr}{p \sinh mr_0} = \frac{r_0}{r}\frac{\sinh\sqrt{\frac{p}{\alpha}}\,r}{p\sinh\sqrt{\frac{p}{\alpha}}\,r_0} \tag{8.66}$$

The use of transform No. 39 from Table 8.1 gives

$$\frac{\theta(r,t)}{\theta_w} = \frac{T(r,t) - T_i}{T_w - T_i} = 1 + \frac{2r_0}{\pi r}\sum_{n=1}^{\infty}\frac{(-1)^n}{n} e^{-n^2\pi^2\alpha t/r_0^2} \sin\frac{n\pi r}{r_0} \tag{8.67}$$

Equation (8.67) is a representation of the unsteady temperature distribution in the sphere in the form of an infinite series. Because of the negative argument of the exponential terms, the convergence of this series is fast for large values of time. However, an alternative expression that converges fast for small values of time can be developed by expanding the transform (8.66) as an asymptotic series in negative exponentials, and then inverting the resulting expansion term by term. The development of this alternative form is left as an exercise (see Problem 8.15).

REFERENCES

1. Hildebrand, F.B., *Advanced Calculus for Application,* 2nd ed., Prentice-Hall, Inc., Englewood Cliffs, New Jersey, 1976.
2. Churchill, R.V., *Operational Mathematics,* 3rd ed., McGraw-Hill Book Co., New York, 1972.
3. Arpacı, V.S., *Conduction Heat Transfer,* Addison-Wesley Publishing Co., Inc., Reading, Massachusetts, 1966.
4. Özışık, M.N., *Heat Conduction,* John Wiley and Sons, Inc., New York, 1980.
5. Luikov, A.V., *Analytical Heat Diffusion Theory,* Academic Press, New York, 1968.
6. Myers, G.E., *Analytical Methods in Conduction Heat Transfer,* McGraw-Hill Book Co., New York, 1971.
7. Carslaw, H.S., and Jaeger, J.C., *Conduction of Heat in Solids,* 2nd ed., Oxford, Clarendon Press, 1959.
8. Speigel, M.R., *Laplace Transforms,* Schaum's Outline Series, McGraw-Hill Book Co., New York, 1965.

PROBLEMS

8.1 Prove that $\mathcal{L}\{\sin \omega t\} = \dfrac{\omega}{p^2 + \omega^2}$ and $\mathcal{L}\{\cos \omega t\} = \dfrac{p}{p^2 + \omega^2}$.

8.2 Prove that if $\mathcal{L}\{f(t)\} = \overline{f}(p)$ then $\mathcal{L}\left\{\dfrac{df}{dt}\right\} = p\overline{f}(p) - f(0)$.

8.3 Prove that if $\mathcal{L}\{f(t)\} = \overline{f}(p)$ then $\mathcal{L}\left\{\displaystyle\int_0^t f(t')dt'\right\} = \dfrac{1}{p}\,\overline{f}(p)$.

8.4 Prove that $g*h = h*g$.

8.5 Find the inverse of $\dfrac{\cosh\sqrt{\frac{p}{\alpha}}\,x}{p\,\cosh\sqrt{\frac{p}{\alpha}}\,L}$ by the Heaviside's expansion theorem and obtain the expression (8.36).

8.6 A semi-infinite solid, $x \geq 0$, is initially at a uniform temperature T_i. Assuming that the thermal conductivity k and the thermal diffusivity α of the solid are constant, obtain an expression for the unsteady temperature distribution $T(x,t)$ in the solid, if the surface at $x = 0$ is maintained for times $t \geq 0$ at temperature $T(0,t) = T_0 + (\Delta T)_0 \sin \omega t$.

8.7 The semi-infinite solid shown in Fig. 8.7 is initially at a uniform temperature T_i and has constant thermo-physical properties. For times $t \geq 0$, the surface at $x = 0$ is exposed to a fluid maintained at a constant temperature T_∞ ($\neq T_i$). Assuming that the heat transfer coefficient h is constant, obtain an expression for the unsteady temperature distribution $T(x,t)$ in the solid for times $t > 0$.

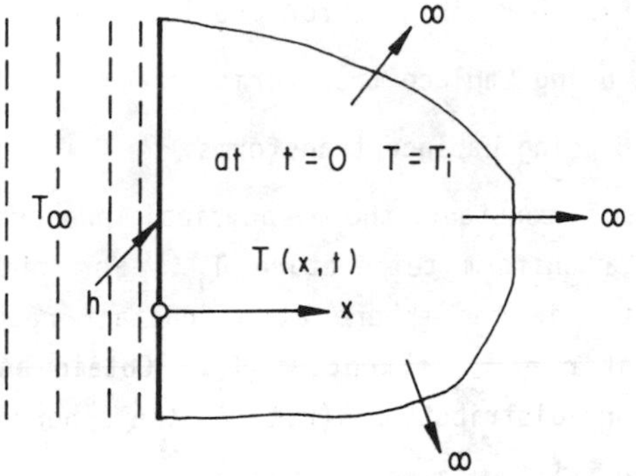

Fig. 8.7 Figure for Problem 8.7

8.8 A semi-infinite solid $x \geq 0$ is initially at a uniform temperature T_i. For times $t \geq 0$, a constant heat flux q_w is applied to the surface at $x = 0$. Assuming that the thermal conductivity k and the thermal diffusivity α of the solid are constant, obtain an expression for the temperature distribution in the solid for times $t > 0$.

8.9 A plane wall of thickness L is initially at a uniform temperature T_i. For times $t \geq 0$, the temperature of the surface at $x = 0$ is varied according to $T(0,t) = F(t)$, where $F(t)$ is a prescribed function of time, and the surface at $x = L$ is kept at T_i. Assuming that the thermo-physical properties of the material of the wall are constant, obtain an expression for the temperature distribution in the wall for $t > 0$.

8.10 Two semi-infinite solids are initially at uniform temperatures T_1 and T_2. The thermo-physical properties of the solids are constant.

(a) Find the temperature distributions in the solids for times $t > 0$ after they have been put, as illustrated in Fig. 8.8, into contact at $t = 0$. Assume perfect thermal contact at the interface.

(b) Obtain an expression for the variation of the interface temperature with time.

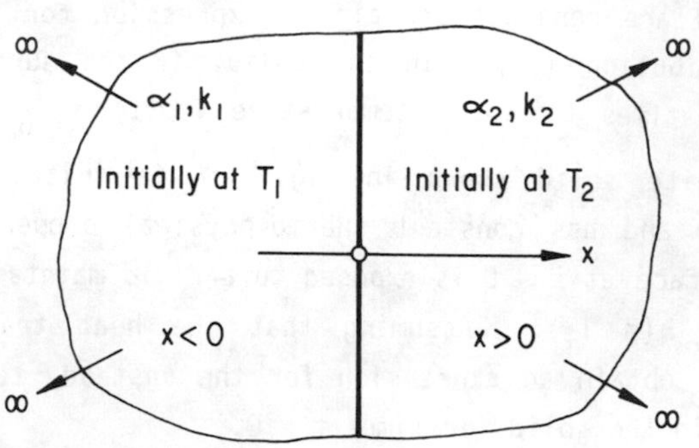

Fig. 8.8 Figure for Problem 8.10.

8.11 Solve Problem 6.1 using Laplace transforms.

8.12 Solve Problem 6.18 using Laplace transforms.

8.13 A solid sphere of constant thermo-physical properties and radius r_o is initially at a uniform temperature T_i. For times $t \geq 0$, internal energy is generated in the sphere at a constant rate $\dot{q}$ per unit volume and the surface at $r = r_o$ is kept at T_i. Obtain an expression for the unsteady temperature distribution $T(r,t)$ in the sphere.

8.14 Solve Problem 6.20 using Laplace transforms.

8.15 A solid sphere of radius r_o is initially at a uniform temperature T_i. The temperature of the surface at $r = r_o$ is maintained for times $t \geq 0$, at a constant temperature T_w ($\neq T_i$). Assume that the thermo-physical properties are constant. Obtain an expression for the unsteady temperature distribution $T(r,t)$ in the sphere that will converge fast for small values of time.

9

Numerical Solutions

9.1 INTRODUCTION

In the preceeding four chapters, three analytical methods to solve linear heat conduction problems have been given for relatively simple geometric shapes in rectangular, cylindrical and spherical coordinates under certain straightforward type boundary conditions. The vast majority of problems encountered in practice, however, cannot be solved analytically as they usually involve irregular geometries with mathematically inconvenient mixed type boundary conditions. In such cases, numerical and/or graphical methods often provide the answer. For example, in a turbine blade the temperature distribution cannot be determined exactly by means of analytical methods, because the boundary of the blade is not parallel to the coordinate surfaces of an orthogonal system. The geometry may be simplified for an analytical solution, but the results will not be accurate enough for practical applications.

With the development of high-speed digital and analog computers, numerical techniques have been developed and extended to handle almost any problem of any degree of complexity. Of the numerical methods available, the *finite-difference* method is the more frequently used one. Accordingly, in this chapter we shall concentrate on the solution of both steady- and unsteady-state heat conduction problems by the finite-difference method.

The essence of the finite-difference method consists of replacing the pertinent differential equation and the boundary conditions by a set of algebraic equations. Our treatment of the method is not intended to be exhaustive in its mathematical rigor, but we shall present the fundamentals of the method and the solutions of finite-differential equations by numerical and graphical means. For more detailed discussion of the applications of the finite-difference and other numerical methods to heat conduction problems, the reader is referred to References [1-15]. Computer programs for the solution of several heat conduction problems can be found in Reference [12].

The finite-element method is another numerical method of solution. This relatively new method is not as straight-forward, conceptually, as the finite-difference method, but it has several advantages in solving conduction problems over the finite-difference method, particularly for problems with complex geometries. Fundamentals of this method are discussed in References [4,13].

9.2 FINITE-DIFFERENCE APPROXIMATION OF DERIVATIVES

In replacing a differential equation or a boundary condition by a set of algebraic equations, the fundamental operation is to approximate the derivatives by finite differences. Let us consider the function T(x) shown in Fig. 9.1. The definition of the derivative of T(x) at x_m is

$$\left.\frac{dT}{dx}\right|_{x_m} = \lim_{\Delta x \to 0} \frac{T(x_m + \Delta x) - T(x_m)}{\Delta x} \tag{9.1}$$

As an approximation let us not go through the limiting process. Then we can write

$$\left.\frac{dT}{dx}\right|_{x_m} \simeq \frac{T(x_m + \Delta x) - T(x_m)}{\Delta x} = \frac{T_{m+1} - T_m}{\Delta x} \tag{9.2}$$

This is an approximate expression for the derivative at x_m and is called the *forward-difference* form of the first derivative. Another approximate expression similar to the forward-difference form (9.2) can also be written

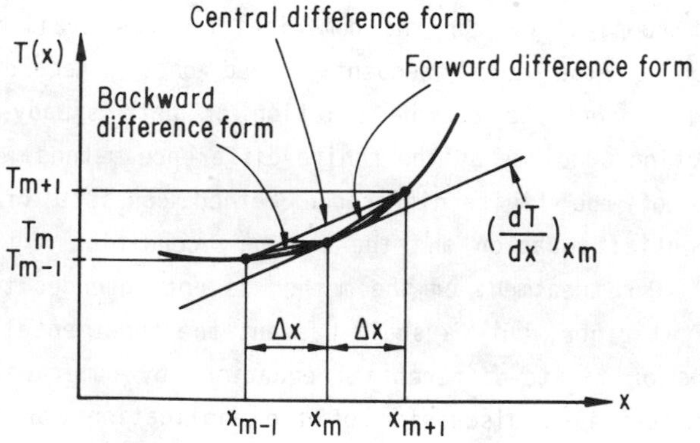

Fig. 9.1 Finite-difference approximation of derivatives.

as

$$\left.\frac{dT}{dx}\right|_{x_m} \simeq \frac{T_m - T_{m-1}}{\Delta x} \tag{9.3}$$

which is known as the *backward-difference form* of the first derivative at x_m.

A more accurate approximate expression than either the forward-difference form or the backward-difference form can clearly be written as

$$\left.\frac{dT}{dx}\right|_{x_m} \simeq \frac{T_{m+1} - T_{m-1}}{2\Delta x} \tag{9.4}$$

which is called the *central-difference form* of the first derivative.

As we shall discuss in Section 9.10, both the forward-difference and the backward-difference forms have a *truncation error* (or *discretetation error*) of the order of magnitude of Δx, and the central-difference form has a truncation error of the order of magnitude of $(\Delta x)^2$.

The second derivative of T(x) can be approximated by the central-difference form as

$$\left.\frac{d^2T}{dx^2}\right|_{x_m} \simeq \frac{\left.\frac{dT}{dx}\right|_{x_{m+\frac{\Delta x}{2}}} - \left.\frac{dT}{dx}\right|_{x_{m-\frac{\Delta x}{2}}}}{\Delta x} \tag{9.5a}$$

When the central-difference forms of $\left.\frac{dT}{dx}\right|_{x_{m+\frac{\Delta x}{2}}}$ and $\left.\frac{dT}{dx}\right|_{x_{m-\frac{\Delta x}{2}}}$ are substituted into Eq. (9.5a) we get

$$\left.\frac{d^2T}{dx^2}\right|_{x_m} \simeq \frac{\frac{T_{m+1} - T_m}{\Delta x} - \frac{T_m - T_{m-1}}{\Delta x}}{\Delta x} \tag{9.5b}$$

which reduces to

$$\left.\frac{d^2T}{dx^2}\right|_{x_m} \simeq \frac{T_{m+1} + T_{m-1} - 2T_m}{(\Delta x)^2} \tag{9.5c}$$

and also has a truncation error of order $(\Delta x)^2$.

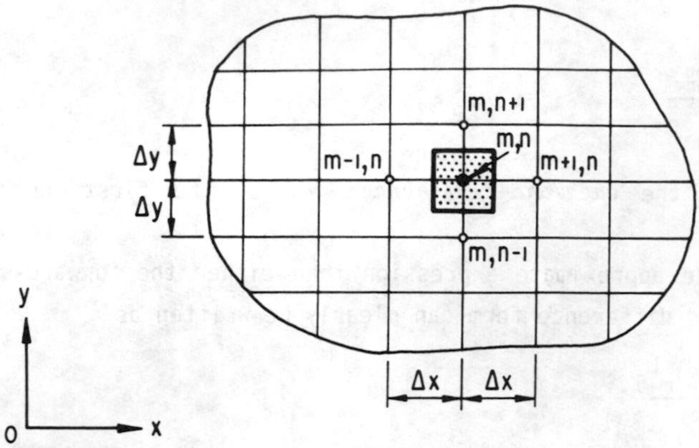

Fig. 9.2 Network of grid points in two-dimensional numerical analysis of heat conduction.

9.3 FINITE-DIFFERENCE FORMULATION OF STEADY-STATE PROBLEMS IN RECTANGULAR COORDINATES

Consider a two-dimensional solid body in rectangular coordinates as shown in Fig. 9.2. The results that we shall obtain can readily be extended to three-dimensional problems in rectangular coordinates and to problems in other geometries as well. Let us assume that the thermal conductivity of the solid is constant and there are no internal energy sources or sinks. Under steady-state conditions the temperature distribution satisfies Laplace's equation

$$\frac{\partial^2 T}{\partial x^2} + \frac{\partial^2 T}{\partial y^2} = 0 \tag{9.6}$$

everywhere within the solid. In other words, the solution of Eq. (9.6), together with the use of the boundary conditions, gives a continuous temperature distribution within the solid. When the differential equation (9.6) and the boundary conditions of the solid are expressed in finite-difference forms, the resulting equations are satisfied only at certain points.

Let us divide the solid body in both x and y directions forming the so-called *network of grid* (or *nodal*) *points* as shown in Fig. 9.2. The grid points are identified by two subscripts, say m and n, m being the number of x-increments and n the number of y-increments. We now approximate Eq. (9.6) at each grid point in the solid by replacing the derivatives with their approximate finite-difference equivalents. For the grid point (m,n) we can apply Eq. (9.5c) to each second-order derivative giving the approximation

$$\frac{T_{m+1,n} + T_{m-1,n} - 2T_{m,n}}{(\Delta x)^2} + \frac{T_{m,n+1} + T_{m,n-1} - 2T_{m,n}}{(\Delta y)^2} = 0 \tag{9.7}$$

If the step lengths in both x and y directions are taken to be the same; that is, $\Delta x = \Delta y$, then this approximation reduces to the algebraic equation

$$T_{m+1,n} + T_{m-1,n} + T_{m,n+1} + T_{m,n-1} - 4T_{m,n} = 0 \tag{9.8a}$$

or

$$T_{m,n} = \frac{1}{4}[T_{m+1,n} + T_{m-1,n} + T_{m,n+1} + T_{m,n-1}] \tag{9.8b}$$

which represents the temperature at the grid point (m,n) in terms of the temperatures at the neighboring points (m+1,n), (m-1,n), (m,n+1) and (m,n-1).

The finite-difference equation (9.8a) is applicable for interior points. If the temperature is known over the whole boundary, then the application of Eq. (9.8a) to each interior grid point is sufficient to allow the temperatures at these grid points to be found. If there are N number of interior grid points, this process gives N simultaneous algebraic equations, and we know from mathematics, at least in principle, how to solve these equations.

If small step lengths are used, then the temperature distribution is more closely approximated. On the other hand, the number of grid points and therefore the number of equations becomes very large. This is a restriction only if the equations are solved by hand. High-speed, large capacity computers eliminate this restriction.

We have given the derivation of the finite-difference equation (9.8a) from mathematical point of view. That is, we have converted the differential equation (9.6) into the finite-difference form by approximating the derivatives at each grid point. The energy balance concept and rate equations can also be used directly to arrive at the same finite-difference equation. To explain this alternative method of deriving the finite-difference equation, reconsider the two-dimensional solid in Fig. 9.2, and define a system around the grid point as shown in Fig. 9.3.

Let us consider the system around the grid point (m,n). The first law of thermodynamics as applied to this system requires that, under steady-state conditions, the net rate of heat transferred to the system be zero. That is,

$$\dot{Q}_{m-1,n} + \dot{Q}_{m,n-1} + \dot{Q}_{m+1,n} + \dot{Q}_{m,n+1} = 0 \tag{9.9}$$

The Fourier's law of heat conduction, for unit depth, gives

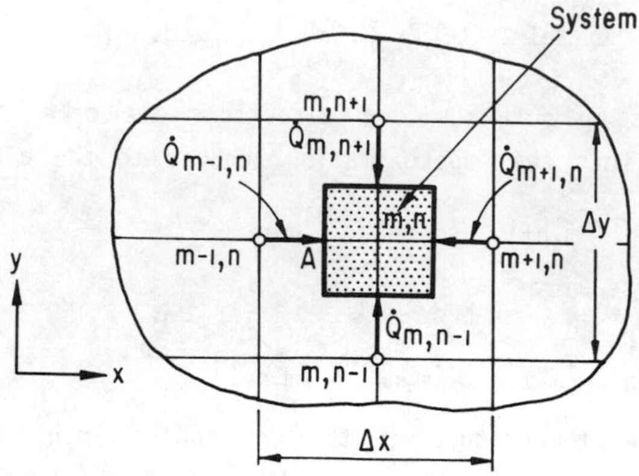

Fig. 9.3 System around the grid point (m,n) for the derivation of the finite-difference equation.

$$\dot{Q}_{m-1,n} = -k\ \Delta y\left(\frac{\partial T}{\partial x}\right)_A$$

which can be approximated as

$$\dot{Q}_{m-1,n} \simeq k\ \Delta y\ \frac{T_{m-1,n} - T_{m,n}}{\Delta x} \tag{9.10a}$$

Similar expressions for $\dot{Q}_{m+1,n}$, $\dot{Q}_{m,n-1}$ and $\dot{Q}_{m,n+1}$ can be written as

$$\dot{Q}_{m+1,n} \simeq k\ \Delta y\ \frac{T_{m+1,n} - T_{m,n}}{\Delta x} \tag{9.10b}$$

$$\dot{Q}_{m,n-1} \simeq k\ \Delta x\ \frac{T_{m,n-1} - T_{m,n}}{\Delta y} \tag{9.10c}$$

and

$$\dot{Q}_{m,n+1} \simeq k\ \Delta x\ \frac{T_{m,n+1} - T_{m,n}}{\Delta y} \tag{9.10d}$$

Substitution of Eqs. (9.10a,b,c,d) into Eq. (9.9) yields

$$k\ \Delta y\ \frac{T_{m-1,n} - T_{m,n}}{\Delta x} + k\ \Delta y\ \frac{T_{m+1,n} - T_{m,n}}{\Delta x} + k\ \Delta x\ \frac{T_{m,n-1} - T_{m,n}}{\Delta y}$$

$$+ k\ \Delta x\ \frac{T_{m,n+1} - T_{m,n}}{\Delta y} = 0 \tag{9.11}$$

If the step lengths Δx and Δy are taken to be equal, then Eq. (9.11) reduces to

$$T_{m+1,n} + T_{m-1,n} + T_{m,n+1} + T_{m,n-1} - 4T_{m,n} = 0$$

which is the same finite-difference relation (9.8a) we have obtained previously.

9.4 FINITE-DIFFERENCE APPROXIMATION OF BOUNDARY CONDITIONS

Finite-difference formulation of a heat conduction problem will be complete if the boundary conditions are also written in finite-difference form. When the temperature on the whole boundary is specified, the known value enters into the finite-difference equations so that each equation for grid points adjacent to the boundary would have a prescribed term in it. If the condition on the whole boundary, or a part of it, is other than a temperature boundary condition, then new finite-difference equations should be derived for the grid points on the boundary. Let us now consider the development of such equations for the grid points on the boundary for different kinds of boundary conditions.

9.4.1 Boundary with Convective Heat Transfer into a Medium at a Prescribed Temperature

We now consider a grid point (m,n) on a boundary which is exposed to a fluid at temperature T_∞ with a constant heat transfer coefficient h as shown in Fig. 9.4. Application of the first law of thermodynamics to the system shown yields

$$k\,\Delta y\,\frac{T_{m-1,n} - T_{m,n}}{\Delta x} + h\,\Delta y(T_\infty - T_{m,n}) + k\,\frac{\Delta x}{2}\,\frac{T_{m,n-1} - T_{m,n}}{\Delta y}$$

$$+\,k\,\frac{\Delta x}{2}\,\frac{T_{m,n+1} - T_{m,n}}{\Delta y} = 0 \tag{9.12}$$

If the step lengths Δx and Δy are taken to be equal then Eq. (9.12) reduces to

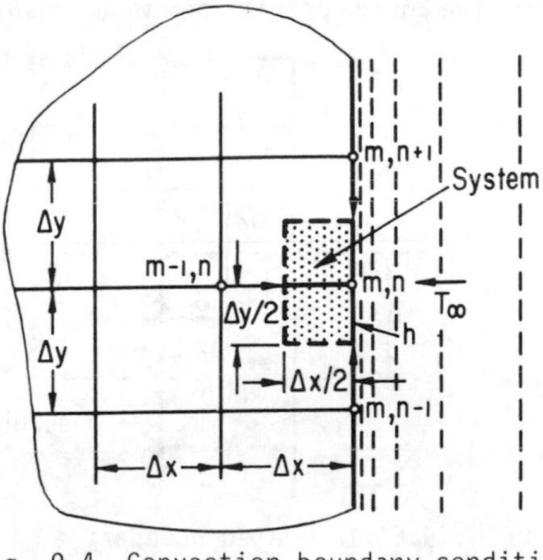

Fig. 9.4 Convection boundary condition.

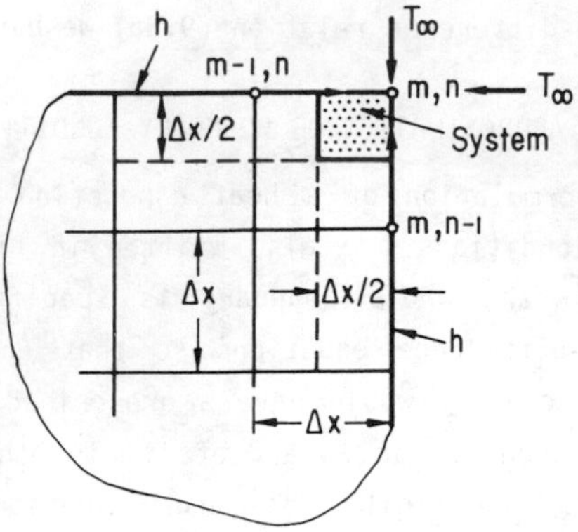

Fig. 9.5 Grid point on a corner with convection boundary condition.

$$\frac{1}{2}(2T_{m-1,n} + T_{m,n+1} + T_{m,n-1}) + \frac{h\ \Delta x}{k}T_\infty - (2 + \frac{h\ \Delta x}{k})T_{m,n} = 0 \tag{9.13}$$

Thus, when a convection boundary condition is present as in Fig. 9.4, the finite-difference equation (9.13) is used for the grid points on the boundary and the finite-difference equation (9.8a) for the interior grid points. If the grid point happens to be on a corner such as in Fig. 9.5, then Eq. (9.13) is not applicable. Let us now consider the corner section in Fig. 9.5. Application of the first law of thermodynamics to the system shown yields

$$\frac{1}{2}(T_{m-1,n} + T_{m,n-1}) + \frac{h\ \Delta x}{k}T_\infty - (1 + \frac{h\ \Delta x}{k})T_{m,n} = 0 \tag{9.14}$$

9.4.2 Insulated Boundary

When part of the boundary is insulated as shown in Fig. 9.6, finite-difference equations for the grid points on this insulated section can be

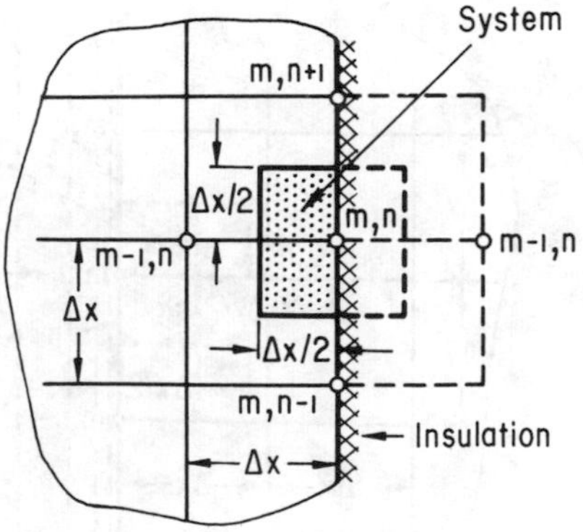

Fig. 9.6 Insulated boundary.

derived as in the previous case. Application of the first law of thermodynamics to the system shown yields

$$T_{m,n+1} + T_{m,n-1} + 2T_{m-1,n} - 4T_{m,n} = 0 \tag{9.15}$$

Since $\partial T/\partial x = 0$ on the boundary, Eq. (9.15) can also be directly written from Eq. (9.8a) by setting $T_{m+1,n} = T_{m-1,n}$ (why?).

9.5 IRREGULAR BOUNDARIES

In problems with simple geometries, we can arrange for certain grid points to lie on the boundaries. However, in cases in which the boundary does not fall on regular grid points, the boundary is considered to be *irregular*. In many practical problems, the boundaries are usually irregular and the numerical methods are the only means for solving such problems. Let us consider the irregular boundary shown in Fig. 9.7 and assume that the temperature of the boundary is known. The finite-difference equation (9.8a) cannot be used, for example, for the grid point (m,n) next to the boundary.

Application of the first law of thermodynamics to the system shown yields

$$k\,\frac{(1+b)\Delta y}{2}\,\frac{T_{m-1,n} - T_{m,n}}{a\Delta x} + k\,\frac{(1+b)\Delta y}{2}\,\frac{T_{m+1,n} - T_{m,n}}{\Delta x}$$

$$+ k\,\frac{(1+a)\Delta x}{2}\,\frac{T_{m,n-1} - T_{m,n}}{b\Delta y} + k\,\frac{(1+a)\Delta x}{2}\,\frac{T_{m,n+1} - T_{m,n}}{\Delta y} = 0 \tag{9.16}$$

If the step lengths are taken to be equal; that is, $\Delta x = \Delta y$, then Eq. (9.16) reduces to

$$\frac{1}{1+a}\,T_{m+1,n} + \frac{1}{a(1+a)}\,T_{m-1,n} + \frac{1}{1+b}\,T_{m,n+1} + \frac{1}{b(1+b)}\,T_{m,n-1}$$

$$- \left[\frac{1}{a} + \frac{1}{b}\right] T_{m,n} = 0 \tag{9.17}$$

where a and b are defined in Fig. 9.7. We note that the grid point (m,n) is not located at the geometric center of the system. If $a = b = 1$, then Eq. (9.17) reduces to Eq. (9.8a).

9.6 SOLUTION OF FINITE-DIFFERENCE EQUATIONS

The finite-difference representation of the differential equation and the boundary conditions of steady-state problems results in a system of algebraic equations. Except for the problems in which thermal conductivity depends on temperature, these equations will be linear. The resultant system of finite-difference equations must be solved to find the temperatures at the

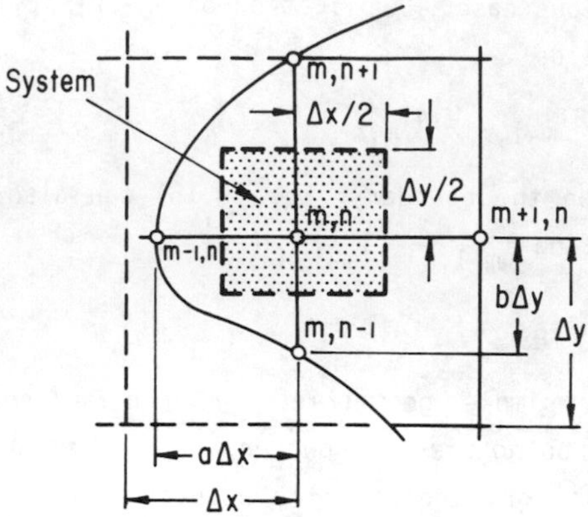

Fig. 9.7 Irregular boundary.

various grid points. If the number of grid points, and therefore the number of equations, are not large, then the system of finite-difference equations can be solved by hand or by the use of a desk calculator using the so-called *relaxation* method which was first introduced by Southwell [14]. If the number of grid points are large, then the equations can be solved with a large-capacity digital computer.

9.6.1 Relaxation Method

When the temperature of the whole boundary is specified, this method for two-dimensional steady-state problems is carried out according to the following steps:

(1) Set the right-hand side of Eq. (9.8a) equal to some residual $R_{m,n}$;

$$T_{m+1,n} + T_{m-1,n} + T_{m,n+1} + T_{m,n-1} - 4T_{m,n} = R_{m,n} \tag{9.18}$$

which we wish to relax to zero.

(2) Guess values for the temperatures of the interior grid points.

(3) Calculate the residual $R_{m,n}$ at each grid point using the assumed values. If the assumed values are the true values of the temperatures at each grid, then the residuals will, of course, be zero. In general, they will not be zero.

(4) Select the largest residual and try to make it, at least approximately, zero by changing the assumed temperature of the corresponding grid point, while holding the temperatures of the other grid points constant.

(5) Compute new residuals. Note that only the residuals of the grid

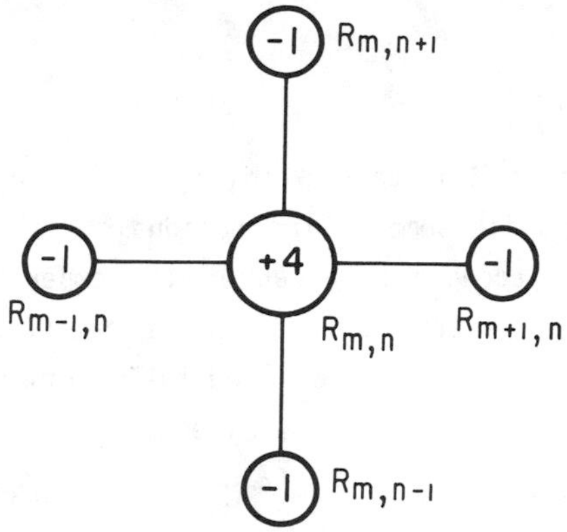

Fig. 9.8 Two-dimensional relaxation pattern; effect of unit decrease in $T_{m,n}$ on $R_{m,n}$'s.

points adjacent to the readjusted grid point are affected. The computation of new residuals may be accomplished by using the relaxation pattern shown in Fig. 9.8. Since $dR_{m,n} = -4dT_{m,n}$ when all other temperatures are held constant, a unit decrease in $T_{m,n}$ results in an increase in $R_{m,n}$ by 4, and a unit decrease in the residuals of the adjacent grid points.

(6) Continue to relax the residuals until they are all as close to zero as desired.

The relaxation method will now be demonstrated by an example.

Example 9.1: Let us find the temperature distribution and the rate of heat flow through a two-dimensional solid with the boundary conditions as shown in Fig. 9.9.

Solution: We divide the solid into increments as shown ($\Delta x = \Delta y$). The residual equations for grid points A, B, C, and D are

$$1000 + T_B - 4T_A = R_A$$

$$600 + T_A + T_C - 4T_B = R_B \tag{9.19a,b}$$

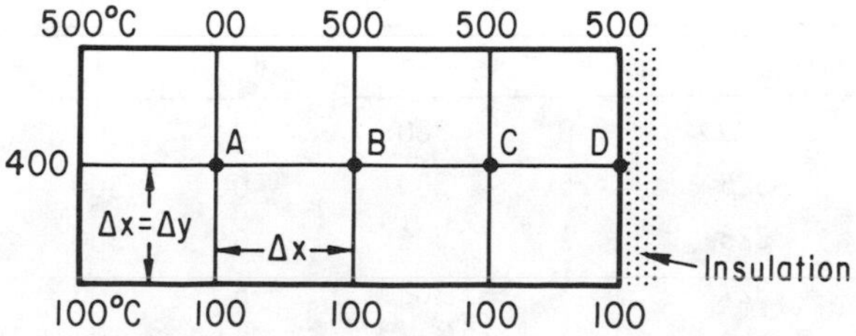

Fig. 9.9 Nomenclature for Example 9.1.

$$600 + T_B + T_D - 4T_C = R_C$$

$$600 + 2T_C - 4T_D = R_D \tag{9.19c,d}$$

The relaxation calculations for this problem are given in Table 9.1. The process has been stopped while some of the residuals still have nonzero values. However, the accuracy is acceptable because all temperatures are within 1°C of their correct value. Notice that we have overrelaxed or underrelaxed the residuals during the calculations to speed the solution.

The rate of heat transfer at 100°C is given by

$$\dot{Q} = kb\,[\tfrac{1}{2}(400 - 100) + (327 - 100) + (307 - 100) + (302 - 100) + \tfrac{1}{2}(300 - 100)] = 886\ kb$$

where k is the thermal conductivity and b is the depth of the solid. If a sufficiently fine grid network were used, a more accurate value of the heat transfer rate would be obtained. The rate of heat transfer at 500°C face or at 400°C face can be obtained in the same way.

9.6.2 Matrix Inversion Method

A system of N number of finite-difference equations is a set of algebraic equations involving the unknown temperatures implicitly. This system of finite-difference equations can be written in matrix notation as

$$\begin{bmatrix} A_{11} & A_{12} & \cdots & A_{1N} \\ A_{21} & A_{22} & \cdots & A_{2N} \\ \cdot & & & \cdot \\ \cdot & & & \cdot \\ \cdot & & & \cdot \\ A_{N1} & A_{N2} & \cdots & A_{NN} \end{bmatrix} \begin{bmatrix} T_1 \\ T_2 \\ \\ \\ \\ T_N \end{bmatrix} = \begin{bmatrix} C_1 \\ C_2 \\ \\ \\ \\ C_N \end{bmatrix} \tag{9.20a}$$

Table 9.1 Relaxation Table for Example 9.1.*

A		B		C		D	
R	T	R	T	R	T	R	T
100	300	0	300	0	300	0	300
0	325	25	300	0	300	0	300
6	325	1	306	6	300	0	300
-2	327	3	306	6	300	0	300
-2	327	5	306	-2	302	2	300
-1	327	1	307	-1	302	2	300

*T in °C

or

$$\mathbf{A} \cdot \mathbf{T} = \mathbf{C} \tag{9.20b}$$

where **A** is the coefficient matrix, **T** is the vector of unknown temperatures, and **C** is a column vector of known constants. In heat conduction problems, most of the off-diagonal elements of **A** would be zero. Multiplication of Eq. (9.20b) from the left with the inverse $\mathbf{A}^{-1}$ of the coefficient matrix yields the unknown vector **T** as

$$\mathbf{T} = \mathbf{A}^{-1} \cdot \mathbf{C} \tag{9.21}$$

whose elements are the unknown temperatures at the grid points. The problem, then, reduces to inverting the matrix **A**, but the inverse can be found efficiently by using a digital computer. We shall now solve an example problem to illustrate the method.

Example 9.2: Let us re-solve the problem in Example 9.1 using the matrix inversion method.

Solution: The finite-difference equations at the grid points A, B, C, and D can be written as

$$\begin{aligned} 4T_A - T_B \qquad\qquad &= 1000 \\ -T_A + 4T_B - T_C \qquad &= 600 \\ -T_B + 4T_C - T_D &= 600 \\ -2T_C + 4T_D &= 600 \end{aligned} \tag{9.22a,b,c,d}$$

In matrix notation these equations would be

$$\mathbf{A} \cdot \mathbf{T} = \mathbf{C}$$

where

$$\mathbf{A} = \begin{bmatrix} 4 & -1 & 0 & 0 \\ -1 & 4 & -1 & 0 \\ 0 & -1 & 4 & -1 \\ 0 & 0 & -2 & 4 \end{bmatrix}, \quad \mathbf{T} = \begin{bmatrix} T_A \\ T_B \\ T_C \\ T_D \end{bmatrix}, \text{ and } \mathbf{C} = \begin{bmatrix} 1000 \\ 600 \\ 600 \\ 600 \end{bmatrix}$$

Now, a digital computer may be employed and standard subroutines may be used to find the inverse $\mathbf{A}^{-1}$. In this special case, $\mathbf{A}^{-1}$ can also be obtained by hand. Here we shall give the result and interested readers may refer to books on linear algebra such as the References [17,19] for matrix inversions. The inverse $\mathbf{A}^{-1}$ is found to be

$$A^{-1} = \begin{bmatrix} 0.268 & 0.072 & 0.021 & 0.005 \\ 0.072 & 0.289 & 0.082 & 0.021 \\ 0.021 & 0.082 & 0.309 & 0.077 \\ 0.010 & 0.041 & 0.155 & 0.289 \end{bmatrix}$$

Therefore,

$$\begin{bmatrix} T_A \\ T_B \\ T_C \\ T_D \end{bmatrix} = \begin{bmatrix} 0.268 & 0.072 & 0.021 & 0.005 \\ 0.072 & 0.289 & 0.082 & 0.021 \\ 0.021 & 0.082 & 0.309 & 0.077 \\ 0.010 & 0.041 & 0.155 & 0.289 \end{bmatrix} \begin{bmatrix} 1000 \\ 600 \\ 600 \\ 600 \end{bmatrix}$$

which yields

$$T_A = 0.268 \times 1000 + (0.072 + 0.021 + 0.005) \times 600 = 326.8°C$$

$$T_B = 0.072 \times 1000 + (0.289 + 0.082 + 0.021) \times 600 = 307.2°C$$

$$T_C = 0.021 \times 1000 + (0.082 + 0.309 + 0.077) \times 600 = 301.8°C$$

$$T_D = 0.010 \times 1000 + (0.041 + 0.155 + 0.289) \times 600 = 301.0°C$$

Clearly for a large number of grid points, the use of a computer is a necessity.

9.6.3 Gaussian Elimination Method

This is probably one of the most commonly used methods for solving a coupled system of linear algebraic equations. The basic idea behind this method is back-substitution. We shall explain the Gaussian elimination method by means of the following example.

Example 9.3: Re-solve the problem in Example 9.1 using the Gaussian elimination method.

Solution: Referring to Eqs. (9.22a,b,c,d), we divide the first equation by 4 (the coefficient of T_A) to obtain an expression for T_A and use the results to eliminate T_A in all other equations. We then divide the second equation by the now modified coefficient of T_B to obtain an expression for T_B which can then be eliminated in the remaining equations, and so on through the rest of the system. We eventually obtain an equivalent system

$$\begin{aligned} 4T_A - T_B \quad &= 1000 \\ 3.75T_B - T_C \quad &= 850 \\ 3.733T_C - T_D &= 826.7 \\ 3.464T_D &= 1042.9 \end{aligned} \qquad (9.23a,b,c,d)$$

The unknown T_D in the system is now found to be

$T_D = 301°C$

and the rest follows by back-substitution: Eq. (9.23c) gives

$T_C = 302.1°C$

and Eq. (9.23b) gives

$T_B = 307.2°C$

and Eq. (9.23a) gives

$T_A = 326.8°C$

The last two methods we have seen are examples of *direct methods*. There are also *iterative methods* of solution such as the *Gauss-Seidel method* or the *successive over relaxation method*. A brief summary of available numerical methods can be found in References [3,5,7].

9.7 FINITE-DIFFERENCE FORMULATION OF ONE-DIMENSIONAL UNSTEADY-STATE PROBLEMS IN RECTANGULAR COORDINATES

In one-dimensional unsteady-state problems in rectangular coordinates, the differential equation which governs the temperature distribution for constant thermo-physical properties is given by

$$\frac{\partial^2 T}{\partial x^2} = \frac{1}{\alpha}\frac{\partial T}{\partial t} \tag{9.24}$$

which can also be expressed in finite-difference form by approximating the derivatives. Let us form a network of grid points by dividing x and t domains into small intervals of Δx and Δt as shown in Fig. 9.10, so that T_n^p represents

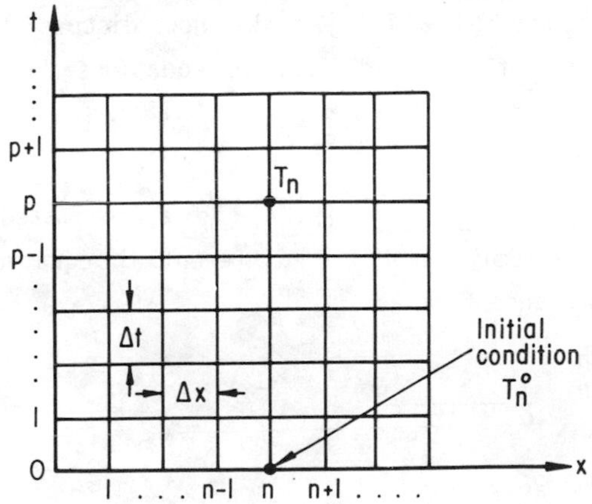

Fig. 9.10 Network of grid points for one-dimensional unsteady-state problems.

the temperature at location $x = n\Delta x$ at time $t = p\Delta t$. At time $t(=p\Delta t)$, the second-order partial derivative in Eq. (9.24) can be represented as

$$\left.\frac{\partial^2 T}{\partial x^2}\right|_t \simeq \frac{T_{n+1}^p + T_{n-1}^p - 2T_n^p}{(\Delta x)^2} \tag{9.25}$$

The time derivative in Eq. (9.24) may be approximated in terms of central, backward, or forward differences as

$$\left.\frac{\partial T}{\partial t}\right|_x \simeq \frac{T^{p+1} - T^{p-1}}{2\,\Delta t} \tag{9.26a}$$

$$\left.\frac{\partial T}{\partial t}\right|_x \simeq \frac{T^p - T^{p-1}}{\Delta t} \tag{9.26b}$$

$$\left.\frac{\partial T}{\partial t}\right|_x \simeq \frac{T^{p+1} - T^p}{\Delta t} \tag{9.26c}$$

These three approximations have different error levels and also have different stability properties as we shall discuss. In other word, each approximation has its own advantages and disadvantages. For example, finite-difference equations resulting from forward-difference approximation of the time derivative are uncoupled and therefore easy to solve, but their solutions are not always stable. On the other hand, use of backward-difference approximation yields finite-difference equations which are coupled and therefore difficult to solve, but their solutions are always stable. The use of the more accurate central-difference approximation of the time derivative, although it is apparently very attractive, results in a set of finite-difference equations, whose solutions are always unstable [7,16]. We now discuss the representation of Eq. (9.24) in terms of finite-difference equations by different methods.

9.7.1 Explicit Method

If the time derivative in Eq. (9.24) is approximated by the forward-difference approach (9.26c), then the differential equation (9.24) can be written in finite-difference form as

$$\frac{T_{n+1}^p + T_{n-1}^p - 2T_n^p}{(\Delta x)^2} = \frac{1}{\alpha}\,\frac{T_n^{p+1} - T_n^p}{\Delta t} \tag{9.27a}$$

which can be rearranged as

$$T_n^{p+1} = \left[1 - \frac{2\alpha\Delta t}{(\Delta x)^2}\right]T_n^p + \frac{\alpha\Delta t}{(\Delta x)^2}(T_{n+1}^p + T_{n-1}^p) \tag{9.27b}$$

This formulation is called *explicit* because it is possible to write the temperature T_n^{p+1} (at location n, at time t + Δt) explicitly in terms of the temperatures at time t and at locations n-1, n, and n+1. Thus, if the temperatures of various grid locations are known at any particular time, the temperatures after a time increment Δt may be calculated by writing an equation like Eq. (9.27b) for each grid location and obtaining values of T_n^{p+1}. The calculation proceeds directly from one time increment to the next until the desired temperature distribution is obtained.

The energy balance concept and rate equations can also be used directly to arrive at the finite-difference equation (9.27b). To explain this alternative method, consider a one-dimensional solid and define a system as shown in Fig. 9.11. The first law of thermodynamics as applied to this system gives

$$kA\,\frac{T_{n-1}^p - T_n^p}{\Delta x} + kA\,\frac{T_{n+1}^p - T_n^p}{\Delta x} = \rho c A \Delta x\,\frac{T_n^{p+1} - T_n^p}{\Delta t}$$

which yields Eq. (9.27a) when rearranged.

If the space increment Δx and time increment Δt are chosen such that

$$\frac{\alpha\Delta t}{(\Delta x)^2} = \frac{1}{2}$$

then Eq. (9.27b) reduces to

$$T_n^{p+1} = \frac{1}{2}(T_{n+1}^p + T_{n-1}^p) \tag{9.28}$$

Therefore, the temperature of the location n after one time increment is

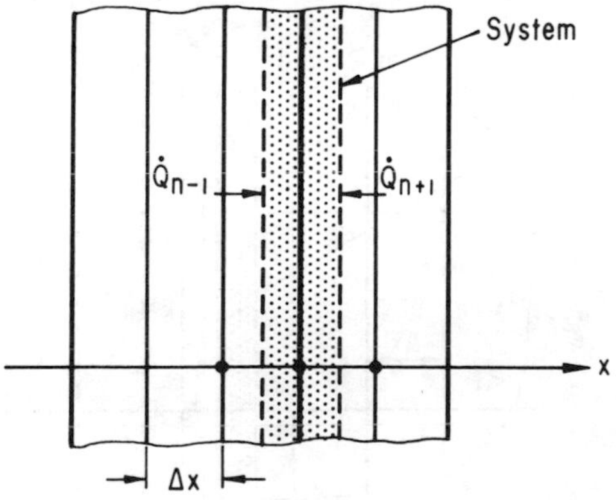

Fig. 9.11 One-dimensional solid.

given by the arithmetic average of the temperatures of the adjacent locations at the beginning of the time increment.

Depending on the values of Δx and Δt, the coefficient of T_n^p in Eq. (9.27b) may be negative, zero, or positive. As we shall show in Section 9.11, if this coefficient is negative, a condition is generated such that the second law of thermodynamics is violated and the solution becomes unstable. One might think that small space increments could be used for greater accuracy in combination with large time increments to speed the calculation. This is usually not the case, because it may result in a negative coefficient for T_n^p. To provide some insight into the physical significance of the stability, suppose that the temperatures T_{n+1}^p and T_{n-1}^p are zero, and the temperature T_n^p is positive. Then, if the coefficient of T_n^p is negative, the temperature T_n^{p+1} at the location n becomes negative. This is impossible because this is against the second law of thermodynamics as heat cannot flow in the direction of a positive temperature gradient. Therefore, for stable solutions we should have

$$\frac{\alpha \Delta t}{(\Delta x)^2} \leq \frac{1}{2}$$

Once Δx is established, this restriction automatically limits our choice of Δt.

If the temperatures of the surfaces of the solid are specified, then the difference equation (9.27b) is used to determine the temperatures of the internal grid locations as a function of time. If convection boundary condition prevails on the whole boundary, or part of it, then the above relations are no longer applicable for the boundary. Therefore, such a boundary must be handled separately. For the one-dimensional system shown in Fig. 9.12,

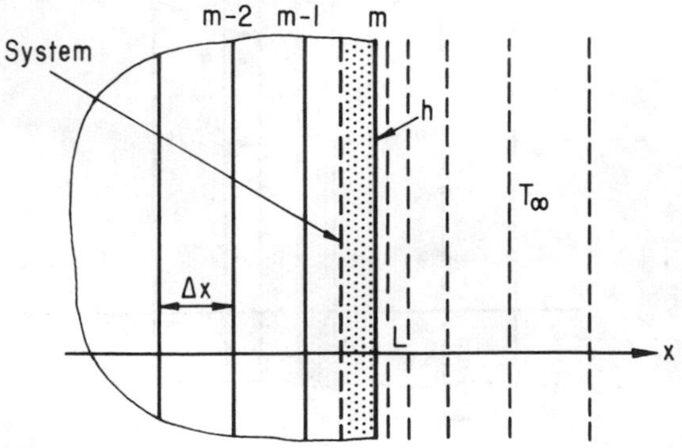

Fig. 9.12 Boundary at x = L.

the boundary condition at the surface at $x = L$ is

$$-k \left.\frac{\partial T}{\partial x}\right|_{x=L} = h[T(L) - T_\infty] \tag{9.29}$$

The finite-difference approximation of the boundary condition (9.29) can be written as

$$k \frac{T_{m-1}^{p+1} - T_m^{p+1}}{\Delta x} = h[T_m^{p+1} - T_\infty] \tag{9.30a}$$

or

$$T_m^{p+1} = \frac{1}{1 + \frac{h\Delta x}{k}} [T_{m-1}^{p+1} + \frac{h\Delta x}{k} T_\infty] \tag{9.30b}$$

Substitution of T_{m-1}^{p+1} from Eq. (9.27b) into Eq. (9.30b) gives

$$T_m^{p+1} = \frac{1}{1 + \frac{h\Delta x}{k}} \left\{ [1 - \frac{2\alpha\Delta t}{(\Delta x)^2}]T_{m-1}^p + \frac{\alpha\Delta t}{(\Delta x)^2} (T_m^p + T_{m-2}^p) + \frac{h\Delta x}{k} T_\infty \right\} \tag{9.31}$$

Equation (9.31) does not include the effect of heat capacity of the element of the wall at the boundary. If small Δx is used, this approximation works fairly well simply because the heat capacity of the element becomes negligible. A better approximation could be obtained by taking the heat capacity of the system shown into consideration. Application of the first law of thermodynamics to the system yields

$$k \frac{T_{m-1}^p - T_m^p}{\Delta x} + h(T_\infty - T_m^p) = \rho c \frac{\Delta x}{2} \frac{T_m^{p+1} - T_m^p}{\Delta t} \tag{9.32a}$$

which can be rearranged as

$$T_m^{p+1} = \frac{\alpha\Delta t}{(\Delta x)^2} \left\{ \left[\frac{(\Delta x)^2}{\alpha\Delta t} - 2\frac{h\Delta x}{k} - 2\right] T_m^p + 2T_{m-1}^p + 2\frac{h\Delta x}{k} T_\infty \right\} \tag{9.32b}$$

For convection boundary condition, the stability of the temperatures of the grid points on the surface must also be insured and therefore the selection of the parameter $(\Delta x)^2/\alpha\Delta t$ is not so simple as it is for interior points. This parameter can be so chosen that the coefficients of T_m^p for both interior and surface grid points become either positive or zero; that is,

$$\frac{(\Delta x)^2}{\alpha\Delta t} \geq 2(\frac{h\Delta x}{k} + 1)$$

Schneider [1] and Price and Slack [15] discuss the stability of boundary conditions and numerical solutions in some detail. For the stability problem and the convergence of the solutions the interested reader may also consult References [3,6,7].

Example 9.4: The initial temperature distribution inside a homogeneous flat plate of thickness 60 cm is as shown in Fig. 9.13. The plate is cooled from both sides by a coolant at 100°C. The surface heat transfer coefficient h on both sides is constant and has a value of 15 W/m²·K. Let us find the temperatures at the grid locations as a function of time. The properties of the solid are k = 3 W/m·K, ρ = 1500 kg/m³, and c = 0.9 kJ/kg·K.

Solution: The temperatures of the interior grid locations (1), (2), (3), (4) and (5) can be calculated using Eq. (9.27b) as

$$T_n^{p+1} = [1 - \frac{2\alpha\Delta t}{(\Delta x)^2}]T_n^p + \frac{\alpha\Delta t}{(\Delta x)^2}(T_{n+1}^p + T_{n-1}^p), \qquad n = 1,2,3,4 \text{ and } 5$$

and the surface temperatures can be found using Eq. (9.32b),

$$T_0^{p+1} = \frac{\alpha\Delta t}{(\Delta x)^2}\left\{\left[\frac{(\Delta x)^2}{\alpha\Delta t} - 2\frac{h\Delta x}{k} - 2\right]T_0^p + 2T_1^p + 2\frac{h\Delta x}{k}T_\infty\right\}$$

$$T_6^{p+1} = \frac{\alpha\Delta t}{(\Delta x)^2}\left\{\left[\frac{(\Delta x)^2}{\alpha\Delta t} - 2\frac{h\Delta x}{k} - 2\right]T_6^p + 2T_5^p + 2\frac{h\Delta x}{k}T_\infty\right\}$$

Since convection boundary condition prevails on the surrounding faces, for a stable solution

$$\frac{(\Delta x)^2}{\alpha\Delta t} \geq 2(\frac{h\Delta x}{k} + 1)$$

On the other hand,

$$\frac{h\Delta x}{k} = 0.5$$

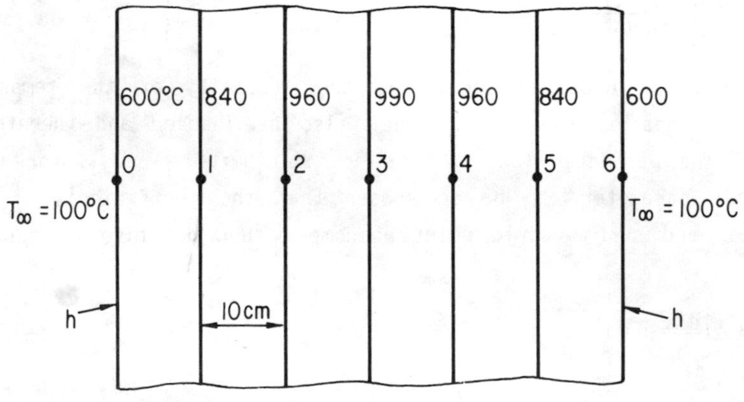

Fig. 9.13 Nomenclature for Example 9.4.

Table 9.2 Results of Example 9.4.*

p	t(min)	T_0	T_1	T_2	T_3	T_4	T_5	T_6
0	0	600	840	960	990	960	840	600
1	25	593	800	930	970	930	800	593
2	50	566	774	900	943	900	774	566
3	75	549	747	872	914	872	947	549
4	100	531	723	844	886	844	723	531

*Temperatures are in °C.

therefore

$$\frac{(\Delta x)^2}{\alpha \Delta t} \geq 3$$

If $(\Delta x)^2/\alpha\Delta t = 3$ is selected, then we can write

$$T_n^{p+1} = \frac{1}{3}(T_n^p + T_{n+1}^p + T_{n-1}^p), \qquad n = 1,2,3,4 \text{ and } 5$$

and

$$T_0^{p+1} = \frac{1}{3}[2T_1^p + T_\infty]$$

$$T_6^{p+1} = \frac{1}{3}[2T_5 + T_\infty]$$

Furthermore,

$$\Delta t = \frac{(\Delta x)^2}{3\alpha}$$

$$= \frac{(0.1)^2 \text{ m}^2 \text{ x}(1500)\text{kg/m}^3 \text{ x}(0.9)\text{kJ/kg K}}{3 \text{ x}(3) \text{ W/m.K } (10^{-3}) \text{ kJ/J}}$$

$$= 1500 \text{ sec.} = 25 \text{ min.}$$

Now, the temperatures of the grid locations can be calculated using the above relations as a function of time with 25 min. time intervals. Results and the calculations are given in Table 9.2.

9.7.2 Implicit Method

In the explicit method discussed in the previous section, we have seen that the requirements, such as $\alpha\Delta t/(\Delta x)^2 \leq \frac{1}{2}$ for interior grid locations, place an undesirable restriction on the time increment Δt. For problems extending over large values of times, this could result in excessive amounts

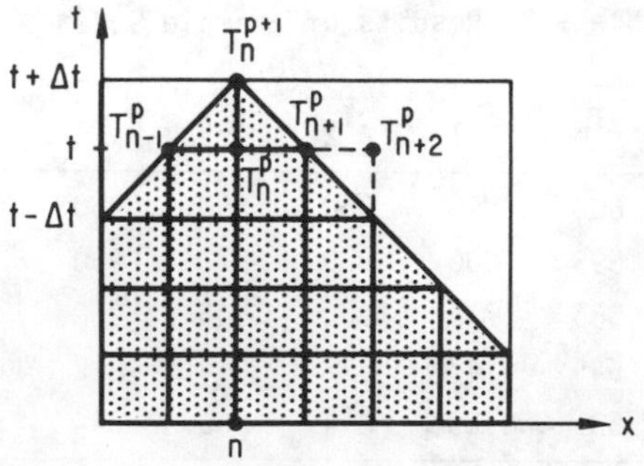

Fig. 9.14 Limitation of the explicit method.

of computation. Furthermore, in the explicit method, T_n^{p+1} depends only on T_n^p, T_{n+1}^p and T_{n-1}^p as depicted in Fig. 9.14. It is natural, however, to expect that T_n^{p+1} should also depend on, for example, T_{n+2}^p.

The implicit method, to be discussed now, overcomes both of these difficulties at the expense of a somewhat more complicated calculational procedure. In this method, $\partial^2 T/\partial x^2$ is replaced by a finite-difference form evaluated at t + Δt and $\partial T/\partial t$ by backward finite-difference form. Then, the differential Equation (9.24) becomes

$$\frac{T_{n+1}^{p+1} + T_{n-1}^{p+1} - 2T_n^{p+1}}{(\Delta x)^2} = \frac{1}{\alpha}\frac{T_n^{p+1} - T_n^p}{\Delta t} \tag{9.33a}$$

which can be written as

$$T_n^p = \left[1 + \frac{2\alpha\Delta t}{(\Delta x)^2}\right] T_n^{p+1} - \frac{\alpha\Delta t}{(\Delta x)^2}\left[T_{n+1}^{p+1} + T_{n-1}^{p+1}\right] \tag{9.33b}$$

We note that this backward-difference formulation does not permit the explicit calculation of T_n^{p+1} in terms of T_n^p. Rather, at any one time level, Eq. (9.33b) will be written once for each grid point, resulting in a system of algebraic equations, which must be solved simultaneously to determine the temperatures T^{p+1}. Solution of such coupled equations, of course, is much more difficult than the solution of uncoupled ones which result in the explicit method. On the other hand, no restriction is imposed on the step size Δx and time increment Δt because of the stability conditions. This means that larger time increments can be selected to speed the calculation. Solution of the finite-difference equations can be performed with the methods discussed in Section 9.6.

9.7.3 Crank-Nicolson Method

In this method of finite-difference formulation, the arithmetic average of Eq. (9.27a) and Eq. (9.33a) is taken. This yields

$$\frac{1}{2}\left[\frac{T_{n+1}^{p} + T_{n-1}^{p} - 2T_{n}^{p}}{(\Delta x)^2} + \frac{T_{n+1}^{p+1} + T_{n-1}^{p+1} - 2T_{n}^{p+1}}{(\Delta x)^2}\right] = \frac{1}{\alpha}\,\frac{T_{n}^{p+1} - T_{n}^{p}}{\Delta t} \tag{9.34a}$$

which can be written as

$$2\left[1 + \frac{\alpha\Delta t}{(\Delta x)^2}\right] T_{n}^{p+1} - \frac{\alpha\Delta t}{(\Delta x)^2}\left[T_{n+1}^{p+1} + T_{n-1}^{p+1}\right]$$

$$= 2\left[1 - \frac{\alpha\Delta t}{(\Delta x)^2}\right] T_{n}^{p} + \frac{\alpha\Delta t}{(\Delta x)^2}\left[T_{n+1}^{p} + T_{n-1}^{p}\right] \tag{9.34b}$$

This result is similar to Eq. (9.33b) for the implicit method. It may be shown that Crank-Nicolson method is stable for all values of $\alpha\Delta t/(\Delta x)^2$ and converges, as we shall discuss in Section 9.10, with the truncation error of the order of $[(\Delta t)^2 + (\Delta x)^2]$. This is a distinct improvement over the previous two methods because both the explicit and the implicit methods lead to a truncation error of the order of magnitude of $[\Delta t + (\Delta x)^2]$. On the other hand, the finite-difference equations in this case are slightly more complicated than the equations for the implicit method.

In taking the average of Eq. (9.27a) and Eq. (9.33a), a weighted average can be taken. In literature, there are other methods for finite differencing obtained in this way. Interested readers can refer to References [5,7] for details of the other methods.

9.8 FINITE-DIFFERENCE FORMULATION OF TWO-DIMENSIONAL UNSTEADY-STATE PROBLEMS IN RECTANGULAR COORDINATES

Consider a two-dimensional solid as shown in Fig. 9.15. Within this solid, the temperature distribution satisfies

$$\frac{\partial^2 T}{\partial x^2} + \frac{\partial^2 T}{\partial y^2} = \frac{1}{\alpha}\frac{\partial T}{\partial t} \tag{9.35}$$

assuming constant thermo-physical properties. The initial temperature distribution at $t = 0$ is specified at all points on and within the boundary, and the subsequent conditions on the boundary are usually either of the first, the second or the third kind. Divide this solid into increments as we did before.

As in the one-dimensional unsteady-state case, the finite-difference approximation of Eq. (9.35) can be written in a number of forms. With the

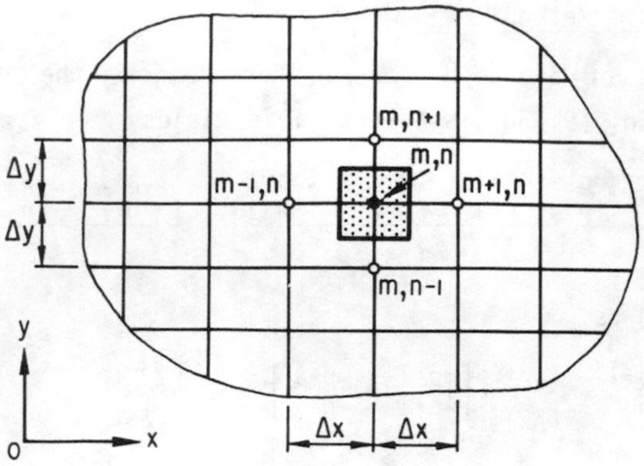

Fig. 9.15 Nomenclature for numerical solution of two-dimensional unsteady-state conduction in rectangular coordinates.

notation of Fig. 9.15, the explicit finite-difference representation of Eq. (9.35) is

$$\frac{T^p_{m+1,n} + T^p_{m-1,n} - 2T^p_{m,n}}{(\Delta x)^2} + \frac{T^p_{m,n+1} + T^p_{m,n-1} - 2T^p_{m,n}}{(\Delta y)^2} = \frac{1}{\alpha}\frac{T^{p+1}_{m,n} - T^p_{m,n}}{\Delta t} \tag{9.36a}$$

Thus, at each grid point, the temperature at a given time is represented in terms of the known values at the previous time steps. If the step lengths of space coordinates are chosen to be the same; that is $\Delta x = \Delta y$, then Eq. (9.36a) can be written as

$$T^{p+1}_{m,n} = \frac{\alpha\Delta t}{(\Delta x)^2}\left[T^p_{m+1,n} + T^p_{m-1,n} + T^p_{m,n+1} + T^p_{m,n-1}\right] + \left[1 - \frac{4\alpha\Delta t}{(\Delta x)^2}\right]T^p_{m,n} \tag{9.36b}$$

As in the case of one-dimensional, unsteady-state problems, this explicit method is stable for certain values of Δx and Δt. For problems with prescribed boundary temperatures for $t \geq 0$, the limit of stability is given by

$$\frac{\alpha\Delta t}{(\Delta x)^2} \leq \frac{1}{4}$$

The finite-difference equation (9.36b) is useful for determining the temperatures of the interior grid points. When the boundary condition is of the third kind, then this relation no longer applies for the grid points on the boundary. A relation for such grid points can be obtained as we did in the one-dimensional case. Thus, for a grid point on a boundary where the condition is of the third kind, the finite-difference relation is given by

$$T_{m,n}^{p+1} = \frac{\alpha\Delta t}{(\Delta x)^2}\left\{2\frac{h\Delta x}{k}T_\infty + 2T_{m-1,n}^{p} + T_{m,n+1}^{p} + T_{m,n-1}^{p} + \left[\frac{(\Delta x)^2}{\alpha\Delta t} - 2\frac{h\Delta x}{k} - 4\right]T_{m,n}^{p}\right\} \qquad (9.37)$$

In this case, to ensure the stability of the solution of the finite-difference equations $(\Delta x)^2/\alpha\Delta t$ should be restricted as

$$\frac{(\Delta x)^2}{\alpha\Delta t} \geq 2\left[\frac{h\Delta x}{k} + 2\right]$$

An implicit method of representing Eq. (9.35) in terms of finite-difference equations can also be given as

$$\frac{T_{m+1,n}^{p+1} + T_{m-1,n}^{p+1} - 2T_{m,n}^{p+1}}{(\Delta x)^2} + \frac{T_{m,n+1}^{p+1} + T_{m,n-1}^{p+1} - 2T_{m,n}^{p+1}}{(\Delta y)^2} = \frac{1}{\alpha}\,\frac{T_{m,n}^{p+1} - T_{m,n}^{p}}{\Delta t} \qquad (9.38a)$$

If Δx is taken to be equal to Δy, then Eq. (9.38a) can be rewritten as

$$T_{m,n}^{p} = \left[1 + \frac{4\alpha\Delta t}{(\Delta x)^2}\right]T_{m,n}^{p+1} - \frac{\alpha\Delta t}{(\Delta x)^2}\left[T_{m+1,n}^{p+1} + T_{m-1,n}^{p+1} + T_{m,n+1}^{p+1} + T_{m,n-1}^{p+1}\right] \qquad (9.38b)$$

This formulation, although is stable for all values of Δx and Δt, does not permit the calculation of T^{p+1} explicitly in terms of T^{p}. Rather, it requires the solution of a large number of simultaneous algebraic equations at each time step.

Example 9.5: Consider the solid body shown in Fig. 9.16, which is initially at a uniform temperature of 100°C. For times $t \geq 0$, the boundary temperatures are maintained at the values given in the figure. Find the temperatures of the grid points A, B, C and D as a function of time.

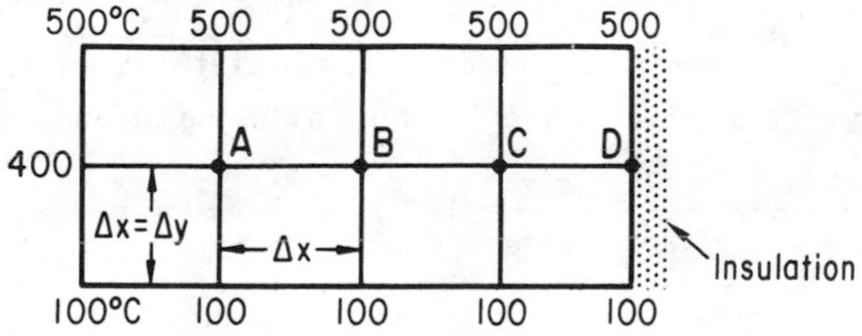

Fig. 9.16 Nomenclature for Example 9.5.

Table 9.3 Results of Example 9.5.*

p	t	T_A	T_B	T_C	T_D
0	0	100	100	100	100
1	$t = \frac{(\Delta x)^2}{4\alpha}$	275	200	200	200
2	$2\Delta t$	300	269	250	250
3	$3\Delta t$	317	288	280	275
4	$4\Delta t$	322	299	291	289
.	.	.	.	.	.
.	.	.	.	.	.
∞	∞	327	307	302	300

*Temperatures are in °C.

Solution: If $\alpha\Delta t/(\Delta x)^2 = \frac{1}{4}$ is selected, then the temperatures at the points A, B, C, and D can be calculated from

$$T_A^{p+1} = \frac{1}{4}[1000 + T_B^p]$$

$$T_B^{p+1} = \frac{1}{4}[600 + T_A^p + T_C^p]$$

$$T_C^{p+1} = \frac{1}{4}[600 + T_B^p + T_D^p]$$

$$T_D^{p+1} = \frac{1}{4}[600 + 2T_C^p]$$

We have tabulated the temperatures of these points as a function of time in Table 9.3.

9.9 FINITE-DIFFERENCE FORMULATION OF PROBLEMS IN CYLINDRICAL COORDINATES

We now consider the finite-difference formulation of problems in cylindrical coordinates. The concepts that we use here can readily be extended to problems in spherical coordinates. The temperature distribution in problems in cylindrical coordinates, under steady-state conditions without internal energy sources and sinks, satisfies the following Laplace equation

$$\frac{\partial^2 T}{\partial r^2} + \frac{1}{r}\frac{\partial T}{\partial r} + \frac{1}{r^2}\frac{\partial^2 T}{\partial \phi^2} + \frac{\partial^2 T}{\partial z^2} = 0 \tag{9.39}$$

if the thermo-physical properties are constants. For problems of the type $T(r,\phi)$, where there is no variation in T along the z-axis, this equation reduces to

$$\frac{\partial^2 T}{\partial r^2} + \frac{1}{r}\frac{\partial T}{\partial r} + \frac{1}{r^2}\frac{\partial^2 T}{\partial \phi^2} = 0 \tag{9.40}$$

This equation may be represented in finite-difference form at the grid point (m,n) in Fig. 9.17 as

$$\frac{T_{m+1,n} + T_{m-1,n} - 2T_{m,n}}{(\Delta r)^2} + \frac{1}{r_m}\frac{T_{m+1,n} - T_{m-1,n}}{2\Delta r} + \frac{1}{r_m^2}\frac{T_{m,n+1} + T_{m,n-1} - 2T_{m,n}}{(\Delta\phi)^2} = 0 \tag{9.41}$$

where $r_m = m\Delta r$. At $r = 0$, Eq. (9.41) reduces to

$$T_0 = T_m$$

where T_0 is the temperature at $r = 0$ and T_m is the mean of the temperatures of the grid points which surround the point $r = 0$ [16]. When Eq. (9.41) is applied to every grid point, together with the boundary conditions, it results in a system of simultaneous linear algebraic equations similar to those we studied previously in rectangular coordinates.

For problems of the type $T(r,z)$, where there is no variation in T along the ϕ-direction, Eq. (9.39) reduces to

$$\frac{\partial^2 T}{\partial r^2} + \frac{1}{r}\frac{\partial T}{\partial r} + \frac{\partial^2 T}{\partial z^2} = 0 \tag{9.42}$$

which may be represented in finite-difference form as

$$\frac{T_{m+1,n} + T_{m-1,n} - 2T_{m,n}}{(\Delta r)^2} + \frac{1}{r_m}\frac{T_{m+1,n} - T_{m-1,n}}{2\Delta r} + \frac{T_{m,n+1} + T_{m,n-1} - 2T_{m,n}}{(\Delta z)^2} = 0 \tag{9.43}$$

where n represents the number of increments in the z-direction and m the number of increments in the r-direction. At $r = 0$, it can be shown that the Laplace equation (9.42) is equivalent to

$$2\frac{\partial^2 T}{\partial r^2} + \frac{\partial^2 T}{\partial z^2} = 0 \tag{9.44}$$

which can easily be represented in finite-difference form.

Similar approaches are, of course, necessary for the treatment of unsteady-state problems in cylindrical coordinates.

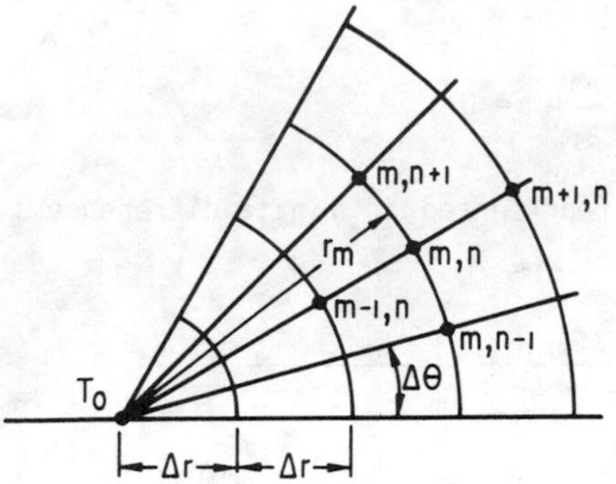

Fig. 9.17 Nomenclature for finite-difference formulation in cylindrical polar coordinates.

9.10 ERRORS IN FINITE-DIFFERENCE SOLUTIONS

Let T(x) and its derivatives be single-valued, finite, and continuous functions of x. Taylor series expansion of T(x) about the point x can be written as

$$T(x + \Delta x) = T(x) + \frac{dT}{dx}\bigg|_x \Delta x + \frac{1}{2!}\frac{d^2T}{dx^2}\bigg|_x (\Delta x)^2 + \frac{1}{3!}\frac{d^3T}{dx^3}\bigg|_x (\Delta x)^3 + \frac{1}{4!}\frac{d^4T}{dx^4}\bigg|_x (\Delta x)^4 + \ldots \tag{9.45a}$$

and

$$T(x - \Delta x) = T(x) - \frac{dT}{dx}\bigg|_x \Delta x + \frac{1}{2!}\frac{d^2T}{dx^2}\bigg|_x (\Delta x)^2 - \frac{1}{3!}\frac{d^3T}{dx^3}\bigg| (\Delta x)^3 + \frac{1}{4!}\frac{d^4T}{dx^4}\bigg|_x (\Delta x)^4 + \ldots \tag{9.45b}$$

By adding Eq. (9.45a) and Eq. (9.45b) we obtain

$$T(x + \Delta x) + T(x - \Delta x) = 2T(x) + \frac{d^2T}{dx^2}\bigg|_x (\Delta x)^2 + O[(\Delta x)^4] \tag{9.46}$$

where $O[(\Delta x)^4]$ denotes terms containing fourth and higher powers of Δx. We can rewrite Eq. (9.46) in the form

$$\frac{d^2T}{dx^2}\bigg|_x = \frac{T(x + \Delta x) + T(x - \Delta x) - 2T(x)}{(\Delta x)^2} + O[(\Delta x)^2] \tag{9.47}$$

Comparison of Eq. (9.5c) and Eq. (9.47) reveals that the finite-difference approximation of d^2T/dx^2 given by Eq. (9.5c) has a truncation error of the order of magnitude of $(\Delta x)^2$, which implies that doubling Δx quadruples the truncation error that is caused by treating the problem in discrete steps.

Similarly, subtracting Eq. (9.45a) from Eq. (9.45b) and rearranging the resulting expression we get

$$\left.\frac{dT}{dx}\right|_x = \frac{T(x + \Delta x) - T(x - \Delta x)}{2\Delta x} + O[(\Delta x)^2] \tag{9.48}$$

Comparison of Eq. (9.4) and Eq. (9.48) shows that the central-difference approximation of the first derivative of T(x) has a truncation error of the order of magnitude $(\Delta x)^2$.

From Eq. (9.45a) we can write

$$\left.\frac{dT}{dx}\right|_x = \frac{T(x + \Delta x) - T(x)}{\Delta x} + O[\Delta x] \tag{9.49}$$

and from Eq. (9.45b)

$$\left.\frac{dT}{dx}\right|_x = \frac{T(x) - T(x - \Delta x)}{\Delta x} + O[\Delta x] \tag{9.50}$$

Thus, we see that both the forward and backward representations of the first derivative of T(x) have truncation error of the order of magnitude of Δx. Therefore, the central-difference form is a more accurate approximation than both the forward- and backward-difference approximations.

When a differential equation is expressed in finite difference form, the so-called *truncation* error is introduced into the resulting algebraic equations, because of the truncation of the Taylor series expansion of the derivatives. This is the result of the technique used; that is, the approximation which is inherent in the method and it cannot be avoided. It is also independent of the characteristics of the computing equipment. It may be reduced only by selecting a finer grid; that is, smaller sizes for space and time increments.

The finite-difference equations are algebraic equations which are solved numerically, and the numerical calculations are carried out only to a finite number of decimal places. In each numerical calculation during the solution, the results are rounded-off, and because of this, errors called *round-off* errors are introduced. Since the round-off errors may accumulate, they may cause a large cumulative error. It is, actually, difficult to determine the order of the magnitude of the cumulative round-off errors or cumulative departure of the solution from the true value due to round-off errors. The use of small step sizes increases the accumulation of round-off errors, although they are desired for less truncation error; that is, for better approximation.

The round-off errors are determined by the characteristics of the machine which does the calculation.

9.11 CONVERGENCE AND STABILITY

The accuracy of the results of the finite-difference equations is hard to estimate. It is known, however, that if the numerical method satisfies two criteria termed "convergence" and "stability", the accuracy is determined by the step sizes used and increased accuracy may be obtained at the expense of increased labor.

If the approximate numerical solution approaches to the exact solution as the grid spacings in time and space are reduced to zero, then the solution is said to be *convergent*. The numerical method is unsatisfactory, unless the numerical solution converges to the exact solution in the limit, and a numerical method which converges in the limit to the exact solution is said to fulfill the convergence criterion.

In replacing the differential equation and the related boundary conditions by finite-difference equations, finite step sizes in both time and space are used, and only a finite number of significant figures can be carried in the calculations. These are practical limitations and they introduce both truncation and round-off errors into the analysis as discussed in the previous section. These errors may not be serious unless they grow in magnitude as the solution proceeds. If they grow as a solution proceeds and this growth is unbounded, the solution is said to be *unstable*. A method which prevents the growth of errors is said to fulfill the stability criterion. Stability is, in fact, necessary for convergence. Instability also results if errors grow at a faster rate than that at which convergence is approached. Price and Slack [15] discuss the stability and accuracy of the numerical solutions of heat conduction problems. The following is a brief outline of the treatment to determine the stability of finite-difference methods. This procedure was first given by O'Brien, Hyman and Kaplan [18], and is also discussed in References [3,7].

To explain the procedure, let us consider the explicit finite-difference form (9.27b):

$$T_n^{p+1} = \left[1 - \frac{2\alpha\Delta t}{(\Delta x)^2}\right]T_n^p + \frac{\alpha\Delta t}{(\Delta x)^2}\left(T_{n+1}^p + T_{n-1}^p\right) \tag{9.51}$$

Solution of Eq. (9.24) at time t can be expanded into a Fourier series and a typical term in the expansion, neglecting the constant coefficient, will be $\phi(t)e^{i\beta x}$. By substituting this term into the finite-difference equation (9.51), the form of $\phi(t)$ can be found, and a criterion thereby can be established as to whether it remains bounded as t becomes large. Substituting $T_n^p = \phi(t)e^{i\beta x}$, we have

$$\phi(t + \Delta t)e^{i\beta x} = \left[1 - \frac{2\alpha\Delta t}{(\Delta x)^2}\right]\phi(t)e^{i\beta x} + \frac{\alpha\Delta t}{(\Delta x)^2}\left[e^{i\beta(x+\Delta x)} + e^{i\beta(x-\Delta x)}\right]\phi(t) \tag{9.52}$$

Rearranging this result we obtain

$$\frac{\phi(t + \Delta t)}{\Delta(t)} = 1 - 4\frac{\alpha\Delta t}{(\Delta x)^2}\sin^2\frac{\beta\Delta x}{2} \tag{9.53}$$

For stability, $\phi(t)$ must remain bounded as Δt, and thus Δx, approaches zero. Clearly, this requires

$$\text{Max }\left|1 - 4\frac{\alpha\Delta t}{(\Delta x)^2}\sin^2\frac{\beta\Delta x}{2}\right| \leq 1 \tag{9.54}$$

for all values of β. In fact, components of all frequencies of β may be present; if they are not present in the initial condition, or not introduced by the boundary conditions, they can be introduced by the round-off errors. To satisfy the condition (9.54), we see that $\alpha\Delta t/(\Delta x)^2$ can be at most 1/2. That is, for the explicit method

$$\frac{\alpha\Delta t}{(\Delta x)^2} \leq \frac{1}{2}$$

is a necessary (and sufficient) condition for stability.

9.12 GRAPHICAL SOLUTIONS

One-dimensional unsteady-state heat conduction problems may also be solved by the graphical method known as *Schmidt-plot*. As we have seen in Section 9.7.1, if Δx and Δt satisfy the condition

$$\frac{\alpha\Delta t}{(\Delta x)^2} = \frac{1}{2}$$

then the temperature at the location m at the time $(p + 1)\Delta t$ will be equal to the arithmetic average of the temperatures at the locations $(m + 1)$ and $(m - 1)$ at the time $p\Delta t$; that is,

$$T_m^{p+1} = \frac{1}{2}\left(T_{m+1}^p + T_{m-1}^p\right)$$

As shown in Fig. 9.18, this arithmetic average can easily be constructed graphically.

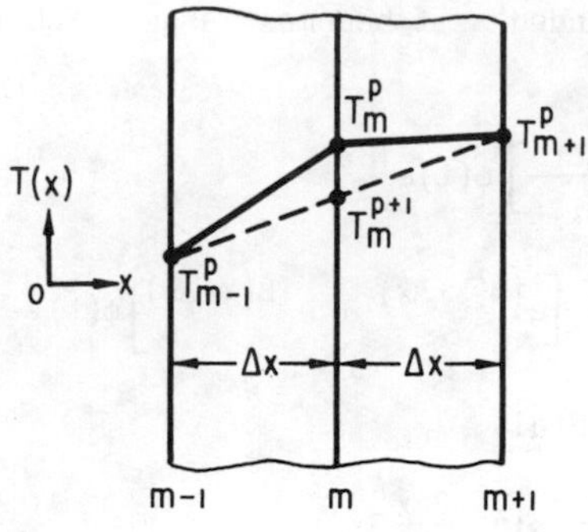

Fig. 9.18 Graphical construction.

For graphical solution of a one-dimensional heat conduction problem, we first draw the initial condition and divide the body into increments of Δx as shown in Fig. 9.19.

The temperature T_m^1 at the location m at time $\Delta t[=\frac{1}{2}(\Delta x)^2/\alpha]$ is simply the intersection of the line drawn between T_{m-1}^0 and T_{m+1}^0 with the grid line at location m. This graphical construction, which is known as Schmidt-plot, is repeated until the desired temperature distribution is obtained.

Example 9.6: A plane wall of thickness 60cm is initially at a uniform temperature of 35°C. At t = 0 its surface temperatures are suddenly raised to 150°C and 350°C and kept constant at these values for time $t \geq 0$. Find the temperature at a depth of 15 cm from the surface maintained at 150°C after 15 hours have passed. The thermal diffusivity of the material of the wall is given to be $\alpha = 0.52 \times 10^{-6}\ m^2/s$.

Solution: To start with the graphical method, let us divide the wall into

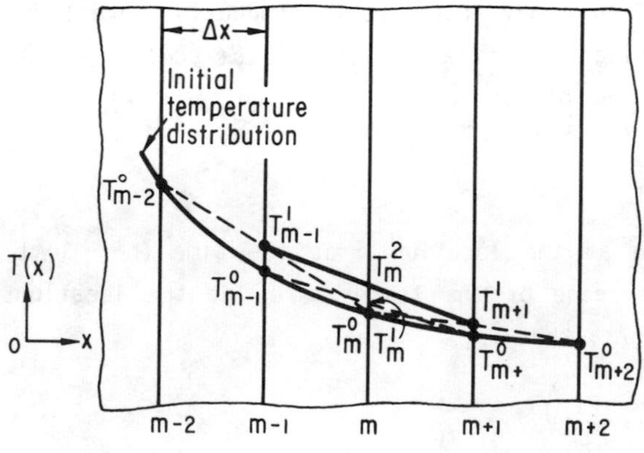

Fig. 9.19 Schmidt-Plot technique.

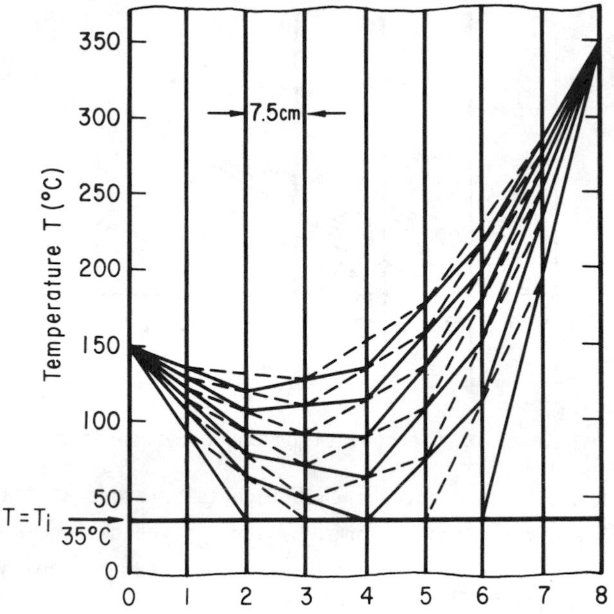

Fig. 9.20 Schmidt-Plot for Example 9.6.

increments of Δx = 7.5 cm as shown in Fig. 9.20. The time increment Δt is then given by

$$\Delta t = \frac{(\Delta x)^2}{2\alpha} = \frac{(0.075)^2 \text{ m}^2}{2 \times (0.52 \times 10^{-6}) \text{ m}^2/\text{s} \times (3600) \text{ s/hr}} = 1.502 \text{ hr}$$

The temperature distributions in the wall at various Δt's are constructed approximately by the Schmidt-plot. This construction can easily be followed in Fig. 9.20, where the temperature distributions are depicted after 12 time increments. The temperature at a depth of 15 cm from the surface kept at 150°C after 15 hrs, which is approximately equal to 10 time increments, is found to be ~ 130°C.

The graphical solution with a boundary condition of the third kind; that is, heat transfer by convection to an ambient fluid at a prescribed temperature with a prescribed heat transfer coefficient, may be constructed in the following way: Equation (9.29) can be written as

$$-\left.\frac{\partial T}{\partial x}\right|_{x=L} = \frac{T_w - T_\infty}{k/h} \tag{9.55}$$

which is the basis for obtaining the surface temperature by Schmidt-plot as illustrated in Fig. 9.21. If a line is drawn between T_∞ and T_m^p, the intersec-

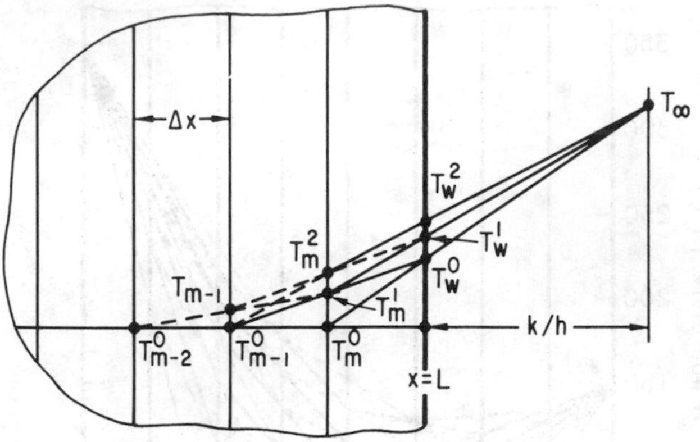

Fig. 9.21 Schmidt-plot for convection boundary condition.

tion of this line with the surface gives the surface temperature at that particular time. If T_∞ and h are functions of time, by varying the distances which correspond to T_∞ and k/h, the above procedure can still be used to find the surface temperature.

REFERENCES

1. Schneider, P.J., *Conduction Heat Transfer*, Addison-Wesley Publishing Co., Inc., Reading, Massachusetts, 1955.
2. Dusinberre, G.M., *Heat Transfer Calculations by Finite Differences*, International Textbook Co., Scranton, Pennsylvania, 1961.
3. Özışık, M.N., *Boundary Value Problems of Heat Conduction*, International Textbook Co., Scranton, Pennsylvania, 1968.
4. Myers, G.E., *Conduction Heat Transfer*, McGraw-Hill Book Co., New York, 1972.
5. Bayley, F.J., Owen, J.M., and Turner, A.B., *Heat Transfer*, Thomas Nelson and Sons Ltd., London, 1972.
6. Arpacı, V.S., *Conduction Heat Transfer*, Addison-Wesley Publishing Co., Inc., Reading Massachusetts, 1966.
7. Carnahan, R., Luther, H.A., and Wilkes, J.O., *Applied Numerical Methods*, John Wiley and Sons, Inc., New York, 1969.
8. Richtmeyer, R.D., *Difference Methods for Initial Value Problems*, Interscience Publishers, Inc., New York, 1957.
9. Crank, J., and Nicolson, P., "A Practical Method for Numerical Evaluation of Solutions of P.D.E. of the Heat Conduction Type", *Proc. Camb. Phil.*

Soc., Vol. 43, pp. 50-67, 1947.

10. Barakat, H.Z., and Clark, J.A., "On the Solution of Diffusion Equation by Numerical Methods", *J. Heat Transfer*, Vol. 88C, pp. 421-427, 1966.
11. Holman, J.P., *Heat Transfer*, 5th ed., McGraw-Hill Book Co., New York, 1981.
12. Schenk, H., *Fortran Methods in Heat Flow*, The Ronald Press Co., New York, 1963.
13. Wilson, E.L., and Nickell, R.E., "Application of the Finite Element Method to Heat Conduction Analysis", *Nucl. Eng. Des.*, Vol. 4, pp. 276-286, 1966.
14. Southwell, R.V., *Relaxation Methods in Engineering Science*, Oxford University Press, London, 1940.
15. Price, P.H., and Slack, M.R., "Stability and Accuracy of Numerical Solutions of Heat Flow Equation", *British J. of Appl. Physics*, Vol. 3, p. 379, 1952.
16. Smith, G.D., *Numerical Solution of Partial Differential Equations*, Oxford University Press, London, 1965.
17. Noble, B., *Applied Linear Algebra*, Prentice-Hall, Inc., Englewood Cliffs, New Jersey, 1969.
18. O'Brien, G., Hyman, M., and Kaplan, S., "A Study of the Numerical Solution of Partial Differential Equations", *J. Math. Phys.*, Vol. 29, pp. 233-251, 1951.
19. Hildebrand, F.B., *Methods of Applied Mathematics*, 2nd ed., Prentice-Hall, Inc., Englewood Cliffs, New Jersey, 1976.

PROBLEMS

9.1 Derive Eq.(9.14).

9.2 Derive Eq.(9.15).

9.3 Obtain an expression in finite-difference form for the temperature of the corner grid point (m,n) of the two-dimensional solid shown in Fig. 9.5 under unsteady-state conditions.

9.4 Consider the one-dimensional fin shown in Fig. 9.22. The base temperature T_b, the surrounding fluid temperature T_∞, and the heat transfer coefficient h are known constants. Divide the fin into six segments as shown, and formulate the problem in terms of finite differences to calculate the temperatures at the six points shown under steady-state conditions.

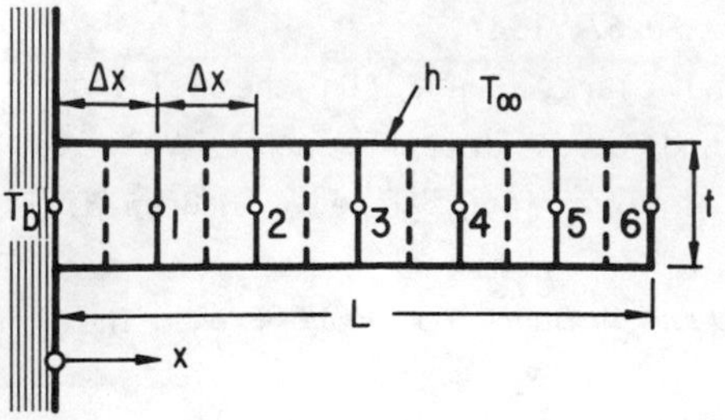

Fig. 9.22 Figure for Problem 9.4.

9.5 A copper bar of length 15 cm and 0.5 cm in diameter is braced to a steel bar with equal length and diameter. This composite bar is placed in an air flow at 0°C. The heat transfer coefficient is 125 $W/m^2 \cdot K$. The copper-side end temperature of the combined bar is 100°C and the steel-side end temperature is 50°C. Find the temperature distribution along the combined bar using relaxation method.

9.6 The fin of Problem 9.4 is made of brass (k=111 W/m·K) of length 2.5 cm with a 0.1 cm x 30 cm cross-section. The base temperature of the fin is 90°C, The surrounding fluid temperature is 20°C, and the heat transfer coefficient is 15 $W/m^2 \cdot K$. Estimate the temperature distribution along the fin and the rate of heat transfer to the surrounding fluid.

9.7 Obtain an expression for the residual $R_{m,n}$ at the insulated boundary of Fig. 9.6, if heat is generated within the solid at a rate of $\dot{q}$ per unit volume.

9.8 Figure 9.23 shows the cross-section of a long steel bar. The temperatures of the surfaces are as indicated in the figure. Estimate the values of temperatures at points A, B, C, D, E and F.

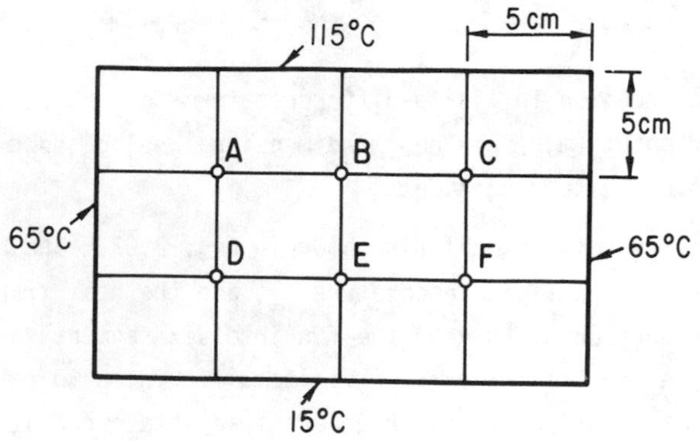

Fig. 9.23 Figure for Problem 9.8.

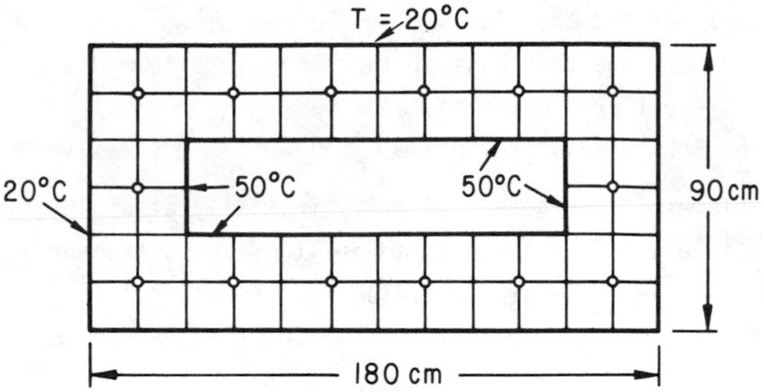

Fig. 9.24 Figure for Problem 9.9.

9.9 The cross-section of a chimney is shown in Fig. 9.24. The inside and outside surface temperatures are as indicated in the figure. Estimate the values of temperatures at the points shown.

9.10 The cross-section of a long solid bar is shown in Fig. 9.25. The boundaries are insulated except two, which are kept at constant temperatures as indicated in the figure. Estimate the temperature distribution in the solid and the heat transfer rates at the top and bottom surfaces by the finite-difference method.

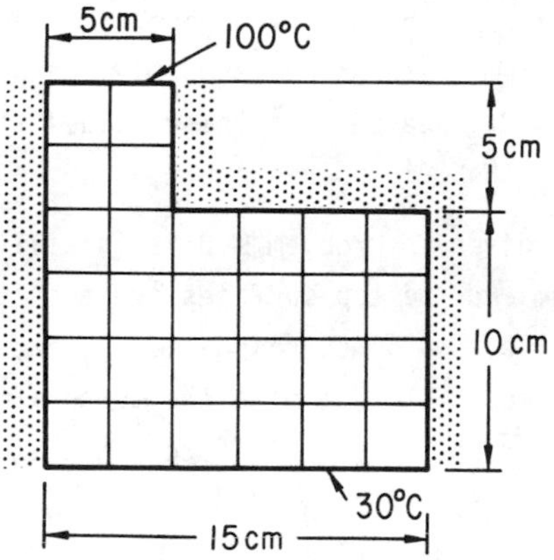

Fig. 9.25 Figure for Problem 9.10.

9.11 A dry wooden plate is initially at 0°C. It is suddenly exposed to air at 30°C. The plate is 10 cm in thickness and the other dimensions are very large. The heat transfer coefficient on both sides of the plate is 10 $W/m^2 \cdot K$, and the properties of the plate material are k=0.109 $W/m \cdot K$, ρ=417 kg/m^3, and c=2720 $J/kg \cdot K$. After two hours of heating, find the temperatures of the surfaces and the midplane of the plate,

a) by dividing the thickness of the plate into four divisions and applying a finite-difference method, and

b) by the Schmidt-plot technique.

9.12 A steel plate of thickness 2.56 cm is initially at 650°C. It is suddenly immersed into a cold water. As a result of this, the surface temperatures drop to 94°C, and stay at this value for the rest of the cooling process. Estimate the time elapsed for the midplane temperature to cool down to 450°C. For steel α=1.16 X 10^{-5} m^2/s may be used.

9.13 A thin steel plate of thickness 1.27 cm comes out of a mill at 540°C with a velocity of 2.44 m/s. The temperature of the surfaces decreases linearly as 110°C/m. Assume that the heat flow is in the normal direction to the surfaces of the plate. By dividing the plate thickness into eight, estimate the temperature distribution across the thickness of the plate and the total surface heat flux at a distance 2.74 cm from the mill. For steel k = 43.3 W/m·K and α=0.98 x 10^{-5} m^2/s may be used.

9.14 While the fin given in Problem 9.4 is at a uniform temperature T_i at t=0, it is exposed to an ambient temperature T_∞ for times t>0. Formulate the problem in terms of finite-differences.

9.15 Assume that the chimney cross-section given in Problem 9.9 is initially at 20°C. The surface temperatures are suddenly changed at t=0 to the values indicated in Fig. 9.24, and kept at these values for times t>0. Estimate the temperatures of the grid points shown as a function of time.

9.16 The long steel bar given in Problem 9.8 is initially at 65°C. The temperatures of the bottom and the top surfaces are then suddenly changed at t=0 to the values of 15°C and 115°C, respectively, and are kept constant at these values for times t>0. Estimate the temperatures at the points shown at various times, and obtain steady-state temperatures at these points. For steel α=1.16 x 10^{-5} m^2/s may be used.

10

Further Methods of Solution

10.1 INTRODUCTION

So far we have discussed various analytical (separation of variables, integral transforms, Laplace transforms) and numerical (finite-difference) methods of solution of heat conduction problems. We investigated the applications of these techniques to several linear problems (except for some non-linear one-dimensional steady-state problems in Chapter 3). In this chapter we present further methods of solution including Duhamel's method, the integral method, and the variational method, and also discuss some phase-change problems.

Among the techniques introduced in this chapter, the integral method is especially important as it can easily be applied to non-linear problems. Such problems need not be linearized, because the method is elastic enough to encompass all sorts of non-linearities associated with heat conduction problems.

In the following sections, we introduce the basic concepts of each method and the solution procedure in terms of some representative problems.

10.2 DUHAMEL'S METHOD

Duhamel's method is a superposition technique which can be used to obtain the solution of a linear heat conduction problem with time-dependent boundary conditions and/or time-dependent internal energy generation from the solution of the same problem with time-independent boundary conditions and/or time-independent internal energy generation.

In order to introduce those concepts which are basic to Duhamel's method, we use the method, in this section, to obtain the solution to a simple heat conduction problem. A detailed treatment of the method as applied to linear heat conduction problems can be found in Reference [1].

Consider a semi-infinite solid, $0 \leq x < \infty$, initially at zero temperature. The surface temperature at $x = 0$ is specified as a prescribed function of time, $f(t)$, for $t > 0$ (see Fig. 8.5). Assuming constant thermo-physical properties, the formulation of the problem for the unsteady temperature distribution $T(x,t)$ is given by

$$\frac{\partial^2 T}{\partial x^2} = \frac{1}{\alpha}\frac{\partial T}{\partial t} \tag{10.1a}$$

$$T(x,0) = 0 \tag{10.1b}$$

$$T(0,t) = f(t) \quad \text{and} \quad \lim_{x\to\infty} T(x,t) \to 0 \tag{10.1c,d}$$

where α is the thermal diffusivity of the solid.

Noting the linearity of the problem, it may be possible to construct a solution for $T(x,t)$ by merely breaking the surface temperature variation $f(t)$ up into a number of constant temperature steps and then superposing the constant surface temperature solutions for each step. Thus, let the surface temperature variation over a time interval $(0,t)$ be approximated as

$$f(t) \cong f(0)\, H(0) + \sum_{i=1}^{n} [f(\tau_i) - f(\tau_{i-1})]\, H(t-\tau_i) \tag{10.2}$$

where $\tau_0 = 0 < \tau_1 < \ldots < \tau_{n-1} < \tau_n = t$, and $H(t)$ is the so-called *Heaviside step function*, defined by

$$H(t) = \begin{cases} 1 & t > 0 \\ 0 & t < 0 \end{cases} \tag{10.3}$$

Then the solution to the problem (10.1) at any time t can be written in the form

$$T(x,t) \cong f(0)\, \phi(x,t) + \sum_{i=1}^{n} [f(\tau_i) - f(\tau_{i-1})]\phi(x,t-\tau_i) \tag{10.4}$$

where $\phi(x,t)$ is the solution of the following auxiliary problem:

$$\frac{\partial^2 \phi}{\partial x^2} = \frac{1}{\alpha}\frac{\partial \phi}{\partial t} \tag{10.5a}$$

$$\phi(x,0) = 0 \tag{10.5b}$$

$$\phi(0,t) = 1 \quad \text{and} \quad \lim_{x\to\infty} \phi(x,t) \to 0 \tag{10.5c,d}$$

In fact, $\phi(x,t)$ can be readily obtained from Eq. (8.46) as

$$\phi(x,t) = \operatorname{erfc}\left(\frac{x}{2\sqrt{\alpha t}}\right) \tag{10.6}$$

Equation (10.4) can be rewritten as

$$T(x,t) \cong f(0)\phi(x,t) + \sum_{i=1}^{n} \frac{\Delta f_i}{\Delta \tau_i} \phi(x,t-\tau_i)\Delta\tau_i \tag{10.7}$$

where

$$\Delta f_i = f(\tau_i) - f(\tau_{i-1})$$

and

$$\Delta\tau_i = \tau_i - \tau_{i-1}$$

Thus, in the limit as $n \to \infty$, Eq. (10.7) takes the form

$$T(x,t) = f(0)\phi(x,t) + \int_0^t \frac{df}{d\tau}\phi(x,t-\tau)d\tau \tag{10.8}$$

As $n \to \infty$ the approximation (10.4) is an exact representation and the summation term in Eq. (10.7), by definition, becomes the integral term in Eq. (10.8).

Equation (10.8) is known as *Duhamel's superposition integral.* Integrating the second term on the right-hand side of this result by parts yields

$$T(x,t) = -\int_0^t f(\tau)\frac{\partial\phi(x,t-\tau)}{\partial\tau}d\tau \tag{10.9}$$

However, since

$$\frac{\partial\phi(x,t-\tau)}{\partial\tau} = -\frac{\partial\phi(x,t-\tau)}{\partial t}$$

Eq. (10.9) can also be written as

$$T(x,t) = \int_0^t f(\tau)\frac{\partial\phi(x,t-\tau)}{\partial t}d\tau \tag{10.10}$$

On the other hand, from Eq. (10.6) we have

$$\frac{\partial\phi(x,t-\tau)}{\partial t} = \frac{1}{2\sqrt{\pi\alpha}}\frac{x}{(t-\tau)^{3/2}}\exp\left[-\frac{x^2}{4\alpha(t-\tau)}\right] \tag{10.11}$$

Thus, introducing Eq. (10.11) into Eq. (10.10) the solution becomes

$$T(x,t) = \frac{x}{2\sqrt{\pi\alpha}}\int_0^t \frac{f(\tau)}{(t-\tau)^{3/2}}\exp\left[-\frac{x^2}{4\alpha(t-\tau)}\right]d\tau \tag{10.12a}$$

which can also be written as

$$T(x,t) = \frac{x}{2\sqrt{\pi\alpha}} \int_0^t \frac{f(t - t')}{\tau^{3/2}} \exp(-\frac{x^2}{4\alpha t'})dt' \tag{10.12b}$$

This result is identical to Eq. (8.52) as expected.

All the fundamental concepts basic to Duhamel's method appear in the example presented above. For further applications of the method, the reader is referred to Reference [1].

10.3 INTEGRAL METHOD OF FORMULATION AND SOLUTION

The objective of this section is to introduce the *integral method*, by which approximate analytical solutions to linear and non-linear heat conduction problems can be obtained. The integral method was first used by von Karman [2] and Pohlhausen [3] to solve boundary layer momentum and energy equations approximately. The method is, however, equally attractive for solving any problem governed by a diffusion type equation, such as unsteady heat conduction problems in solids. Landahl [4], in 1953, used the integral method for the first time to solve the diffusion equation in the field of biophysics in connection with the spread of a concentrate. Later on, this technique was used by several investigators to solve various linear and non-linear heat conduction problems, including transient heat conduction problems, melting and solidification problems, and melting and extrusion of polymers. Review of the literature on the application of the integral method can be found in References [1,5].

In this section, we introduce the basic concepts of the integral method and the solution procedure by solving a simple heat conduction problem.

Consider a semi-infinite solid, $0 \leq x < \infty$, initially at a uniform temperature T_i (see Fig. 8.4). The surface temperature at $x = 0$ is changed to T_w at $t = 0$ and maintained constant at this value for times $t > 0$. Assuming constant thermo-physical properties, the formulation of the problem for the unsteady temperature distribution in the solid is given by

$$\frac{\partial^2 T}{\partial x^2} = \frac{1}{\alpha}\frac{\partial T}{\partial t} \tag{10.13a}$$

$$T(x,0) = T_i \tag{10.13b}$$

$$T(0,t) = T_w \quad \text{and} \quad \lim_{x\to\infty} T(x,t) \to T_i \tag{10.13c,d}$$

where α is the thermal diffusivity of the solid.

We now define a quantity $\delta(t)$ called the *penetration depth* or *thermal layer* such that, for all practical purposes,

$$T(\delta,t) = T_i \qquad \text{and} \qquad \frac{\partial T(\delta,t)}{\partial x} = 0 \tag{10.14a,b}$$

That is, for $x > \delta(t)$ the solid is at the initial temperature and there is no heat transfer beyond this point in the solid. Next we integrate the differential equation (10.13a) over x from $x = 0$ to $x = \delta(t)$:

$$\int_0^{\delta(t)} \frac{\partial^2 T}{\partial x^2}\, dx = \frac{1}{\alpha} \int_0^{\delta(t)} \frac{\partial T}{\partial t}\, dx \tag{10.15}$$

The term on the left reduces to

$$\int_0^{\delta(t)} \frac{\partial^2 T}{\partial x^2}\, dx = \frac{\partial T}{\partial x}\Big|_{x=\delta} - \frac{\partial T}{\partial x}\Big|_{x=0} = -\frac{\partial T}{\partial x}\Big|_{x=0}$$

The term on the right-hand side of Eq. (10.15) can be rewritten, using Leibnitz rule, as

$$\int_0^{\delta(t)} \frac{\partial T}{\partial t}\, dx = \frac{d}{dt} \int_0^{\delta(t)} T(x,t)dx - T(\delta,t)\frac{d\delta}{dx}$$

$$= \frac{d}{dt} \int_0^{\delta(t)} T(\delta,t)dx - T_i \frac{d\delta}{dx}$$

Thus, Eq. (10.15) becomes

$$-\frac{\partial T}{\partial x}\Big|_{x=0} = \frac{1}{\alpha}\left[\frac{d}{dt} \int_0^{\delta(t)} T(x,t)dx - T_i \frac{d\delta}{dx} \right] \tag{10.16}$$

This result can also be written as

$$\frac{d}{dt} \int_0^{\delta(t)} [T(x,t) - T_i]dx = -\alpha \frac{\partial T}{\partial x}\Big|_{x=0} \tag{10.17}$$

which is called the *heat-balance integral* or *energy integral equation.*

Let us assume that the temperature distribution $T(x,t)$ can be represented over $0 \le x \le \delta(t)$ by a second-degree polynomial in the form:

$$T(x,t) = a + bx + cx^2 \tag{10.18}$$

where the coefficients a, b, and c may depend upon t. Applying Eqs.(10.13c) and (10.14a,b) these constants are found to be

$$a = T_w , \qquad b = 2\,\frac{T_i - T_w}{\delta} , \qquad \text{and} \qquad c = -\,\frac{T_i - T_w}{\delta^2}$$

Thus the assumed profile (10.18) becomes

$$\frac{T(x,t) - T_i}{T_w - T_i} = 1 - 2\left(\frac{x}{\delta}\right) + \left(\frac{x}{\delta}\right)^2 \tag{10.19}$$

Substituting this result into the energy integral equation (10.17) and then performing the integration we obtain

$$\delta \frac{d\delta}{dt} = 6\,\alpha \tag{10.20}$$

Since $\delta(0) = 0$, the solution of Eq. (10.20) gives

$$\delta = \sqrt{12\alpha t} \tag{10.21}$$

Knowing $\delta(t)$, the temperature distribution $T(x,t)$ is determined from Eq. (10.19).

The heat flux q_w at the boundary surface is a quantity of interest and it is given by

$$q_w(t) = -k\left(\frac{\partial T}{\partial x}\right)_{x=0} = \frac{1}{\sqrt{3}}\,\frac{k(T_w - T_i)}{\sqrt{\alpha t}} \tag{10.22}$$

where k is the thermal conductivity of the solid.

The same problem was solved by Laplace transforms in Chapter 8 (see Section 8.8) and the exact result for the temperature distribution is given, from Eq. (8.46), by

$$\frac{T(x,t) - T_i}{T_w - T_i} = \operatorname{erfc}\left(\frac{x}{2\sqrt{\alpha t}}\right) \tag{10.23}$$

This exact result gives

$$q_w(t) = \frac{1}{\sqrt{\pi}}\,\frac{k(T_w - T_i)}{\sqrt{\alpha t}} \quad \text{(exact)} \tag{10.24}$$

By comparing Eqs. (10.22) and (10.24), it is seen that the results for the surface heat flux are of the same form, differing only by a numerical factor. The approximate result (10.22) has an error of about 2.3%.

For a better approximation for the temperature distribution, for example, a third-degree polynomial can be used in the form:

$$T(x,t) = a + bx + cx^2 + dx^3 \tag{10.25}$$

where there are four coefficients to be determined. Equations (10.13c) and (10.14a,b) give only three conditions, which are called the *natural conditions*. An additional *derived* condition may be obtained, for example, by evaluating the differential equation (10.13a) at $y = \delta$ where $T = T_i$ = constant. The resulting condition is

$$\left.\frac{\partial^2 T}{\partial x^2}\right|_{x=\delta} = 0 \tag{10.26}$$

This condition is sometimes called the *smoothing condition* because it tends to make the assumed profile go smoothly into the undisturbed initial temperature at $x = \delta$. With the natural conditions, (10.13c) and (10.14a,b), and the derived condition (10.26) the third degree profile (10.25) takes the form

$$\frac{T(x,t) - T_i}{T_w - T_i} = \left(1 - \frac{x}{\delta}\right)^3 \tag{10.27}$$

Substituting this result into Eq. (10.17) we are led to a differential equation for $\delta(t)$, whose solution, with the condition $\delta(0) = 0$, yields

$$\delta = \sqrt{24\alpha t} \tag{10.28}$$

The third-degree polynomial approximation (10.25) then gives the surface heat flux as

$$q_w(t) = \sqrt{\frac{3}{8}}\,\frac{k(T_w - T_i)}{\sqrt{\alpha t}} \tag{10.29}$$

which is again identical to the exact result (10.24), except the numerical factor $\left(\frac{3}{8}\right)^{\frac{1}{2}}$, and the error in the wall heat flux is about 8.5%.

Evaluating the differential equation (10.13a) at $x = 0$ where $T = T_w$ = constant, another derived condition can also be obtained as

$$\left.\frac{\partial^2 T}{\partial x^2}\right|_{x=0} = 0 \tag{10.30}$$

If this condition is used together with the natural conditions, the third-degree polynomial (10.25) becomes

$$\frac{T(x,t) - T_i}{T_w - T_i} = 1 - \frac{3}{2}\left(\frac{x}{\delta}\right) + \frac{1}{2}\left(\frac{x}{\delta}\right)^3 \tag{10.31}$$

where

$$\delta = \sqrt{8\alpha t} \tag{10.32}$$

For the surface heat flux Eq. (10.31) gives

$$q_w(t) = \frac{3}{4\sqrt{2}} \frac{k(T_w - T_i)}{\sqrt{\alpha t}} \tag{10.33}$$

The error in this case is about 6%.

It is also possible that with the five conditions (10.13c), (10.14a,b), (10.26) and (10.30) a fourth-degree polynomial can be assumed for the temperature profile in the form

$$T(x,y) = a + by + cy^2 + dy^3 + ey^4 \tag{10.34}$$

Solving the problem with this profile by the same procedures as in the above cases gives

$$\frac{T(x,t) - T_i}{T_w - T_i} = 1 - 2\left(\frac{x}{\delta}\right) + 2\left(\frac{x}{\delta}\right)^3 - \left(\frac{x}{\delta}\right)^4 \tag{10.35}$$

where

$$\delta = \sqrt{\frac{40}{3}\alpha t} \tag{10.36}$$

The profile (10.35) gives the surface heat flux as

$$q_w(t) = \sqrt{\frac{3}{10}} \frac{k(T_w - T_\infty)}{\sqrt{\alpha t}} \tag{10.37}$$

with an error of about 3%.

As illustrated, the choice of the profile is never unique, and the error in the final solution depends, to a large extent, on the selection of the profile. Among the profiles considered above, although the second-degree polynomial (10.19) gives the least error in the surface heat flux, the fourth-degree polynomial (10.35) is the best approximation to the exact temperature distribution (10.23). However, there is no significant improvement in the accuracy of the solution if a polynomial greater than fourth-degree is used [1].

An approximate solution for the semi-infinite solid with arbitrary time-dependent surface temperature $T(0,t) = f(t)$ can also be developed from the approximate solutions obtained above by the use of Duhamel's method discussed in the previous section. In that case, the use of Duhamel's superposition integral (10.8) gives

$$T(x,t) - T_i = [f(0) - T_i]\phi(x,t) + \int_0^t \frac{df}{d\tau}\phi(x,t-\tau)d\tau \tag{10.38}$$

where, for the fourth-degree polynomial approximation, we would have

$$\phi(x,t) = 1 - 2\left(\frac{x}{\delta}\right) + 2\left(\frac{x}{\delta}\right)^3 - \left(\frac{x}{\delta}\right)^4 \tag{10.39}$$

with δ given by Eq. (10.36).

10.3.1 Problems with Temperature-Dependent Thermal Conductivity

When the thermal conductivity depends on temperature the general heat conduction equation is given by

$$\vec{\nabla} \cdot [k(T)\vec{\nabla}T] + \dot{q} = \rho c \frac{\partial T}{\partial t} \tag{10.40}$$

Because of the dependence of thermal conductivity k on temperature T, Eq. (10.40) is a non-linear differential equation. It can, however, be reduced to a linear form by introducing a new temperature function $\theta(\vec{r},t)$ by means of the Kirchhoff transformation introduced in Section 2.4 by Eq. (2.38). If the boundary conditions can also be transformed, then the problem may be solved exactly for $\theta(\vec{r},t)$ by the use of the techniques introduced in the previous chapters. But inverting the solution back to the temperature distribution $T(r,t)$ may not be an easy task in most of the cases.

Non-linear problems involving temperature-dependent thermal conductivity may, however, be solved approximately by the integral method. Following is an example on the use of the integral method to solve such problems.

Consider a semi-infinite solid, $0 \leq x < \infty$, initially at a uniform temperature T_i. For times $t \geq 0$, a constant heat flux q_w is applied to the surface at $x = 0$. Assuming that the thermal conductivity k, specific heat c, and the density ρ of the solid are all temperature dependent, the formulation of the problem for the temperature distribution $T(x,t)$ is given by

$$\frac{\partial}{\partial x}\left(k \frac{\partial T}{\partial x}\right) = \rho c \frac{\partial T}{\partial t} \tag{10.41a}$$

$$T(x,0) = T_i \tag{10.41b}$$

$$-k_w \frac{\partial T(0,t)}{\partial x} = q_w \quad \text{and} \quad \lim_{x\to\infty} T(x,t) \to T_i \tag{10.41c,d}$$

where $k_w = k(T_w)$ with $T_w = T(0,t)$.

At this point, as suggested by Goodman [6], we transform the dependent variable $T(x,t)$ according to

$$\theta(x,t) = \int_{T_i}^{T(x,t)} \rho c\,dT \tag{10.42}$$

Since

$$\frac{\partial \theta}{\partial t} = \rho c \frac{\partial T}{\partial t} \tag{10.43a}$$

and

$$\frac{\partial \theta}{\partial x} = \rho c \frac{\partial T}{\partial x} \tag{10.43b}$$

the system (10.41) becomes

$$\frac{\partial}{\partial x}\left(\alpha \frac{\partial \theta}{\partial x}\right) = \frac{\partial \theta}{\partial t} \tag{10.44a}$$

$$\theta(x,0) = 0 \tag{10.44b}$$

$$-\alpha_w \frac{\partial \theta(0,t)}{\partial x} = q_w \quad \text{and} \quad \lim_{x\to\infty} \theta(x,t) \to 0 \tag{10.44c,d}$$

where $\alpha = \frac{k}{\rho c}$ and $\alpha_w = \alpha(T_w)$.

Define now a penetration depth $\delta(t)$ such that

$$T(\delta,t) = T_i \quad \text{and} \quad \frac{\partial T(\delta,t)}{\partial x} = 0 \tag{10.45a,b}$$

By means of the transformation (10.42) these conditions can also be written as

$$\theta(\delta,t) = 0 \quad \text{and} \quad \frac{\partial \theta(\delta,t)}{\partial x} = 0 \tag{10.46a,b}$$

Upon integrating Eq. (10.44a) from $x = 0$ to $x = \delta(t)$, and applying Eqs. (10.44c) and (10.46a,b), we obtain the following energy integral equation:

$$\frac{d}{dt}\int_0^\delta \theta(x,t)dx = q_w \tag{10.47}$$

We adopt a third-degree polynomial representation* for $\theta(x,t)$ as

$$\theta(x,t) = a + bx + cx^2 + dx^3 \tag{10.48}$$

In addition to the three natural conditions, Eqs. (10.44c) and (10.46a,b), we may obtain an additional derived condition by evaluating the differential equation (10.44a) at $y = \delta$ as

$$\left.\frac{\partial^2 \theta}{\partial x^2}\right|_{x=\delta} = 0 \tag{10.49}$$

With the natural conditions (10.44c) and (10.46a,b) and the derived condition (10.49), the assumed profile (10.48) takes the form

*The polynomial representation is adequate provided that the thermo-physical properties of the material do not vary rapidly with temperature. This characteristic applies to most materials.

$$\theta(x,t) = \frac{q_w \delta}{3\alpha_w}\left(1 - \frac{x}{\delta}\right)^3 \tag{10.50}$$

Substituting this profile into Eq. (10.47) and performing the indicated operations, we obtain the following differential equation for $\delta(t)$:

$$\frac{d\delta^2}{dt} = 12\alpha_w \tag{10.51}$$

Notice that only the thermo-physical properties at $x = 0$ are involved when the problem is cast in terms of the transformed variable $\theta(x,t)$. The solution of Eq. (10.51) subject to the initial condition $\delta(0) = 0$, is

$$\delta = \sqrt{12\alpha_w t} \tag{10.52}$$

This equation cannot yet be used to calculate the penetration depth δ directly, because α_w is the thermal diffusivity to be evaluated at $T(0,t)$, which is also not yet known. However, by setting $x = 0$ in Eq. (10.50), and eliminating δ between the resulting equation and Eq. (10.52), the following transcendental equation for $\theta(0,t)$ is obtained:

$$\sqrt{\alpha_w}\,\theta(0,t) = \sqrt{\frac{4}{3}}\, q_w \sqrt{t} \tag{10.53}$$

Since α_w is a function of $T(0,t)$ or $\theta(0,t)$ Eq. (10.53) enables us to determine $\theta(0,t)$ and therefore α_w as a function of time. Once α_w is known, δ is obtained from Eq. (10.52) as a function of time. Equation (10.50) then gives the transformed temperature function $\theta(x,t)$. The actual temperature distribution $T(x,t)$ is obtained through the transformation (10.42).

As a special case, suppose that ρc is constant. Then,

$$\theta(x,t) = \rho c[T(x,t) - T_i] \tag{10.54a}$$

Furthermore, let $k = k_i[1 + \beta(T - T_i)]$ represent the variation of thermal conductivity, where $k_i = k(T_i)$. Then, Eq. (10.50) becomes

$$T(x,t) - T_i = \frac{q_w \delta}{3k_w}\left(1 - \frac{x}{\delta}\right)^3 \tag{10.54b}$$

where k_w is given by

$$k_w = k_i[1 + \beta(T_w - T_i)] \tag{10.55a}$$

By substituting $T_w - T_i$ from Eq. (10.54b) and δ from Eq. (10.52), Eq. (10.55a) can be rewritten as

$$k_w(k_w - k_i)^2 = \frac{4}{3}\,\frac{(k_i \beta q_w)^2}{\rho c}\, t \tag{10.55b}$$

from which $k_w(t)$ is obtained directly. Once $k_w(t)$ is available, Eq.(10.52) will give $\delta(t)$. The temperature distribution $T(x,t)$ is then given by Eq. (10.54b).

If k, ρ and c are all constants then the temperature distribution is given from Eq. (10.54b) by

$$T(x,t) - T_i = \frac{q_w \delta}{3k}\left(1 - \frac{x}{\delta}\right)^3 \tag{10.56}$$

with

$$\delta = \sqrt{12\alpha t} \tag{10.57}$$

where $\alpha = k/\rho c$. In this case Eqs. (10.56a,b) will give the temperature distribution in the solid directly.

10.3.2 Non-Linear Boundary Conditions

The integral method can also be used to obtain approximate solutions to problems with non-linear boundary conditions. In the following example we illustrate the use of the method to solve a heat conduction problem in a semi-infinite solid with a non-linear boundary condition.

Consider a semi-infinite solid, $0 \leq x < \infty$, initially at zero temperature. For times $t \geq 0$, the surface at $x = 0$ is subjected to a heat flux q_w which is a prescribed function of the surface temperature T_w. Assuming constant thermo-physical properties, the formulation of the problem for the unsteady temperature distribution $T(x,t)$ for $t > 0$ is given by

$$\frac{\partial^2 T}{\partial x^2} = \frac{1}{\alpha}\frac{\partial T}{\partial t} \tag{10.58a}$$

$$T(x,0) = 0 \tag{10.58b}$$

$$-k\,\frac{\partial T(0,t)}{\partial x} = q_w(T_w) \qquad \text{and} \qquad \lim_{x\to\infty} T(x,t) \to 0 \tag{10.58c,d}$$

where $T_w = T(0,t)$.

Define a penetration depth $\delta(t)$ such that

$$T(\delta,t) = 0 \qquad \text{and} \qquad \frac{\partial T(\delta,t)}{\partial x} = 0 \tag{10.59a,b}$$

Upon integrating Eq. (10.58a) from $x = 0$ to $x = \delta$, and applying Eqs. (10.58c) and (10.59a,b), we obtain

$$\frac{d}{dx}\int_0^{\delta} T(x,t)dx = \frac{q_w(T_w)}{\rho c} \tag{10.60}$$

The temperature profile will be taken to be a cubic polynomial as

$$T(x,t) = a + bx + cx^2 + dx^3 \tag{10.61}$$

With the natural conditions (10.58c) and (10.59a,b), and the derived condition

$$\frac{\partial^2 T}{\partial x^2}\bigg|_{x=\delta} = 0 \tag{10.62}$$

which is obtained by evaluating the differential equation (10.58a) at $y = \delta$, the assumed profile (10.61) takes the form

$$T(x,t) = \frac{q_w(T_w)}{k}\frac{\delta}{3}\left(1 - \frac{x}{\delta}\right)^3 \tag{10.63}$$

which can also be written as

$$\frac{T(x,t)}{T_w} = \left(1 - \frac{x}{\delta}\right)^3 \tag{10.64}$$

where

$$T_w = \frac{q_w(T_w)}{k}\frac{\delta}{3} \tag{10.65}$$

Equation (10.65) is a relationship between the surface temperature T_w and the penetration depth δ. As a consequence, there is really only one unknown function of time in Eq. (10.64).

By substituting the profile (10.64) into the energy integral equation (10.60), performing the indicated operations, and then eliminating δ from the resulting expression by using Eq. (10.65), we obtain

$$\frac{d}{dt}\left(\frac{T_w^2}{q_w}\right) = \frac{4}{3}\frac{q_w}{\rho ck} \tag{10.66}$$

which is, in fact, an ordinary differential equation for T_w. With the initial condition $T_w(0) = T_i$, Eq. (10.66) can be integrated analytically. The result is

$$\int_0^{T_w}\left[2q_w(T_w) - T_w\frac{dq_w}{dT_w}\right]\frac{T_w}{[q_w(T_w)]^3}\,dT_w = \frac{4}{3}\frac{t}{\rho ck} \tag{10.67}$$

Equation (10.67) expresses a relationship between the surface temperature T_w and time t. Once $T_w(t)$ is available, then Eqs. (10.64) and (10.65) yield the temperature distribution.

10.3.3 Plane Wall

In the examples considered so far, we have applied the integral method to several heat conduction problems in the semi-infinite region. In this section we apply the method to a plane wall problem. In such a problem the penetration depth $\delta(t)$ loses its significance after a certain time limit as it cannot increase indefinitely.

Consider a plane wall of thickness L in the x-direction and initially at a uniform temperature T_i. For times $t \geq 0$, a constant heat flux q_w is applied to the surface at $x = 0$ while the surface at $x = L$ is maintained at the initial temperature T_i. Assuming constant thermo-physical properties, the formulation of the problem for the unsteady temperature distribution $T(x,t)$ in the wall for $t > 0$ can be stated as

$$\frac{\partial^2 T}{\partial x^2} = \frac{1}{\alpha}\frac{\partial T}{\partial t} \tag{10.68a}$$

$$T(x,0) = T_i \tag{10.68b}$$

$$-k\,\frac{\partial T(0,t)}{\partial x} = q_w \qquad \text{and} \qquad T(L,t) = T_i \tag{10.68c,d}$$

Initially, when the penetration depth $\delta(t)$ is less than the thickness L, the effect of the boundary condition (10.68c) is not felt at $x = L$, and the plane wall behaves as if it were a semi-infinite solid. As soon as $\delta = L$, the concept of the penetration depth loses its significance and a different type of analysis is then required. Therefore, we separate the analysis into two stages:

Solution for $\delta < L$: As discussed above when $\delta < L$ the plane wall behaves as if it were a semi-infinite solid. If we adopt a third-degree polynomial representation for the temperature distribution, then the results given by Eqs. (10.56) and (10.57) would be applicable. That is,

$$T(x,t) - T_i = \frac{q_w\delta}{3k}\left(1 - \frac{x}{\delta}\right)^3 \tag{10.69}$$

with

$$\delta = \sqrt{12\alpha t} \tag{10.70}$$

The solution (10.69) is valid for $\delta \leq L$, or, therefore, for $t \leq L^2/12\alpha$.

Solution for $t > L^2/12\alpha$: In this case the penetration depth has no meaning. Again assume that the temperature profile is given by a third-degree polynomial as

$$T(x,t) = a + bx + cx^2 + dx^3, \qquad t > \frac{L^2}{12\alpha} \tag{10.71}$$

where there are four coefficients to be determined. We, therefore, need four independent conditions to determine these coefficients. Two of these are the two natural conditions (10.68c,d). The third condition can be derived by evaluating the differential equation (10.68a) at $x = L$. The result is

$$\left.\frac{\partial^2 T}{\partial x^2}\right|_{x=L} = 0 \tag{10.72}$$

In terms of these three conditions and the unknown surface temperature $T(0,t)$ the profile (10.71) takes the form

$$T(x,t) - T_i = \frac{1}{2}\left(3\frac{\theta_w}{L} - \frac{q_w}{k}\right)(L - x) - \frac{1}{2L^2}\left(\frac{\theta_w}{L} - \frac{q_w}{k}\right)(L - x)^3 \tag{10.73}$$

where $\theta_w(t) = T(0,t) - T_i$ and is an unknown function of time. However, it can be determined as follows: Integrating the differential equation (10.68a) from $x = 0$ to $x = L$, and making use of the condition (10.68c) we obtain

$$\left.\frac{\partial T}{\partial x}\right|_{x=L} + \frac{q_w}{k} = \frac{1}{\alpha}\frac{d}{dt}\int_0^L T(x,t)dx \tag{10.74}$$

Substituting the profile (10.73) into Eq. (10.74) yields the following differential equation for the unknown function $\theta_w(t)$:

$$\frac{d\theta_w}{dt} = \frac{12}{5}\frac{\alpha}{L}\left(\frac{q_w}{k} - \frac{\theta_w}{L}\right) \tag{10.75}$$

The initial condition at $t = L^2/12\alpha$ needed to solve this differential equation is determined by setting $x = 0$ and $\delta = L$ in Eq. (10.69). That is,

$$\theta_w\left(\frac{L^2}{12\alpha}\right) = \frac{q_w L}{3k} \tag{10.76}$$

The solution of Eq. (10.75), with the condition (10.76), is given by

$$\theta_w(t) = \frac{q_w L}{k}\left[1 - 0.814\exp\left(-\frac{12}{5}\frac{\alpha t}{L^2}\right)\right], \qquad t > \frac{L^2}{12\alpha} \tag{10.77}$$

Thus, Eq. (10.73) with $\theta_w(t)$ as given by Eq. (10.75) represents the temperature distribution in the slab for times $t > L^2/12\alpha$.

10.3.4 Problems with Cylindrical and Spherical Symmetry

So far we have considered some representative problems in the rectangular coordinate system, where the use of polynomial representation for the temperature distribution gives reasonably good results. However, Lardner and Pohle [7] have demonstrated that polynomial representation will be inappropriate in problems involving cylindrical and spherical symmetry. This is due to the fact that the volume into which heat diffuses does not remain the same for equal increments of r in the cylindrical and spherical coordinate systems. They suggest the following representations:

Cylindrical symmetry: $T = (\text{polynomial in } r)\ \ln r$ (10.78a)

Spherical symmetry: $T = (\text{polynomial in } r)/r$ (10.78b)

For further details, the reader is referred to Lardner and Pohle's paper.

10.4 VARIATIONAL FORMULATION AND SOLUTION

In this section we first introduce the basics of variational calculus and then apply the results to formulate a heat conduction problem in variational form. Finally, the Ritz method is discussed as an approximate technique to solve the variational forms of heat conduction problems.

10.4.1 Basics of Variational Calculus

Let us start with the problem of finding the function $y(x)$, defined over $x_1 \le x \le x_2$, and passing through the given end points $y(x_1) = y_1$ and $y(x_2) = y_2$, such that

$$I = \int_{x_1}^{x_2} f(x, y, \frac{dy}{dx})\,dx = \text{extremum (minimum or maximum)} \tag{10.79}$$

A definite integral such as Eq. (10.79) is called a *functional*, because the value of it depends on the choice of $y(x)$.

We can reduce the problem of extremization (*i.e.*, maximization or minimization) of the functional (10.79) to the extremization of a *function* as follows. Denoting the solution (as yet unknown) as $y(x)$, let us consider a one-parameter family of "comparison functions" $\tilde{y}(x)$, defined by

$$\tilde{y}(x) = y(x) + \varepsilon\eta(x) \tag{10.80}$$

where $\eta(x)$ is an arbitrary, continuously differentiable function satisfying the end conditions

$$\eta(x_1) = \eta(x_2) = 0 \tag{10.81}$$

and ε is a parameter. Note that no matter how $\eta(x)$ is defined, the solution

$y(x)$ will certainly be a member of the $\tilde{y}$-family, namely, for $\varepsilon = 0$. Note also that $\tilde{y}(x_1) = y_1$ and $\tilde{y}(x_2) = y_2$; thus all members of the family satisfy the end conditions of $y(x)$.

Next consider the quantity

$$I(\varepsilon) = \int_{x_1}^{x_2} f(x, \tilde{y}, \frac{d\tilde{y}}{dx})dx \tag{10.82}$$

which is a function of the parameter ε. Note that

$$\lim_{\varepsilon \to 0} I(\varepsilon) = I \tag{10.83}$$

Thus the value of $I(\varepsilon)$ is an extrenum at $\varepsilon = 0$. According to differential calculus for $I(\varepsilon)$ to be an extrenum at this point, the necessary condition is

$$\frac{dI(0)}{d\varepsilon} = 0 \tag{10.84}$$

Since, from Eq. (10.82),

$$\frac{dI(\varepsilon)}{d\varepsilon} = \int_{x_1}^{x_2} \left(\frac{\partial f}{\partial \tilde{y}} \eta + \frac{\partial f}{\partial \tilde{y}'} \eta'\right) dx$$

$$= \int_{x_1}^{x_2} \left[\frac{\partial f}{\partial (y + \varepsilon\eta)} \eta + \frac{\partial f}{\partial (y' + \varepsilon\eta')} \eta'\right] dx \tag{10.85}$$

the condition (10.84) reduces to

$$\frac{dI(0)}{d\varepsilon} = \int_{x_1}^{x_2} \left(\frac{\partial f}{\partial y} \eta + \frac{\partial f}{\partial y'} \eta'\right) dx = 0 \tag{10.86a}$$

where primes are used to denote differentiation with respect to x. Integrating the second term under the integral sign by parts we obtain

$$\frac{dI(0)}{d\varepsilon} = \int_{x_1}^{x_2} \left[\frac{\partial f}{\partial y} - \frac{d}{dx} \left(\frac{\partial f}{\partial y'}\right)\right] \eta(x)dx + \frac{\partial f}{\partial y'} \eta\Big|_{x_1}^{x_2} = 0 \tag{10.86b}$$

Since by definition $\eta(x_1) = \eta(x_2) = 0$, Eq. (10.86b) reduces to

$$\frac{dI(0)}{d\varepsilon} = \int_{x_1}^{x_2} \left[\frac{\partial f}{\partial y} - \frac{d}{dx} \left(\frac{\partial f}{\partial y'}\right)\right] \eta(x)dx = 0 \tag{10.86c}$$

Equation (10.86c) must hold for any $\eta(x)$ [which satisfy the conditions (10.80) and are continuously differentiable]. We therefore conclude that*

$$\frac{\partial f}{\partial y} - \frac{d}{dx}\left(\frac{\partial f}{\partial y'}\right) = 0 \tag{10.87}$$

Equation (10.87) is the so-called *Euler equation* associated with the functional (10.79). Thus the condition necessary for $y(x)$ to minimize (or maximize) the functional (10.79) is that $f(x,y,dy/dx)$ must satisfy the corresponding Euler equation. Note that the second term in Eq. (10.87) is

$$\frac{d}{dx}\left(\frac{\partial f}{\partial y'}\right) = \frac{\partial^2 f}{\partial x \partial y'} + \frac{\partial^2 f}{\partial y \partial y'}\frac{dy}{dx} + \frac{\partial^2 f}{\partial y'^2}\frac{d^2 y}{dx^2} \tag{10.88}$$

In general, the Euler equation (10.87) will be a non-linear second-order differential equation in the unknown $y(x)$.

Equation (10.86b) holds for all (continuously differentiable) η's. If $y(x_1)$ and $y(x_2)$ are prescribed then $\eta(x_1) = \eta(x_2) = 0$ and Eq. (10.86b) reduces to Eq. (10.86c). If, on the other hand, $y(x_1)$ and $y(x_2)$ are unspecified, then $\eta(x_1)$ and $\eta(x_2)$ will be arbitrary. However, since Eq. (10.86b) must hold for all η, it must obviously hold for the subclass of η's that vanish at x_1 abd x_2 as well. Then, the boundary terms in Eq. (10.86b) drop out and as before it leads to the Euler equation

$$\frac{\partial f}{\partial y} - \frac{d}{dx}\left(\frac{\partial f}{\partial y'}\right) = 0 \tag{10.89}$$

So, Eq. (10.86b) reduces to

$$\frac{\partial f}{\partial y'}\eta\Big|_{x_1}^{x_2} = \frac{\partial f}{\partial y'}\Big|_{x_2}\eta(x_2) - \frac{\partial f}{\partial y'}\Big|_{x_1}\eta(x_1) = 0 \tag{10.90}$$

Thus, if $y(x_1)$ and/or $y(x_2)$ are unspecified, then $\eta(x_1)$ and/or $\eta(x_2)$ are arbitrary, and it follows from Eq. (10.90) that

$$\frac{\partial f}{\partial y'}\Big|_{x_1} = 0 \qquad \text{and/or} \qquad \frac{\partial f}{\partial y'}\Big|_{x_2} = 0 \tag{10.91a,b}$$

These are referred to as *natural boundary conditions* in variational problems.

---

*If $F(x)$ is continuous and

$$\int_{x_1}^{x_2} F(x)\eta(x)\,dx = 0$$

for all continuously differentiable functions $\eta(x)$ for which $\eta(x_1) = \eta(x_2) = 0$, then it must be true that $F(x) = 0$ for all $x_1 \le x \le x_2$ [8].

10.4.2 Variational Formulation of Heat Conduction Problems

The variational formulation of a heat conduction problem may be obtained by considering the differential equation of the problem as the Euler equation associated with its desired variational formulation and working the steps of the previous section in the reverse order. The procedure will now be illustrated in terms of an example below.

Consider a steady-state heat conduction problem formulated as

$$\frac{d}{dx}\left(k\frac{dT}{dx}\right) + \dot{q}(x) = 0 \tag{10.92a}$$

$$T(0) = T_1 \quad \text{and} \quad T(L) = T_2 \tag{10.92b,c}$$

Now take the governing differential equation (10.92a) of the problem as the Euler equation of the variational formulation to be obtained. A look at the derivation of Eq. (10.87) suggests that we proceed as follows:

$$\int_0^L \left[\frac{d}{dx}\left(k\frac{dT}{dx}\right) + \dot{q}(x)\right]\eta(x)\,dx = 0 \tag{10.93}$$

Integrating the first term by parts gives

$$k\frac{dT}{dx}\eta\Big|_0^L - \int_0^L k\frac{dT}{dx}\frac{d\eta}{dx}\,dx + \int_0^L \dot{q}(x)\eta(x)\,dx = 0 \tag{10.94}$$

The first term vanishes because $\eta(0) = \eta(L) = 0$ (T is specified at $x = 0$ and $x = L$). Thus, Eq. (10.94) reduces to

$$\int_0^L \left[\dot{q}(x)\eta(x) - k\frac{dT}{dx}\frac{d\eta}{dx}\right]dx = 0 \tag{10.95}$$

Comparing Eq. (10.95) with Eq. (10.86a) we obtain

$$\frac{\partial f}{\partial T} = \dot{q}(x) \tag{10.96a}$$

and

$$\frac{\partial f}{\partial T'} = -k\frac{dT}{dx} \tag{10.96b}$$

where $T' = dT/dx$. Integrating we get

$$f = \dot{q}(x)T(x) - \frac{1}{2}k\left(\frac{dT}{dx}\right)^2 + F(x) \tag{10.97}$$

where $F(x)$ is an arbitrary function of x. Thus, the variational form of the problem is given by

$$I = \int_0^L [\dot{q}(x)T(x) - \frac{1}{2} k (\frac{dT}{dx})^2]dx + C = \text{extremum} \tag{10.98}$$

where the constant C is given by

$$C = \int_0^L F(x)dx$$

and it can be eliminated from Eq. (10.98) because the function T(x) which minimizes (or maximizes) the functional I defined by Eq. (10.98) also minimizes (or maximizes) the same functional with C = 0. Therefore, the variational form of the problem reduces to

$$I = \int_0^L [\dot{q}(x)T(x) - \frac{1}{2} k (\frac{dT}{dx})^2]dx = \text{extremum} \tag{10.99}$$

Thus, the problem has been reduced to the determination of a function T(x), such that it will pass through the end points $T(0) = T_1$ and $T(L) = T_2$, and will make the functional I defined by Eq. (10.99) an extremum.

10.4.3 Approximate Solutions by the Ritz Method

Once the variational form of a heat conduction problem is available, finding the exact form of the temperature distribution which will make the functional of the formulation an extremum may not be possible. On the other hand, an approximate solution can be obtained by the procedure called the Ritz method.

Let us now reconsider the problem of finding the function y(x), passing through the given end points $y(x_1) = y_1$ and $y(x_2) = y_2$, such that it makes the functional (10.79) an extremum (*i.e.*, a minimum or a maximum). The Ritz method is based on approximating the unknown function y(x) in terms of some known linearly independent functions $\phi_n(x)$, n = 0,1,2,...,N in the form

$$y(x) = \phi_0(x) + \sum_{n=1}^{N} c_n\phi_n(x) \tag{10.100}$$

where the function $\phi_0(x)$ satisfies the boundary conditions and the functions $\phi_n(x)$ for n = 1,2,...,N satisfy the homogeneous part of the boundary conditions. Thus, the expression (10.100) satisfies the boundary conditions. Inserting Eq.

(10.100) into Eq. (10.79) we obtain the functional in the following form

$$I = I(c_0, c_1, c_2, \ldots, c_N) \tag{10.101}$$

On the other hand, the fact that I = extremum requires

$$\frac{\partial I}{\partial c_n} = 0, \qquad n = 0,1,2,\ldots,N \tag{10.102}$$

This procedure results in N algebraic equations for the determination of N unknown coefficients, c_n's. Once c_n's are available, then Eq. (10.100) represents an approximate solution to the problem. The accuracy and the convergence of this solution largely depend on the choice of functions $\phi_n(x)$. A discussion of the error estimation in the Ritz method can be found in Reference [9].

Let us now obtain an approximate solution by the Ritz method to the problem formualted as

$$\frac{d^2T}{dx^2} + \frac{x}{k} = 0 \tag{10.103a}$$

$$T(0) = T_w \quad \text{and} \quad T(1) = T_w \tag{10.103b,c}$$

This is a special case of the problem considered in the previous section. Thus, the variational form is obtainable from Eq. (10.99) as

$$I = \int_0^1 \left[xT(x) - \frac{1}{2}k\left(\frac{dT}{dx}\right)^2\right]dx = \text{extremum} \tag{10.104}$$

Let the trial solution be

$$T(x) \cong T_w + c_1 x(1 - x) \tag{10.105}$$

Clearly, the trial solution $T_w + c_1 x(1 - x)$ satisfies the boundary conditions (10.103b,c). Introducing Eq. (10.105) into Eq. (10.104) we obtain

$$I(c_1) = \frac{1}{2}\left(T_w + \frac{c_1}{6}\right) - \frac{1}{6}kc_1^2 \tag{10.106}$$

The coefficient c_1 is determined according to Eq. (10.102) as

$$\frac{dI}{dc_1} = \frac{1}{12} - \frac{1}{3}kc_1 = 0 \rightarrow c_1 = \frac{1}{4k} \tag{10.107}$$

Then, the approximate solution becomes

$$T(x) = T_w + \frac{1}{4k}x(1 - x) \tag{10.108}$$

which represents a first-order approximation of the temperature distribution.

A second-order approximation can, for example, be obtained by letting

$$T(x) \cong T_w + c_1 x(1 - x) + c_2 x^2(1 - x) \qquad (10.109)$$

Introducing Eq. (10.109) into Eq. (10.104) we obtain $I(c_1, c_2)$. Then, the application of Eq. (10.102) gives two algebraic equations for the determination of the coefficients c_1 and c_2. Solving these equations we get

$$c_1 = \frac{1}{6k} \quad \text{and} \quad c_2 = \frac{1}{6k} \qquad (10.110a,b)$$

Thus, a second-order approximation of the temperature distribution is

$$T(x) \cong T_w + \frac{x}{6k}(1 - x^2) \qquad (10.111)$$

which is, in fact, the exact solution of the problem.

The following is a comparison of the first-order approximation for $k[T(x) - T_w]$ with the exact result for the same term at three different x locations.

x	0.25	0.5	0.75
Exact	0.0391	0.0625	0.0547
1st Order-App.	0.0469	0.0625	0.0469
Error	∿20%	0%	∿14%

In this section the basic principles of the variational formulation of heat conduction problems and the Ritz method for approximate solution of variational formulation have been given. Furthermore, the technique has been demonstrated by means of an example. For further discussions of the methods as applied to heat conduction problems the reader is referred to References [1,10-12].

10.5 PROBLEMS INVOLVING PHASE CHANGE

Heat conduction problems in which melting or solidification occur are generally referred to as *phase-change* or *moving-boundary* problems and can be of two types. For some materials, there is a distinct line of demarcation between the solid and liquid phases, called the *melt line*. That is, the solidification or melting takes place at a discrete temperature, and the two phases are separated by a definite moving interface. Typical of such materials are water and pure metals. Other materials, such as alloys, mixtures and impure materials, do not possess a definite line of demarcation between the liquid and solid phases and melting or solidification takes place over a temperature range. The two phases are separated by a two-phase region, which is distinguished by a gradual change in the physical properties of the material from one phase to the other.

One of the difficulties in handling phase-change problems is that the boundary, either in the form of a melt line or a two-phase region, is moving. In addition, some motion of the liquid phase may be present so that heat transfer by convection has to be considered in the liquid phase. In this section we will deal exclusively with phase-change problems involving a sharp moving interface and assume that heat transfer in the liquid phase is by conduction.

The exact solution of phase-change problems is limited to few idealized situations [1,13]. This is mainly due to the fact that phase change problems are non-linear. Accordingly, in most of the cases, either an approximate method or a numerical technique is used. Among the approximate methods, the integral method is the most commonly used one. It provides a relatively simple and straightforward approach. The basic principles of this method have already been described in Section 10.3. A review of the literature on the solutions of phase-change problems by various approximate methods is given in Reference [1].

10.5.1 Boundary Conditions at a Sharp Moving Interface

Consider now a situation in which melting or solidification takes place at a definite temperature (melting point) and therefore there is a distinct line of demarcation between the two phases. Assume that heat transfer takes place only by conduction in both phases and the thermo-physical properties of each phase are constants. Designate the conditions by a subscript s in the solid phase and by a subscript ℓ in the liquid phase. Let $x = s(t)$ define the location of the interface. If T_m denotes the melting (or solidification) temperature, which is constant for a given substance, then

$$T_s(x,t)\big|_{x=s(t)} = T_\ell(x,t)\big|_{x=s(t)} = T_m \tag{10.112}$$

Assuming that the density does not change during the melting or the solidification process, from an energy balance across the solid-liquid interface, we obtain

$$k_s \frac{\partial T_s}{\partial x} - k_\ell \frac{\partial T_\ell}{\partial x} = \rho L \frac{ds}{dt} \qquad \text{at} \qquad x = s(t) \tag{10.113}$$

where ρ is the density and L is the latent heat of melting or solidification. In obtaining Eq. (10.113), it has been assumed that the solid-liquid interface is moving in the positive x-direction. Equation (10.113) states that the difference between the heat flux entering and leaving the solid-liquid interface is the latent heat absorbed or released. Furthermore, the term ds/dt in Eq. (10.113) is the velocity of the interface in the positive x-direction.

10.5.2 Melting of a Solid with a Constant Surface Temperature-Exact Solution

Consider a solid, $x \geq 0$, initially at the melting temperature T_m. Let the temperature of the surface at x = 0 be raised to T_w ($> T_m$) at t = 0 and maintained at this constant value for times t > 0. Thus melting starts at the surface and the liquid-solid interface moves in the positive x-direction. Being at the uniform temperature T_m throughout, there is no heat transfer in the solid phase. Assuming that heat transfer in the liquid phase is by conduction only, and the thermo-physical properties are constant, the formulation of the problem for the temperature distribution T(x,t) in the liquid phase is given by

$$\frac{\partial^2 T}{\partial x^2} = \frac{1}{\alpha}\frac{\partial T}{\partial t} \tag{10.114a}$$

$$T(0,t) = T_w \quad \text{and} \quad T(x,t)\Big|_{x=s(t)} = T_m \tag{10.114b,c}$$

Furthermore, at the interface we also have

$$-k\,\frac{\partial T}{\partial x}\Big|_{x=s(t)} = \rho L\,\frac{ds}{dt} \tag{10.114d}$$

For simplicity we have dropped the subscript ℓ.

We now assume a solution in the form

$$T(x,t) = A + B\,\mathrm{erfc}\,\frac{x}{2\sqrt{\alpha t}} \tag{10.115}$$

where A and B are two arbitrary constants, and α is the thermal diffusivity of the liquid phase. Equation (10.115) satisfies the heat conduction equation (10.114a) (see Section 8.8). Imposing the boundary conditions (10.114b,c) we obtain

$$A = T_m - (T_w - T_m)\,\frac{\mathrm{erfc}(\gamma)}{\mathrm{erf}(\gamma)} \tag{10.116a}$$

$$B = \frac{T_w - T_m}{\mathrm{erf}(\gamma)} \tag{10.116b}$$

where γ is defined by

$$\gamma = \frac{s(t)}{2\sqrt{\alpha t}} = \text{constant} \tag{10.116c}$$

Introducing Eqs. (10.117a,b) into Eq. (10.115) we get

$$\frac{T(x,t) - T_m}{T_w - T_m} = 1 - \frac{\mathrm{erf}\left(\frac{x}{2\sqrt{\alpha t}}\right)}{\mathrm{erf}(\gamma)} \tag{10.117}$$

where the constant parameter γ is yet to be determined. However, the use of the condition (10.114d) gives

$$\gamma e^{\gamma^2} \operatorname{erf}(\gamma) = \frac{c_p(T_w - T_m)}{L\sqrt{\pi}} \tag{10.118}$$

where c_p is the specific heat of the liquid phase. Equation (10.118) is a transcendental equation for the determination of γ. Knowing γ, the location of the interface s(t) is obtained from

$$s(t) = 2\gamma\sqrt{\alpha t} \tag{10.119}$$

and the temperature distribution in the liquid phase T(x,t) is given by Eq. (10.117).

10.5.3 Melting of a Solid with a Constant Surface Temperature - Approximate Solution by the Integral Method

Let us now obtain an approximate solution by the integral method to the problem solved exactly in the previous section. The formulation of the problem is given by Eq. (10.114). Upon integrating Eq. (10.114a) from $x = 0$ to $x = s(t)$, and applying Eq. (10.114d) we are led to the following energy-integral equation:

$$\frac{d}{dt}\int_0^{s(t)} [T(x,t) - T_m]dx = -\alpha\left[\frac{\rho L}{k}\frac{ds}{dt} + \frac{\partial T(0,t)}{\partial x}\right] \tag{10.120}$$

To obtain an approximate solution let T(x,t) be represented by a second-degree polynomial in x in the form

$$T(x,t) = a + b(x-s) + c(x-s)^2 \tag{10.121}$$

where $s = s(t)$. Two of the three conditions required to determine the coefficients in Eq. (10.121) are Eqs. (10.114b,c). The third condition is Eq. (10.114d); but it is not suitable in it present from, because if it is used, the resulting temperature profile will involve ds/dt. In turn, the energy-integral equation (10.120) would yield a second-order differential equation for s(t), whereas there is only one initial condition for s(t); that is, $s(0) = 0$. To overcome this difficulty we differentiate the condition (10.114c) with respect to t and obtain

$$\frac{\partial T}{\partial x}\frac{ds}{dt} + \frac{\partial T}{\partial t} = 0 \qquad \text{at} \qquad x = s(t) \tag{10.122}$$

Upon eliminating ds/dt between Eqs. (10.114d) and (10.122), it follows that

$$\left(\frac{\partial T}{\partial x}\right)^2 = \frac{\rho L}{k}\frac{\partial T}{\partial t} \qquad \text{at} \qquad x = s(t) \tag{10.123}$$

Furthermore, eliminating $\partial T/\partial t$ between Eqs. (10.114a) and (10.123) we get

$$\left(\frac{\partial T}{\partial x}\right)^2 = \frac{L}{c_p}\frac{\partial^2 T}{\partial x^2} \qquad \text{at} \qquad x = s(t) \tag{10.124}$$

With this condition, the non-linearity of the problem is self-evident. The coefficients a, b and c in Eq. (10.121) can now be determined by the use of Eqs. (10.114b,c) and (10.124). The results are

$$a = T_m \tag{10.125a}$$

$$b = \frac{L}{c_p s}\left[1 - (1 + \mu^2)^{1/2}\right] \tag{10.125b}$$

$$c = \frac{bs + (T_w - T_m)}{s^2} \tag{10.125c}$$

where

$$\mu = \frac{2c_p(T_w - T_m)}{L} \tag{10.126}$$

Substituting the profile (10.121) into the energy-integral equation (10.120) and performing the indicated operations we finally obtain the differential equation

$$s\frac{ds}{dt} = 6\alpha\,\frac{1 - (1 + \mu)^{1/2} + \mu}{5 + (1 + \mu)^{1/2} + \mu} \tag{10.127}$$

The solution of Eq. (10.127), with the condition $s(0) = 0$, leads to

$$s(t) = 2\gamma\sqrt{\alpha t} \tag{10.128}$$

where

$$\gamma = \left[3\,\frac{1 - (1 + \mu)^{1/2} + \mu}{5 + (1 + \mu)^{1/2} + \mu}\right]^{\frac{1}{2}} \tag{10.129}$$

It is to be noted that the approximate solution (10.128) for $s(t)$ is of the same form as the exact solution (10.119); but the parameter γ is given by Eq. (10.129) for the approximate solution, whereas it is the root of the transcendental equation (10.118) for the exact solution. The transcendental equation (10.118) can also be written in terms of μ; that is,

$$\gamma e^{\gamma^2}\operatorname{erf}(\gamma) = \frac{\mu}{2\sqrt{\pi}} \tag{10.130}$$

A graphical comparison of the exact and approximate values of γ as a function of the quantity μ is given in Reference [14]. The agreement between the exact and the approximate results is reasonably good. In fact, the error is about 7% for

$\mu = 2.8$, the largest value of μ considered. For smaller values of μ, the percent error is less. The agreement would be much better if a third-degree polynomial were used [14].

In this section we have outlined the steps in the application of the integral method to a melting problem. For further applications of the method to phase-change problems the reader is referred to References [1,5,6,14-21].

REFERENCES

1. Özışık, M.N., *Heat Conduction*, John Wiley and Sons, Inc., New York, 1980.
2. von Karman, T., Uber Laminare und Turbulente Reibung, Z. *Angew, Math. Mech.*, Vol. 1, p. 233, 1921.
3. Pohlhausen, K., Zur Nahenrungsweisen Integration der Laminaren Reibungschicht, Z. *Angew,Math.Mech.*, Vol 1, p. 252, 1921.
4. Landahl, H.D., *Bull.Math. Biophys.*, Vol. 15, pp. 49-61, 1953.
5. Goodman, T.R., Applications of Integral Methods to Transient Nonlinear Heat Transfer, in *Advances in Heat Transfer*, T.F. Irvine and J.P. Hartnelt (eds.), Vol. 1, pp. 51-122, Academic Press, New York, 1964.
6. Goodman, T.R., *J. Heat Transfer*, Vol. 83c, pp. 83-86, 1961.
7. Lardner T.J., and Pohle, F.V., *J. Appl. Mech.*, Vol. 28, pp. 310-312, 1961.
8. Greenberg, M.D., *Foundations of Applied Mathematics*, Prentice-Hall, Inc., Englewood Cliffs, New Jersey, 1978.
9. Kryloff, N., *Memorial Sci. Math. Paris*, Vol. 49, 1931.
10. Arpacı, V.S., and Vest C.M., Variational Formulation of Transformed Diffusion Problems, ASME Paper No. 67-HT-77.
11. Schechter, R.S., *The Variational Method in Engineering*, McGraw-Hill Book Co., New York, 1967.
12. Arpacı, V.S., *Conduction Heat Transfer*, Addison-Wesley Publishing Co. Inc., Reading, Massachusetts, 1966.
13. Carslaw, H.S., and Jaeger, J.C., *Conduction of Heat in Solids*, 2nd ed., Clarendon Press, London, 1959.
14. Goodman, T.R., *Trans. Am. Soc. Mech. Eng.*, Vol. 80, pp. 335-342, 1958.
15. Goodman, T.R., and Shea, J., *J. Appl. Mech.*, Vol. 32, pp. 16-24, 1960.
16. Poots, G., *Int. J. of Heat Mass Transfer*, Vol. 5, pp. 339-348, 1962.
17. Poots, G., *Int. J. of Heat Mass Transfer*, Vol. 5, p. 525, 1962.
18. Tien, R.H., and Greiger, G.E., *J. Heat Transfer*, Vol. 89c, pp. 230-234, 1967.
19. Tien, R.H., and Greiger, G.E., *J. Heat Transfer*, Vol. 90c, pp. 27-31, 1968.
20. Cho, S.H., and Sunderland, J.E., *J. Heat Transfer*, Vol. 91c, pp. 421-426, 1969.
21. Mody, K., and Özışık, M.N., *Lett. Heat Mass Transfer*, Vol. 2, pp. 487-493, 1975.

PROBLEMS

10.1 A semi-infinite solid, $0 \le x < \infty$, is initially at a uniform temperature T_i. For times $t \ge 0$ a time-dependent heat flux $q_w(t)$ is applied to the surface at $x = 0$. Assuming constant thermo-physical properties, obtain an expression for the unsteady temperature distribution in the solid by Duhamel's method for $t > 0$.

10.2 Solve Problem 8.9 by Duhamel's method.

10.3 A long solid cylinder of constant thermo-physical properties and radius r_0 is initially at a uniform temperature T_i. For times $t \ge 0$, the temperature of the surface at $r=r_0$ is varied according to $T(r_0,t) = f(t)$, where $f(t)$ is a prescribed function of time. Obtain an expression for the unsteady temperature distribution in the solid by Duhamel's method for $t > 0$.

10.4 A long solid cylinder of constant thermo-physical properties and radius r_0 is initially at a uniform temperature T_i. For times $t \ge 0$ internal energy is generated in the cylinder at a rate $\dot{q}(t)$ per unit volume while the peripheral surface at $r=r_0$ is maintained at T_i. Obtain an expression for the unsteady temperature distribution in the cylinder by Duhamel's method for $t > 0$.

10.5 A solid sphere of constant thermo-physical properties and radius r_0 is initially at a uniform temperature T_i. For times $t \ge 0$, internal energy is generated in the sphere at a rate $\dot{q}(t)$ per unit volume while the surface at $r=r_0$ is maintained at T_i. Obtain an expression for the unsteady temperature distribution in the sphere by Duhamel's method for $t > 0$.

10.6 Solve Problem 10.1 using the Integral Method. Approximate the temperature profile by a third-degree polynomial in x.

10.7 A semi-infinite solid, $0 \le x < \infty$, is initially at a uniform temperature T_i. The surface temperature at $x=0$ is changed to T_w at $t=0$ and maintained constant at this value for times $t > 0$. Assuming that the thermal conductivity k, specific heat c, and the density ρ of the solid are all temperature dependent, obtain an expression for the unsteady temperature distribution in the solid by the integral method. Assume that the temperature profile can be represented by a third-degree polynomial.

10.8 Solve Problem 8.7 by the integral method. Approximate the temperature profile in the solid by a third-degree polynomial.

10.9 A plane wall of thickness L in the x-direction is initially at a uniform temperature T_i. For times $t > 0$, the surface at $x=0$ is maintained at a constant temperature T_w and the surface at x=L is kept perfectly insulated. Assuming constant thermo-physical properties, obtain an expression for the unsteady temperature distribution in the wall by the integral method.

10.10 An infinite region, $r_0 \leq r < \infty$, with a cylindrical hole of radius $r=r_0$ in it is initially at a uniform temperature T_i. For times $t \geq 0$, the cylindrical surface at $r=r_0$ is maintained at a constant temperature T_w. Assuming constant thermo-physical properties, obtain an expression for the temperature distribution T(r,t) in the medium by the integral method by assuming the following profiles:

a) $T(r,t) = a + br + cr^2$

b) $T(r,t) = (a + br + cr^2) \cdot \ln r$

Also obtain the exact solution of the problem and compare the two approximate solutions obtained in (a) and (b) with it.

10.11 Obtain the variational form of the heat conduction problem formulated as

$$\frac{d}{dx}\left(k\frac{dT}{dx}\right) + \dot{q}(x) = 0$$

$$\left[-k\frac{dT}{dx} + h_1\, T(x)\right]_{x=0} = h_1 T_{f1}$$

$$\left[k\frac{dT}{dx} + h_2\, T(x)\right]_{x=L} = h_2\, T_{f2}$$

where h_1, h_2 T_{f1} and T_{f2} are constants.

10.12 Obtain an approximate solution by the Ritz method to the problem given by Eqs. (10.103a,b,c) by assuming a trial solution in the form

$$T(x) = T_w + c_1 x(1-x) + c_2\, x\, (1-x)^2$$

10.13 A solid, $x \geq 0$, is initially at the melting temperature T_m. For times $t \geq 0$, a constant heat flux q_w is applied to the surface at x=0. Obtain an expression for the location of the solid-liquid interface as a function of time by the integral method. Assume a second-order polynomial for the temperature distribution in the liquid phase.

10.14 A liquid confined to $x \geq 0$ is initially at a uniform temperature T_i, higher than the melting temperature T_m of the solid phase. At time t=0 the temperature of the boundary at x=0 is lowered to T_w, below T_m, and maintained at this constant value for $t > 0$. Obtain exact expressions for the temperature distributions in both phases and the location of the solid-liquid interface for times $t > 0$.

Appendix A: Thermo-Physical Properties

APPENDIX A

THERMO-PHYSICAL PROPERTIES

Table A.1 Thermo-physical properties of metals.

Metal	Temperature Range T °C	Density ρ g/cm^3	Specific heat c kJ/kg·K	Thermal conductivity k W/m·K	Emissivity ε
Aluminum	0-400	2.72	0.895	204-250	0.04-0.06 (polished) 0.07-0.09 (commercial) 0.2-0.3 (oxidized)
Brass (70% Cu, 30% Zn)	100-300	8.52	0.38	104-147	0.03-0.07 (polished) 0.2-0.25 (commercial) 0.45-0.55 (oxidized)
Bronze (75% Cu, 25% Sn)	0-100	8.67	0.34	26	0.03-0.07 (polished) 0.4-0.5 (oxidized)
Constantan (60% Cu, 40% Ni)	0-100	8.92	0.42	22-26	0.03-0.06 (polished) 0.2-0.4 (oxidized)
Copper	0-600	8.95	0.38	385-350	0.02-0.04 (polished) 0.1-0.2 (commercial)
Iron (C≃4%, cast)	0-1,000	7.26	0.42	52-35	0.2-0.25 (polished) 0.55-0.65 (oxidized) 0.6-0.8 (rusted)
Iron (C≃0.5%, wrought)	0-1,000	7.85	0.46	59-35	0.3-0.35 (polished) 0.9-0.95 (oxidized)
Lead	0-300	11.37	0.13	35-30	0.05-0.08 (polished) 0.3-0.6 (oxidized)

Table A.1 (continued)

Metal	Temperature Range T °C	Density ρ g/cm^3	Specific heat c kJ/kg·K	Thermal conductivity k W/m·K	Emissivity ε
Magnesium	0-300	1.75	1.01	171-157	0.07-0.13 (polished)
Mercury	0-300	13.4	0.125	8-10	0.1-0.12
Molybdenum	0-1,000	10.22	0.251	125-99	0.06-0.10 (polished)
Nickel	0-400	8.9	0.45	93-59	0.05-0.07 (polished) 0.35-0.49 (oxidized)
Platinum	0-1,000	21.4	0.24	70-75	0.05-0.03 (polished) 0.07-0.11 (oxidized)
Silver	0-400	10.52	0.23	410-360	0.01-0.03 (polished) 0.02-0.04 (oxidized)
Steel (C≃1%)	0-1,000	7.80	0.47	43-28	0.07-0.17 (polished)
Steel (Cr≃1%)	0-1,000	7.86	0.46	62-33	0.07-0.17 (polished)
Steel (Cr 18%,Ni 8%)	0-1,000	7.81	0.46	16-26	0.07-0.17 (polished)
Tin	0-200	7.3	0.23	65-57	0.04-0.06 (polished)
Tungsten	0-1,000	19.35	0.13	166-76	0.04-0.08 (polished) 0.1-0.2 (filament)
Zinc	0-400	7.14	0.38	112-93	0.02-0.03 (polished) 0.10-0.11 (oxidized) 0.2-0.3 (galvanized)

Table A.2 Thermo-physical properties of nonmetals.

Material	Temperature Range T °C	Density ρ g/cm^3	Specific heat c kJ/kg·K	Thermal conductivity k W/m·K	Emissivity ε
Asbestos	100-1,000	0.47-0.57	0.816	0.15-0.22	0.93-0.97
Brick, rough red	100-1,000	1.76	0.84	0.38-0.43	0.90-0.95
Clay	0-200	1.46	0.88	1.3	0.91
Concrete	0-200	2.1	0.88	0.81-1.4	0.94
Glass, window	0-600	2.7	0.84	0.78	0.94-0.66
Glass wool	23	0.024	0.7	0.038	
Ice	0	0.91	1.9	2.2	0.97-0.99
Limestone	100-300	2.5	0.9	1.3	0.95-0.80
Marble	0-100	2.60	0.80	2.07-2.94	0.93-0.95
Plasterboard	0-100	1.25	0.84	0.43	0.92
Rubber (hard)	0-100	1.2	1.42	0.15	0.94
Sandstone	0-300	2.24	0.71	1.83	0.83-0.9
Wood (oak)	0-100	0.6-0.8	2.4	0.17-0.21	0.90

REFERENCES

1. Adams, J.A., Rogers, D.F., *Computer-Aided Heat Transfer Analysis*, McGraw-Hill Book Co., New York, 1973.

2. Holman, J.P., *Heat Transfer*, 5th ed., McGraw-Hill Book Co., New York, 1981.

3. Ibele, W., Thermophysical properties, Section 2 in *Handbook of Heat Transfer*, W.M. Rohsenow and J.P. Hartnett, eds., McGraw-Hill Book Co., New York, 1973.

4. Eckert, E.R.G., and Drake, R.M., *Analysis of Heat and Mass Transfer*, McGraw-Hill Book Co., New York, 1972.

Appendix B: Bessel Functions

APPENDIX B

BESSEL FUNCTIONS

A linear second-order ordinary differential equation with variable coefficients of the form

$$x^2 \frac{d^2y}{dx^2} + x\frac{dy}{dx} + (m^2x^2 - \nu^2)y = 0 \tag{B.1}$$

is known as *Bessel's differential equation of order* ν, where m is a parameter and ν is any real number. Since only ν^2 appears in Eq. (B.1), we may also consider ν to be non-negative without loss of generality. The general solution of Eq. (B.1) may be obtained by the use of the method of Frobenius and the result is

$$y(x) = C_1 J_\nu(mx) + C_2 Y_\nu(mx) \tag{B.2}$$

where the functions $J_\nu(mx)$ and $Y_\nu(mx)$ are known as the *Bessel functions* of the *first kind* and of the *second kind*, of *order* ν, respectively. For all values $\nu \geq 0$, the function $J_\nu(mx)$ is defined by

$$J_\nu(mx) = \sum_{k=0}^{\infty} (-1)^k \frac{(mx/2)^{2k+\nu}}{k!\,\Gamma(k+\nu+1)} \tag{B.3}$$

If $\nu \neq n = 0,1,2,\ldots$, the function $Y_\nu(mx)$ is defined by

$$Y_\nu(mx) = \frac{J_\nu(mx)\cos(\nu\pi) - J_{-\nu}(mx)}{\sin(\nu\pi)} \tag{B.4a}$$

and for values of $\nu = n = 0,1,2,\ldots$, it is given by

$$Y_n(mx) = \frac{2}{\pi}\left[\ln\frac{mx}{2} + \gamma\right]J_n(mx) - \frac{1}{\pi}\sum_{k=0}^{n-1}\frac{(n-k-1)!}{k!}\left(\frac{mx}{2}\right)^{2k-n}$$
$$- \frac{1}{\pi}\sum_{k=0}^{\infty}(-1)^k[\phi(k) + \phi(k+n)]\frac{(mx/2)^{2k+n}}{k!(n+k)!} \tag{B.4b}$$

The function $J_{-\nu}(mx)$ in Eq. (B.4a) is obtained by replacing ν by $-\nu$ in Eq. (B.3); that is,

$$J_{-\nu}(mx) = \sum_{k=0}^{\infty} (-1)^k \frac{(mx/2)^{2k-\nu}}{k!\,\Gamma(k-\nu+1)} \tag{B.5}$$

If $\nu = n = 0,1,2,\ldots$, then the functions $J_\nu(mx)$ and $J_{-\nu}(mx)$ are linearly dependent, because they are related to each other in the form

$$J_n(mx) = (-1)^n J_{-n}(mx) \tag{B.6}$$

If, however, $\nu \neq n = 0,1,2,\ldots$, the functions $J_\nu(mx)$ and $J_{-\nu}(mx)$ are linearly independent. Hence, the solution (B.2) can also be written as

$$y(x) = D_1 J_\nu(mx) + D_2 J_{-\nu}(mx), \qquad \nu \neq n = 0,1,2,\ldots \tag{B.7}$$

The *gamma function* appearing in the above equations is defined by the integral

$$\Gamma(\alpha) = \int_0^\infty e^{-t}\, t^{\alpha-1}\, dt, \qquad \alpha > 0 \tag{B.8}$$

Integration by parts gives the following important relation:

$$\Gamma(\alpha + 1) = \alpha\Gamma(\alpha) \tag{B.9}$$

Hence, Eq. (B.9), together with $\Gamma(1) = 1$, yields

$$\Gamma(n + 1) = n!, \qquad n = 0,1,2,\ldots \tag{B.10}$$

It can be shown that for *fractional* numbers the following relation is true

$$\Gamma(\nu)\Gamma(\nu - 1) = \frac{\pi}{\sin(\nu\pi)}, \qquad \Gamma\left(\frac{1}{2}\right) = \sqrt{\pi} \tag{B.11}$$

Furthermore, the expression $\phi(k)$ in Eq. (B.4b) is defined as

$$\phi(k) = \sum_{n=1}^{k} \frac{1}{n}, \qquad \phi(0) = 0 \tag{B.12}$$

and $\gamma = 0.5772156\ldots$ is *Euler's Constant.*

It is difficult to predict the behavior of Bessel functions from their series representations. For integer values of ν, general behavior of these functions are shown in Figs. B.1 and B.2. It is to be noted, however, that Bessel functions of the second kind Y_ν are unbounded at $x = 0$ for all values of $\nu \geq 0$. In Table B.1 we present the numerical values of $J_n(x)$ and $Y_n(x)$ for $n = 0$ and 1. In Table B.2 we give the first forty zeros of $J_0(x)$ and the corresponding values of $J_1(x)$. Furthermore, in Table B.3 we list the first ten zeros of $J_n(x)$ for $n = 1,2,3,4$, and 5.

MODIFIED BESSEL FUNCTIONS

A linear second-order ordinary differential equation with variable coefficients of the form

$$x^2 \frac{d^2y}{dx^2} + x\frac{dy}{dx} - (m^2x^2 + \nu^2)y = 0 \tag{B.13}$$

is known as *modified Bessel's differential equation of order* ν. The general solution of Eq. (B.13) may be written as

$$y(x) = C_1\, I_\nu(mx) + C_2\, K_\nu(mx) \tag{B.14}$$

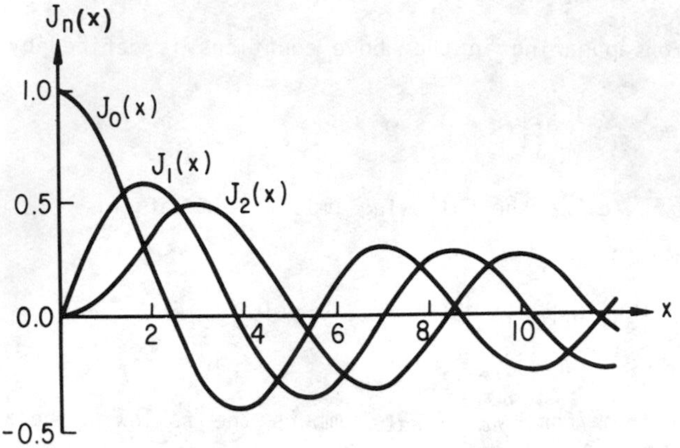

Fig. B.1 Bessel functions of the first kind.

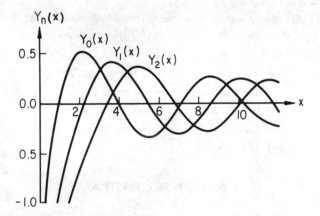

Fig. B.2 Bessel functions of the second kind.

where the functions $I_\nu(mx)$ and $K_\nu(mx)$ are known as the *modified Bessel functions* of the *first kind* and the *second kind*, of *order* ν, respectively. For all values of $\nu \geq 0$, the function $I_\nu(mx)$ is given by

$$I_\nu(mx) = (i)^{-\nu} J_\nu(imx), \qquad i = \sqrt{-1} \tag{B.15}$$

If $\nu \neq n = 0,1,2,\ldots$, the function $K_\nu(mx)$ is defined by

$$K_\nu(mx) = \frac{\pi}{2}\,\frac{I_{-\nu}(mx) - I_\nu(mx)}{\sin(\nu\pi)} \tag{B.16}$$

and for values of $\nu = n = 0,1,2,\ldots$, it is given by

$$\begin{aligned} K_n(mx) = {} & (-1)^{n+1}[\ln \tfrac{mx}{2} + \gamma]I_n(mx) \\ & + \frac{1}{2}\sum_{k=0}^{n-1} (-1)^k \frac{(n-k-1)!}{k!}\left(\frac{mx}{2}\right)^{2k-n} \\ & + \frac{1}{2}(-1)^n \sum_{k=0}^{\infty} [\phi(k) + \phi(k+n)]\,\frac{(mx/2)^{2k+n}}{k!(n+k)!} \end{aligned} \tag{B.17}$$

The function $I_{-\nu}(mx)$ in Eq. (B.16) is obtained by replacing ν by $-\nu$ in Eq. (B.15). If ν is not an integer or zero, the functions $I_\nu(mx)$ and $I_{-\nu}(mx)$ are linearly independent. Hence, the solution (B.14) can also be written as

$$y(x) = D_1\, I_\nu(mx) + D_2\, I_{-\nu}(mx), \qquad \nu \neq n = 0,1,2,\ldots \tag{B.18}$$

If $\nu = n = 0,1,2,\ldots$, then the functions $I_\nu(mx)$ and $I_{-\nu}(mx)$ are related to each other in the form

$$I_n(mx) = I_{-n}(mx) \tag{B.19}$$

The definitions of $\phi(k)$ and γ are as defined before. Graphs of the general behavior of these functions for integer value of ν are shown in Figs. B.3 and B.4. It is to be noted again that the modified Bessel functions of the second kind K_ν are unbounded at $x = 0$ for all $\nu \geq 0$. In the Table B.1 we also present the numerical values of $I_n(x)$ and $K_n(x)$ for $n = 1$ and 2.

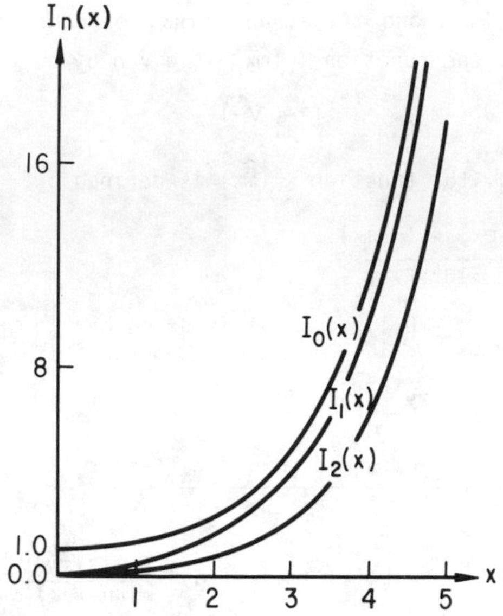

Fig. B.3 Modified Bessel functions of the first kind.

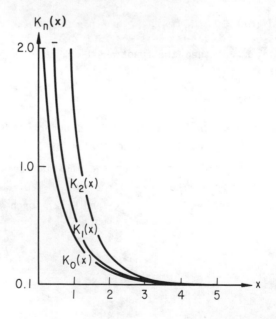

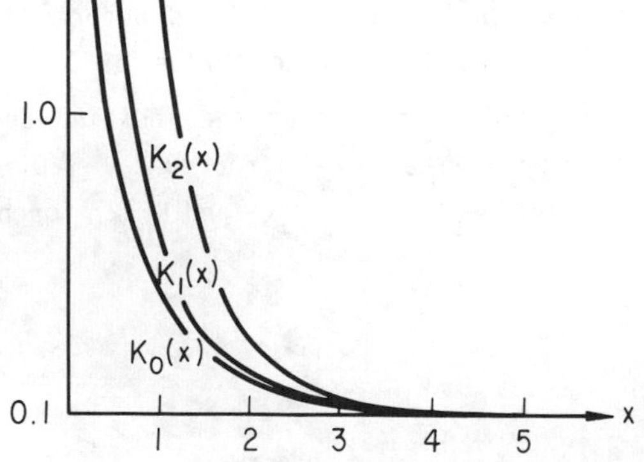

Fig. B.4 Modified Bessel functions of the second kind.

ASYMPTOTIC FORMULAS FOR BESSEL FUNCTIONS

For large values of x we have the following asymptotic formulas:

$$J_\nu(x) = \sqrt{\frac{2}{\pi x}}\ \cos[x - \frac{\pi}{4} - \frac{\nu\pi}{2}] \tag{B.20a}$$

$$Y_\nu(x) = \sqrt{\frac{2}{\pi x}}\ \sin[x - \frac{\pi}{4} - \frac{\nu\pi}{2}] \tag{B.20b}$$

$$I_\nu(x) = \frac{e^x}{\sqrt{2\pi x}} \tag{B.20c}$$

$$K_\nu(x) = \sqrt{\frac{\pi}{2x}}\ e^{-x} \tag{B.20d}$$

It is important that the asymptotic values of $I_\nu(x)$ and $K_\nu(x)$ do not depend on ν.

DERIVATIVES OF BESSEL FUNCTIONS

Some derivatives of Bessel functions are as follows:

$$\frac{d}{dx}[W_\nu(mx)] = \begin{cases} m\,W_{\nu-1}(mx) - (\nu/x)W_\nu(mx), & W = J,Y,I \\ -m\,W_{\nu-1}(mx) - (\nu/x)W_\nu(mx), & W = K \end{cases} \tag{B.21a}$$

$$\frac{d}{dx}[W_\nu(mx] = \begin{cases} -m\,W_{\nu+1}(mx) + (\nu/x)W_\nu(mx), & W = J,Y,K \\ m\,W_{\nu+1}(mx) + (\nu/x)W_\nu(mx), & W = I \end{cases} \tag{B.21b}$$

$$\frac{d}{dx}[x^\nu\,W_\nu(mx)] = \begin{cases} mx^\nu\,W_{\nu-1}(mx), & W = J,Y,I \\ -mx^\nu\,W_{\nu-1}(mx), & W = K \end{cases} \tag{B.21c}$$

$$\frac{d}{dx}[x^{-\nu}\,W_\nu(mx)] = \begin{cases} -mx^{-\nu}\,W_{\nu+1}(mx), & W = J,Y,K \\ mx^{-\nu}\,W_{\nu+1}(mx), & W = I \end{cases} \tag{B.21d}$$

EQUATIONS TRANSFORMABLE INTO BESSEL DIFFERENTIAL EQUATIONS

The solution of the second-order differential equation

$$x^2 \frac{d^2y}{dx^2} + (1 - 2k)x \frac{dy}{dx} + (\alpha x^{2r} + \beta^2)y = 0 \tag{B.22}$$

where k, α, r and β are constants, can be written in the form

$$y(x) = x^k Z_\nu(\frac{\delta}{r} x^r) \tag{B.23}$$

where $\nu = \sqrt{k^2 - \beta^2}/r$. If α is a positive constant then Z_ν is to be interpreted, with $\delta = \sqrt{\alpha}$, by Eq. (B.2), or Eq. (B.7), depending on whether ν is a positive integer (or zero) or not, respectively. On the other hand, if α is a negative constant then Z_ν stands for Eq. (B.14), or Eq. (B.18), with $\delta = \sqrt{-\alpha}$.

If α = 0, then Eq. (B.22) is a Cauchy-Euler equation and has the solution

$$y(x) = x^k(C_1 x^{r\nu} + C_2 x^{-r\nu}) \tag{B.24}$$

TABLES

The numerical values of $J_n(x)$, $Y_n(x)$, $I_n(x)$ and $K_n(x)$ are presented for n = 0 and 1 in Table B.1. The first forty zeros of $J_0(x)$ and the corresponding values of $J_1(x)$ are tabulated in Table B.2. Furthermore, the first ten zeros of $J_n(x)$ for n = 1,2,3,4, and 5 are listed in Table B.3.

Table B.1 Numerical Values of $J_n(x)$, $Y_n(x)$, $I_n(x)$, and $K_n(x)$.

x	$J_0(x)$	$J_1(x)$	$Y_0(x)$	$Y_1(x)$	$I_0(x)$	$I_1(x)$	$K_0(x)$	$K_1(x)$
0.0	0.0000	0.000	$-\infty$	$-\infty$	1.000	0.000	∞	∞
0.1	0.9975	0.0499	-1.5342	-6.4590	1.0025	0.0501	2.4271	9.8538
0.2	0.9900	0.0995	-1.0811	-3.3238	1.0100	0.1005	1.7527	4.7760
0.3	0.9776	0.1483	-0.8073	-2.2931	1.0226	0.1517	1.3725	3.0560
0.4	0.9604	0.1960	-0.6060	-1.7809	1.0404	0.2040	1.1145	2.1844
0.5	0.9385	0.2423	-0.4445	-1.4715	1.0635	2.2579	0.9244	1.6564
0.6	0.9120	0.2867	-0.3085	-1.2604	1.0920	0.3137	0.7775	1.3028
0.7	0.8812	0.3290	-0.1907	-1.1032	1.1263	0.3719	0.6605	1.0503
0.8	0.8463	0.3688	-0.0868	-0.9781	1.1665	0.4329	0.5653	0.8618
0.9	0.8075	0.4059	0.0056	-0.8731	1.2130	0.4971	0.4867	0.7165
1.0	0.7652	0.4401	0.0883	-0.7812	1.2661	0.5652	0.4210	0.6019
1.1	0.7196	0.4709	0.1622	-0.6981	1.3262	0.6375	0.3656	0.5098
1.2	0.6711	0.4983	0.2281	-0.6211	1.3937	0.7147	0.3185	0.4346
1.3	0.6201	0.5520	0.2865	-0.5485	1.4693	0.7973	0.2782	0.3725
1.4	0.5669	0.5419	0.3379	-0.4791	1.5534	0.8861	0.2437	0.3208
1.5	0.5118	0.5579	0.3824	-0.4123	1.6467	0.9817	0.2138	0.2774
1.6	0.4554	0.5699	0.4204	-0.3476	1.7500	1.0848	0.1880	0.2406
1.7	0.3980	0.5778	0.4520	-0.2847	1.8640	1.1963	0.1655	0.2094
1.8	0.3400	0.5815	0.4774	-0.2237	1.9896	1.3172	0.1459	0.1826
1.9	0.2818	0.5812	0.4968	-0.1644	2.1277	1.4482	0.1288	0.1597
2.0	0.2239	0.5767	0.5104	-0.1070	2.2796	1.5906	0.1139	0.1399
2.1	0.1666	0.5683	0.5183	-0.0517	2.4463	1.7455	0.1008	0.1228
2.2	0.1104	0.5560	0.5208	0.0015	2.6291	1.9141	0.0893	0.1079
2.3	0.0555	0.5399	0.5181	0.0523	2.8296	2.0978	0.0791	0.0950
2.4	0.0025	0.5202	0.5104	0.1005	3.0493	2.2981	0.0702	0.0837
2.5	-0.0484	0.4971	0.4981	0.1459	3.2898	2.5167	0.0624	0.0739
2.6	-0.0968	0.4708	0.4813	0.1884	3.5533	2.7554	0.0554	0.0653
2.7	-0.1424	0.4416	0.4605	0.2276	3.8417	3.0161	0.0493	0.0577
2.8	-0.1850	0.4097	0.4359	0.2635	4.1573	3.3011	0.0438	0.0511
2.9	-0.2243	0.3754	0.4079	0.2959	4.5027	3.6126	0.0390	0.0453
3.0	-0.2601	0.3391	0.3769	0.3247	4.8808	3.9534	0.0347	0.0402
3.1	-0.2921	0.3009	0.3431	0.3496	5.2945	4.3262	0.0310	0.0356
3.2	-0.3202	0.2613	0.3071	0.3707	5.7472	4.7343	0.0276	0.0316
3.3	-0.3443	0.2207	0.2691	0.3879	6.2426	5.1810	0.0246	0.0281
3.4	-0.3643	0.1792	0.2296	0.4010	6.7848	5.6701	0.0220	0.0250
3.5	-0.3801	0.1374	0.1896	0.4102	7.3782	6.2058	0.0196	0.0222
3.6	-0.3918	0.0955	0.1477	0.4154	8.0277	6.7927	0.0175	0.0198
3.7	-0.3992	0.0538	0.1061	0.4167	8.7386	7.4357	0.0156	0.0176
3.8	-0.4226	0.0128	0.0645	0.4141	9.5169	8.1404	0.0140	0.0157
3.9	-0.4018	-0.0272	0.0234	0.4078	10.369	8.9128	0.0125	0.0140
4.0	-0.3971	-0.0660	-0.0169	0.3979	11.302	9.7595	0.0011	0.0125

Table B.1 (continued)

x	$J_0(x)$	$J_1(x)$	$Y_0(x)$	$Y_1(x)$	$I_0(x)$	$I_1(x)$	$10^{-2}xK_0(x)$	$10^{-2}xK_1(x)$
4.0	-0.3971	-0.0660	-0.0169	0.3979	11.302	9.7595	1.1160	1.2484
4.2	-0.3766	-0.1386	-0.0938	0.3680	13.442	11.706	0.8927	0.9938
4.4	-0.3423	-0.2028	-0.1633	0.3260	16.010	14.046	0.7149	0.7923
4.6	-0.2961	-0.2566	-0.2235	0.2737	19.093	16.863	0.5730	0.6325
4.8	-0.2404	-0.2985	-0.2723	0.2136	22.794	20.253	0.4597	0.5055
5.0	-0.1776	-0.3276	-0.3085	0.1479	27.240	24.336	0.3691	0.4045
5.2	-0.1103	-0.3432	-0.3313	0.0792	32.584	29.254	0.2966	0.3239
5.4	-0.0412	-0.3453	-0.3402	0.0101	39.009	35.182	0.2385	0.2597
5.6	0.0270	-0.3343	-0.3354	-0.0568	46.738	42.328	0.1918	0.2083
5.8	0.0917	-0.3110	-0.3177	-0.1192	56.038	50.946	0.1544	0.1673
6.0	0.1506	-0.2767	-0.2882	-0.1750	67.234	61.342	0.1244	0.1344
6.2	0.2017	-0.2329	-0.2483	-0.2223	80.718	73.886	0.1003	0.1081
6.4	0.2433	-0.1816	-0.1999	-0.2596	96.962	89.026	0.0808	0.0869
6.6	0.2740	-0.1250	-0.1452	-0.2857	116.54	107.30	0.0652	0.0700
6.8	0.2931	-0.0652	-0.0864	-0.3002	140.14	129.38	0.0526	0.0564
7.0	0.3001	-0.0047	-0.0259	-0.3027	168.59	156.04	0.0425	0.0454
7.2	0.2951	0.0543	0.0339	-0.2934	202.92	188.25	0.0343	0.0366
7.4	0.2786	0.1096	0.0907	-0.2731	244.34	227.17	0.0277	0.0295
7.6	0.2516	0.1592	0.1424	-0.2428	294.33	274.22	0.0224	0.0238
7.8	0.2154	0.2014	0.1872	-0.2039	354.68	331.10	0.0181	0.0192
8.0	0.1717	0.2346	0.2235	-0.1581	427.56	399.87	0.0146	0.0155
8.2	0.1222	0.2580	0.2501	-0.1072	515.59	483.05	0.0118	0.0126
8.4	0.0692	0.2708	0.2662	-0.0535	621.94	583.66	0.0096	0.0101
8.6	0.0146	0.2728	0.2715	0.0011	750.46	705.38	0.0078	0.0082
8.8	-0.0392	0.2641	0.2659	0.0544	905.80	852.66	0.0063	0.0066
9.0	-0.0903	0.2453	0.2499	0.1043	1093.6	1030.9	0.0051	0.0054
9.2	-0.1367	0.2174	0.2245	0.1491	1320.7	1246.7	0.0041	0.0043
9.4	-0.1768	0.1816	0.1907	0.1871	1595.3	1507.9	0.0033	0.0035
9.6	-0.2090	0.1395	0.1502	0.2171	1927.5	1824.1	0.0027	0.0028
9.8	-0.2323	0.0928	0.1045	0.2379	2329.4	2207.1	0.0022	0.0023
10.0	-0.2459	0.0435	0.0557	0.2490				
10.5	-0.2366	-0.0789	0.0675	0.2337				
11.0	-0.1712	-0.1768	-0.1688	0.1637				
11.5	-0.0677	-0.2284	-0.2252	0.0579				
12.0	0.0477	-0.2234	-0.2252	-0.0571				
12.5	0.1469	-0.1655	-0.1712	-0.1538				
13.0	0.2069	-0.0703	-0.0782	-0.2101				
13.5	0.2150	0.0380	-0.0301	-0.2140				
14.0	0.1711	0.1334	0.1272	-0.1666				
14.5	0.0875	0.1934	0.1903	-0.0810				
15.0	-0.0142	0.2051	0.2055	0.0211				
15.5	-0.1092	0.1672	0.1706	0.1148				

Table B.2 Zeros x_n of $J_0(x)$ and the corresponding values of $J_1(x)$.

n	x_n	$J_1(x_n)$	n	x_n	$J_1(x_n)$
1	2.4048	+0.5191	21	65.1900	+0.09882
2	5.5201	-0.3403	22	68.3315	-0.09652
3	8.6537	+0.2715	23	71.4730	+0.09438
4	11.7915	-0.2325	24	74.6145	-0.09237
5	14.9309	+0.2065	25	77.7560	+0.09049
6	18.0711	-0.1877	26	80.8976	-0.08871
7	21.2116	+0.1733	27	84.0391	+0.08704
8	24.3525	-0.1617	28	87.1806	-0.08545
9	27.4935	+0.1522	29	90.3222	+0.08395
10	30.6346	-0.1442	30	93.4637	-0.08253
11	33.7758	+0.1373	31	96.6053	+0.08118
12	36.9171	-0.1313	32	99.7468	-0.07989
13	40.0584	+0.1261	33	102.8884	+0.07866
14	43.1998	-0.1214	34	106.0299	-0.07749
15	46.3412	+0.1172	35	109.1715	+0.07636
16	49.4846	-0.1134	36	112.3131	-0.07529
17	52.6241	+0.1100	37	115.4546	+0.07426
18	55.7655	-0.1068	38	118.5962	-0.07327
19	58.9070	+0.1040	39	121.7377	+0.07232
20	62.0485	-0.1013	40	124.8793	-0.07140

Table B.3 Zeros $x_{n,k}$ of $J_n(x)$.

k	n = 1	n = 2	n = 3	n = 4	n = 5
1	3.832	5.136	6.380	7.588	8.771
2	7.016	8.417	9.761	11.065	12.339
3	10.173	11.620	13.015	14.373	15.700
4	13.324	14.796	16.223	17.616	18.980
5	16.471	17.960	19.409	20.827	22.218
6	19.616	21.117	22.583	24.019	25.430
7	22.760	24.270	25.748	27.199	28.627
8	25.904	27.421	28.908	30.371	31.812
9	29.047	30.569	32.065	33.537	34.989
10	32.190	33.717	35.219	36.699	38.160

REFERENCES

1. Hildebrand, F.B., *Advanced Calculus for Applications*, 2nd ed., Prentice-Hall, Inc., Englewood Cliffs, New Jersey, 1976.
2. Arpacı, V.S., *Conduction Heat Transfer*, Addison-Wesley Pub. Co., Inc., Reading, Massachusetts, 1966.
3. Özışık, M.N., *Heat Conduction*, John Wiley and Sons, Inc., New York, 1980.
4. Janke, E.,, Emde, F., and Losch, F., *Tables of Higher Functions*, McGraw-Hill Book Co., New York, 1960.

Appendix C: Numerical Values of Error Function

APPENDIX C

Table C1 Numerical values of error function

z	erf z	z	erf z	z	erf z	z	erf z
0.00	0.00000	0.40	0.42839	0.80	0.74210	1.20	0.91031
0.01	0.01128	0.41	0.43796	0.81	0.74800	1.21	0.91295
0.02	0.02256	0.42	0.44746	0.82	0.75381	1.22	0.91553
0.03	0.03384	0.43	0.45688	0.83	0.75952	1.23	0.91805
0.04	0.04511	0.44	0.46622	0.84	0.76514	1.24	0.92050
0.05	0.05637	0.45	0.47548	0.85	0.77066	1.25	0.92290
0.06	0.06762	0.46	0.48465	0.86	0.77610	1.26	0.92523
0.07	0.07885	0.47	0.49374	0.87	0.78143	1.27	0.92751
0.08	0.09007	0.48	0.50274	0.88	0.78668	1.28	0.92973
0.09	0.10128	0.49	0.51166	0.89	0.79184	1.29	0.93189
0.10	0.11246	0.50	0.52049	0.90	0.79690	1.30	0.93400
0.11	0.12362	0.51	0.52924	0.91	0.80188	1.31	0.93606
0.12	0.13475	0.52	0.53789	0.92	0.80676	1.32	0.93806
0.13	0.14586	0.53	0.54646	0.93	0.81156	1.33	0.94001
0.14	0.15694	0.54	0.55493	0.94	0.81627	1.34	0.94191
0.15	0.16799	0.55	0.56332	0.95	0.82089	1.35	0.94376
0.16	0.17901	0.56	0.57161	0.96	0.82542	1.36	0.94556
0.17	0.18999	0.57	0.57981	0.97	0.82987	1.37	0.94731
0.18	0.20093	0.58	0.58792	0.98	0.83423	1.38	0.94901
0.19	0.21183	0.59	0.59593	0.99	0.83850	1.39	0.95067
0.20	0.22270	0.60	0.60385	1.00	0.84270	1.40	0.95228
0.21	0.23352	0.61	0.61168	1.01	0.84681	1.41	0.95385
0.22	0.24429	0.62	0.61941	1.02	0.85083	1.42	0.95537
0.23	0.25502	0.63	0.62704	1.03	0.85478	1.43	0.95685
0.24	0.26570	0.64	0.63458	1.04	0.85864	1.44	0.95829
0.25	0.27632	0.65	0.64202	1.05	0.86243	1.45	0.95969
0.26	0.28689	0.66	0.64937	1.06	0.86614	1.46	0.96105
0.27	0.29741	0.67	0.65662	1.07	0.86977	1.47	0.96237
0.28	0.30788	0.68	0.66378	1.08	0.87332	1.48	0.96365
0.29	0.31828	0.69	0.67084	1.09	0.87680	1.49	0.96489
0.30	0.32862	0.70	0.67780	1.10	0.88020	1.50	0.96610
0.31	0.33890	0.71	0.68466	1.11	0.88353	1.51	0.96772
0.32	0.34912	0.72	0.69143	1.12	0.88678	1.52	0.96841
0.33	0.35927	0.73	0.69810	1.13	0.88997	1.53	0.96951
0.34	0.36936	0.74	0.70467	1.14	0.89308	1.54	0.97058
0.35	0.37938	0.75	0.71115	1.15	0.89612	1.55	0.97162
0.36	0.38932	0.76	0.71753	1.16	0.89909	1.56	0.97262
0.37	0.39920	0.77	0.72382	1.17	0.90200	1.57	0.97360
0.38	0.40900	0.78	0.73001	1.18	0.90483	1.58	0.97454
0.39	0.41873	0.79	0.73610	1.19	0.90760	1.59	0.97546

Table C.1 (continued)

z	erf z	z	erf z	z	erf z	z	erf z
1.60	0.97634	1.75	0.98667	1.90	0.99279	3.00	0.999978
1.61	0.97720	1.76	0.98719	1.91	0.99308	3.20	0.999994
1.62	0.97803	1.77	0.98769	1.92	0.99337	3.40	0.999998
1.63	0.97884	1.78	0.98817	1.93	0.99365	3.60	1.000000
1.64	0.97962	1.79	0.98864	1.94	0.99392		
1.65	0.98037	1.80	0.98909	1.95	0.99417		
1.66	0.98110	1.81	0.98952	1.96	0.99442		
1.67	0.98181	1.82	0.98994	1.97	0.99466		
1.68	0.98249	1.83	0.99034	1.98	0.99489		
1.69	0.98315	1.84	0.99073	1.99	0.99511		
1.70	0.98379	1.85	0.99111	2.00	0.99532		
1.71	0.98440	1.86	0.99147	2.20	0.99814		
1.72	0.98500	1.87	0.99182	2.40	0.99931		
1.73	0.98557	1.88	0.99215	2.60	0.99976		
1.74	0.98613	1.89	0.99247	2.80	0.99992		

About the Authors

Sadik Kakaç received his Dipl. Ing. (1955) in Mechanical Engineering from the Technical University of Istanbul, MS(1959) in Mechanical Engineering and MS(1960) in Nuclear Engineering from the Massachusetts Institute of Technology. He received his Ph.D.(1965) in the field of heat transfer from the Victoria University of Manchester, England.

Formerly, Dr. Kakaç was professor of Mechanical Engineering at the Middle East Technical University, Ankara, Turkey. Currently, he is professor of Mechanical Engineering at the University of Miami, Coral Gables, Florida. He is a member of the Honorary Editorial Advisory Board of the International Journal of Heat and Mass Transfer.

Dr. Kakaç has authored three books and about 60 research papers in the field of thermal sciences. He is the co-author of a Convective Heat Transfer book. He edited fourteen volumes in the area of heat and mass transfer.

Yaman Yener received his BS(1968) and MS(1970) in Mechanical Engineering from the Middle East Technical University, Ankara, Turkey and his Ph.D.(1973) in Mechanical and Aerospace Engineering from the North Carolina State University, Raleigh, N.C.

Dr. Yener is the author of 30 research papers in the areas of heat transfer, radiative transfer, neutron transport theory, and the co-author of a Convective Heat Transfer book.

Formerly, Dr. Yener was associate professor of Mechanical Engineering at the Middle East Technical University, Ankara, Turkey. He also taught for two years as a visiting professor at the University of Delaware. Currently, he is associate professor of Mechanical Engineering at Northeastern University, Boston, Massachusetts.

Index